T0135878

Integrated institutional water regimes

Realisation in Greece

vorgelegt von

Eleftheria Kampa

M.Sc. Environmental Change & Management

von der Fakultät VI – Planen Bauen Umwelt

der Technischen Universität Berlin

zur Erlangung des akademischen Grades

Doktorin der Ingenieurwissenschaften

- Dr.-Ing. -

genehmigte Dissertation

Promotionsausschuss:

Vorsitzender: Prof. Dr. S. Heiland

Berichter: Prof. Dr. V. Hartje

Berichter: Prof. Dr. H. Th. A. Bressers

Tag der wissenschaftlichen Aussprache: 25. Mai 2007

Berlin 2007

D 83

Bibliografische Information der Deutschen Nationalbibliothek

Die Deutsche Nationalbibliothek verzeichnet diese Publikation in der Deutschen Nationalbibliografie; detaillierte bibliografische Daten sind im Internet über http://dnb.d-nb.de abrufbar.

ISBN 978-3-8325-1738-0

Logos Verlag Berlin GmbH
Comeniushof, Gubener Str. 47,
10243 Berlin
Tel.: +49 030 42 85 10 90
Fax: +49 030 42 85 10 92
INTERNET: http://www.logos-verlag.de

Acknowledgements

This dissertation project would not have been as an enjoyable challenge, without the help of many people who supported me in its different phases, from the first conceptual set-up, through field work and all the way until its finalisation.

I wish to thank Prof. Volkmar Hartje (Institute for Environmental and Landscape Planning, Technical University of Berlin) for his supervision since the first steps of my work, for always being open to new routes and to modifications in my research and for supporting me in securing research funding. Prof. Hans Bressers (Centre for Clean Environmental Technology, University of Twente) has given me invaluable advice, especially on the theoretical approaches used in this dissertation. I wish to thank him for his guidance and for the time he devoted to share with me his work on the institutional resource regime and the contextual interaction theories. Also Dr. Giorgos Kallis (Fellow at University of California; former Aegean University), Dr. Stefan Kuks (University of Twente) and Dr. Waltina Scheumann (TU Berlin) inspired me through discussions we had on the conceptual focus of this dissertation.

The time which I spent in the field in Greece was exciting, occasionally difficult and in many ways unforgettable. I am very grateful to all individuals whom I interviewed and communicated with, for generously offering me their time and their unique knowledge on the Greek water management "scene". All interview partners and personal communications in the numerous organisations and locations, which I visited in the Vegoritida and Mygdonian basins, as well as at ministries in Athens are listed in the annex of this dissertation. I am also thankful to all those people, whom I repeatedly visited in the field to search through their document and data archives.

Additionally, I wish to thank Dr. Martha Stefouli (Institute of Geology and Mineral Exploitation) and Dr. Thomas Alexandridis (Aristotle University of Thessaloniki) for producing the maps of this dissertation on the Vegoritida and Mygdonian basins.

I would also like to gratefully acknowledge the Bodossaki Foundation in Athens for its financial support in the form of a two-year postgraduate scholarship (2004-2006).

Last but not least, I am greatly indebted to Markus, my family and friends as well as work-colleagues at the Ecologic Institute. Their continuous support and their patience kept me motivated and committed to my dissertation project.

Eleftheria Kampa
Berlin, March 2007

Abstract

Greece faces a great challenge to manage its water resources in a more integrated and sustainable manner, in view of its rising water problems and EU requirements due to the Water Framework Directive. To elaborate on this challenge, this study used the concept of institutional water regimes, aiming at identifying whether there has been any change or attempts to change the Greek institutional water regime towards more integration to date. Secondly, the study aimed at explaining why there has been change (if any) or lack of change towards integration. These aims were pursued empirically first on the national level and secondly in two water basins (Vegoritida and Mygdonian).

Institutional water regimes are defined as a combination of public governance and property rights. Regime changes are qualified by the criteria of the extent of uses considered by the regime and of the coherence within and between public governance and property rights. Furthermore, certain context conditions are defined as potential determinants of the outcome of regime change attempts. The analytical framework, as this was adapted to assess institutional regimes in Greek basins, also distinguishes the process of regime change from the process of its implementation. Explicit emphasis is placed on the influence of actors' motivation, information and resources and actors' interactions on the outcome of these processes.

The research findings show that neither the national regime nor the two basin regimes could change towards more integration so far, influenced by integration-unfavourable context conditions. The context influencing national regime changes was characterised mainly by unbalanced power distribution between different administrative levels and lack of cooperative intersectoral policy style. The context conditions influencing change in the basin regimes involved lack of tradition in actor cooperation, lack of integration-supportive institutional interfaces and lack of joint problem perceptions. Furthermore, key policies and measures adopted during regime change could only be inadequately implemented. The slightly higher tendency of the Mygdonian basin regime towards integration was stimulated mainly by external pressure from the EU & international level for a better implemented wetland protection framework in this basin.

Zusammenfassung

Die Entwicklung und Etablierung eines integrierten Wassermanagements stellt für Griechen-
land angesichts seiner vielfältigen Wasserprobleme sowie der neuen EU Anforderungen
durch die Wasserrahmenrichtlinie eine große Herausforderung dar. Um die Frage nach der
Entwicklung des Wassermanagements im Detail zu beleuchten, wurde in der vorliegenden
Studie das Konzept der institutionellen Wasserregime benutzt. Das erste Ziel bestand darin,
die Entwicklung der Integrationsbemühungen zu analysieren, um bereits erfolgte
Veränderungen innerhalb des griechischen Wasserregimes zu ermitteln. Auch Ansätze und
Versuche für ein höheres Maß an Integration wurden berücksichtigt. Das zweite Ziel der
Arbeit war es, die Ursachen und Gründe für Veränderungen zu mehr Integration oder deren
Ausbleiben zu identifizieren. Um die Ziele der Studie zu erreichen, wurde das institutionelle
Wasserregime zunächst auf nationaler Ebene und anschließend anhand von Fallstudien
über zwei Einzugsgebiete (Vegoritida und Mygdonian) auf regionaler Ebene untersucht.

Die institutionellen Wasserregime werden innerhalb der Studie als eine Kombination aus
öffentlicher Governance sowie existierenden Eigentumsrechten definiert. Im verwendeten
analytischen Rahmen werden Regimeveränderungen durch zwei Kriterien qualifiziert:
Einerseits durch das Ausmaß der vom Regime regulierten Nutzungen und andererseits
durch die Kohärenz des Regimes. Letztere unterteilt sich wiederum in die interne Kohärenz,
die die Konsistenz innerhalb der öffentlichen Governance und den Eigentumsrechten
beschreibt, sowie in die externe Kohärenz, die den wechselseitigen Bezug zwischen
öffentlicher Governance und Eigentumsrechten charakterisiert. Außerdem werden bestimmte
Kontext-Konditionen definiert, die die Entwicklung eines integrierten Regimes fördern
könnten.

Der zur Analyse der griechischen Einzugsgebietsregime angepasste Rahmen wurde
zusätzlich in den Prozess der Regimeveränderung und seiner nachfolgenden Umsetzung
unterteilt. Zudem wurden diese Prozesse erheblich durch die handelnden Akteure geprägt.
Dies wurde durch einen Interaktionsansatz erklärt, der den Fokus auf Motivation, Information
und Ressourcen der Akteure legte.

Die Forschungsergebnisse zeigen, dass die Wasserregime weder auf nationaler noch auf
regionaler Ebene (Vegoritida- und Mygdonian-Einzugsgebiete) ein höheres Integrations-
niveau erreicht haben. Als wichtiger Grund dafür wurde der Mangel an einem integrationsförderndem Kontext identifiziert. Auf das nationale Regime bezogen bestand dieser
hauptsächlich aus der ungleicher Machtverteilung zwischen unterschiedlichen
administrativen Ebenen sowie dem Fehlen einer kooperativen, sektorenübergreifenden
Politik.

Der Mangel an integrationsförderndem Kontext in Bezug auf die Einzugsgebietsregime
bestand im Wesentlichen im Fehlen einer traditionell verankerten Kooperation der Akteure,
einer fehlenden Wahrnehmung für gemeinsame Probleme sowie dem Fehlen
integrationsfördernder institutioneller Schnittstellen. Außerdem wurden beschlossene
Regelungen und Maßnahmen bis jetzt nur teilweise in die Praxis umgesetzt. Die etwas

höhere Integrationstendenz des Mygdonian Einzugsgebietsregimes gründete sich vor allem auf externe Impulse aus der EU- und von internationaler Ebene, die die Umsetzung eines verstärkten Feuchtgebietsschutzes forderten.

Contents

List of figures

List of tables

List of boxes

Abbreviations

CAP	Common Agricultural Policy (of the EU)
CMD	Common Ministerial Decision
COD	Chemical Oxygen Demand
comm.	communication
DEYA	Municipal Enterprises for Water Supply and Sewerage
DWRLR	Division for Water Resources and Land Reclamation (of the Prefecture)
EC	European Commission
ECJ	European Court of Justice
EIA	Environmental Impact Assessment
e-NGO	Environmental Non-Governmental Organisation
EU	European Union
HOS	Hellenic Ornithological Society
IACS	Integrated administration and control system (for crop subsidy payments)
IWBM	Integrated water basin management
IWRM	Integrated water resources management
L.	Law
LVPS	Lake Vegoritida Preservation Society
masl	meters above sea level
MDEV	Ministry of Development
MEPPW	Ministry of Environment, Physical Planning and Public Works
MESNA	Educational and cultural association of the community of Arnissa
MINAGR	Ministry of Agriculture
NGO	Non-Governmental Organisation
PD	Presidential Decree
p.e.	population equivalent
pers.comm.	personal communication
PoF	Prefecture of Florina
PoK	Prefecture of Kozani
PoP	Prefecture of Pella
PoT	Prefecture of Thessaloniki
PPC	Public Power Corporation
Pref.	Prefecture
RBA	River Basin Authority
RBD	River Basin District

RDEPP	Regional Directorate for Environment and Physical Planning
RWD	Regional Water Directorates (of 2003 Water Law)
RWM	Region West Macedonia
RWMD	Regional Water Management Department (of 1987 Water Law)
SES	Specific Environmental Study
Sub-Pref.	Sub-Prefecture
UWWTP	Urban wastewater treatment plant
US	United States (of America)
vs.	versus
WFD	Water Framework Directive
WSSD	World Summit for Sustainable Development
WWF	World Wildlife Fund

1 Introducing the theme and purpose of research

1.1 Integration as a multi-dimensional concept in water resources management

The present dissertation is written in the context of developments in recent international and especially European water policy towards the achievement of more integrated water resources management (IWRM). IWRM for all surface and groundwater within a single water basin is a sub-category of IWRM, referred to as integrated water basin management (IWBM).

On the international level, the concept of IWRM emerged as early as 1992, through the guiding principles set by the Dublin Conference on Water and the Environment. The term "integrated" appeared several times in the Dublin Statement linked to principles like "integrated management of river basins", "integrated planning" and "integrated management plans".

Later on, IWRM and especially IWBM became even more widely accepted in international environmental policy and were embedded in several international agreements and protocols on water protection. This trend in international environmental policy and law was, for instance, expressed in article 25 of the Implementation Plan for sustainable development of the Johannesburg World Summit in September 2002 and in paragraph 8 of the Ministerial Declaration of the 3rd World Water Forum in Kyoto in March 2003 (Karageorgou, 2003).

On the European Union (EU) level, the Water Framework Directive, which since 2000 constitutes the cornerstone of European water policy, also firmly supports integrated water resources management on a basin level, as elaborated in the next section.

In research literature, integration in water resources management appears as a multi-dimensional concept. Van Hofwegen (2001) defined IWRM as the management of surface and groundwater in qualitative, quantitative and ecological terms from a multi-disciplinary perspective, focusing on the water needs and requirements of society at large. Earlier on, Mitchell (1990) considered integrated water management in three ways:

- Consideration of the various dimensions of water, i.e. surface and groundwater, quantity and quality.
- Interactions between water, land and the environment broadening the focus to include also issues such as floodplain management, non-point sources of pollution and the preservation of wetlands.
- Interrelationships between water, social and economic development.

Integrated water management is often also characterised as "holistic" or "ecosystem" integrated management. Marty (2001) argues that whether reference is made to "integrated", "holistic", "comprehensive" or "ecosystem" management, the principal ideas behind those concepts when it comes to water and river management are fundamentally the same.

In the case of integrated water basin management (IWBM), specific emphasis is placed on the spatial dimension of integration, whereby the unit of a water basin is the focus of the integrated model. In the past, the acknowledgement developed that the water basin is the best spatial unit for national and international water management (Teclaff, 1987). A water basin is the geographical area determined by the watershed limits of the system of waters, including surface and underground waters, flowing into a common terminus (cf. Helsinki Rules of the International Law Association, 1966, article II).[1] If the common terminus is a lake, a coastal zone, an estuary etc., this may be regarded as an integrated part of the water basin.[2] In this context, IWBM can be understood as the management of all surface and groundwater of a water basin in its entirety with due attention to water quality, quantity and environmental integrity. A participatory approach is followed, focusing on the integration of natural limitations with all social, economic and environmental interests (Jaspers, 2003).

1.2 The EU Water Framework Directive as driver for integration

The EU Directive 2000/60/EC "establishing a framework for Community action in the field of water policy" (Water Framework Directive, WFD) was adopted in December 2000 and marked the beginning of a new era in European water and environmental policy in general. To overcome the voluminous and fragmented EU legislation on water resources prior to its adoption, the WFD promotes the streamlining of water legislation and the adoption of water policy measures in a coherent manner.

Aim of the WFD is the achievement of "good status" of surface water and groundwater by 2015. For surface water, the objective is to achieve good ecological and good chemical status. For groundwater, the objective is to achieve good chemical and good quantitative status. The definitions of the several aspects of "good status" are given in the technical annexes of the Directive.

To reach the environmental objectives of the WFD, coordinated management plans and programmes of measures must be formulated and implemented in all river basin districts (RBDs)[3] of the EU. More specifically, Member States must initially prepare an inventory of the river basins. Secondly, for each RBD, they must collect information and make an analysis of the characteristics of the basins, a study of impacts of human activities on water, and an economic analysis of the water uses. Moreover, they must put monitoring programmes in place, in order to make a complete water status analysis in each RBD. Finally, in each national or international RBD, a management plan must be elaborated. In international RBDs, Member States shall coordinate their actions with the aim of producing a single RBD management plan. The RBD management plans should include a general description of the characteristics of the waters in the RBD, an analysis of the pressures and impacts of human

---

[1] *Helsinki Rules on the Uses of the Waters of International Rivers* adopted by the International Law Association at its 52nd Conference, Helsinki, 20th August 1966.

[2] The term "water basin" expresses better than the term "river basin", the fact that a hydrological basin may also consist of water elements which are not rivers.

[3] "River basin district" means the area of land and sea, made up of one or more neighbouring river basins together with their associated groundwater and coastal waters, and is identified as the main unit for the management of river basins (from Article 2(15) of the WFD).

activities on water status, a description of the monitoring network, a list of the environmental objectives defined for each water body, a summary of the economic analysis of water uses and a summary of the programme of measures adopted (see WFD Annex VII).

The WFD is considered to be a strong impetus for integration and cooperation in European water management (van der Brugge & Rotmans, 2007). Aubin & Varone (2004a) pointed out several principles of integration, which are reflected in the WFD requirements, starting with the consideration of water as a resource in its entirety. Namely, this refers to the purpose of the WFD to achieve good status in all EU waters, groundwater and surface waters, in terms of quality and quantity and according to the environmental needs of water ecosystems. This objective fulfils integration in the sense of combined management of surface and groundwater, water quantity, quality and environmental integrity. Secondly, the Directive combines water quality objectives with objectives of emissions standards for pollutants. Until its adoption, European policies on water quality followed the two separate lines of water quality objectives and emission standards, which reflect different national approaches. The WFD is an attempt to reconcile the two approaches and, additionally, to integrate (partially) water quantity management aspects. Another element of integration consists in ensuring that administrative arrangements for water management are appropriate for applying the WFD requirements within each RBD, which involves also identifying a competent authority for implementing the WFD in each RBD (Aubin & Varone, 2004a). In this context, the Directive also requires the coordination of planning and measures for water protection within the RBDs. The requirement for coordination is linked to further implicit requirements for multi-dimensional consensus processes between water authorities, across national borders and across regional borders in the RBDs, and even cross-sectoral consensus processes to reach the environmental objectives of the Directive on time (Moss, 2003). Integration is additionally built on the coordinated production of RBD management plans and the integration of national, regional and European legislation in a single programme of measures. Finally, integration is reflected in the public participation requirements of the WFD (Aubin & Varone, 2004a). Specifically, Article 14 of the WFD requires the participation of the organised and non-organised public, in a three-step process for public consultation and information in the formulation of the RBD management plans. Moreover, the active involvement of all interested parties should be encouraged.

A more systematic discussion of the integrative elements of the WFD is made in chapter 3/3.3.3.4, which also explores the links between the WFD and the theory-based interpretation of integration adopted for use in this dissertation.

All in all, the legally binding WFD requirements can be considered as an undisputable driver for more integrated water management in EU Member States. In fact, it has been argued that the Directive has far-reaching implications for the institutional system of water management of Member States and it gives impulses for a new configuration of existing institutional arrangements (Moss, 2003). The progress of the various EU Member States in terms of meeting the WFD requirements, time-wise and content-wise, is expected to be influenced by the existing degree of integration in the water management system and institutional setting of each Member State. In fact, integrated water management will most likely be a greater challenge in those Member States with limited previous experience and structures supportive of integration.

1.3 Problem case of Greece: Greek water management facing a great challenge

In the southern EU Member State of Greece, water resources management has been characterised to date by the lack of coordinated water policies and planning. In 2003, the Greek Committee for Social and Economic Issues concluded that the "country lacks a real water policy framework, thus leading to serious problems in development and environmental protection" (CSEH, 2003a).

The first national Water Management Law was adopted in 1987 (L. 1739/1987). Although this Law was considered a worthwhile policy effort for the harmonisation of the water management framework, it was very partially activated and implemented. According to CSEH (2003a) and WWF (2003), main obstacles in its application were the division of responsibilities for water management into a number of different authorities and ministries as well as the weakness of the public sector to support the Law financially and administratively (see chapter 5 for more details).

In view of the obligations of Greece to comply with the EU WFD, it is considered important to set up an effective system for water resource management on a central and regional level, according to the Directive principles (Karageorgou, 2003). In December 2003, a new national Law on the Management and Protection of Water Resources was adopted (L. 3199/2003), transposing the EU WFD and attempting to reform the existing water management system and administrative structure. However, the prospect of applying this new national water policy framework adequately is also viewed with scepticism.

Given this background, empirical knowledge is needed on the extent to which Greek water management and institutions have fulfilled any principles of integration so far. It is also equally important to explain, why there has been change (if any) or lack of change towards integration to date. From a policy perspective, the realisation of aspects of integrated water management in Greece up until now is worth assessing and explaining, in order to identify factors which may influence this nation's chances of future compliance with the WFD.

1.4 Formulation of the research purpose

The main purpose of this dissertation is to assess whether Greek water management has experienced any changes to date, which have allowed it to fulfil principles of integration. Secondly, the aim is to explain why there has been change (if any) or lack of change towards integration. This research purpose will be pursued first on the national level of Greece and then on a more regional level in specific water basins. This distinction should allow a discussion of integrated water management aspects both in terms of their more abstract conceptual set-up in national policy and in terms of their more operational realisation in specific water management processes within water basins. Moreover, research first on the national and then on the water basin level can provide useful insight into the process and outcome of any attempted decentralisation in the water management sector, which is often closely connected to attempts at integrated water management.

The theory-based analytical framework of this dissertation unfolds in chapters 3 and 4, whereby it becomes obvious that institutional analysis forms an important part of the approach adopted to pursue the research purpose. In specific, the interpretation of institutions is borrowed from institutional resource regime theory and, for analytical purposes, use is made of the concept of institutional resource regimes (rooted in the work of Knoepfel et al. (2003), which was later further developed and specified by Bressers et al. (2004)). The analytical framework based on institutional resource regime theory also assists in the operationalisation of the concept of integration for discussing Greek attempts at integrated water management.

At a first stage, this dissertation aims at characterising and explaining the recent development of the Greek national institutional water regime in terms of integration. Special attention will be given to the 1987 Water Law, which constituted the first national attempt towards partially more integrated water management. The time-frame of investigations of this dissertation also allows a limited discussion of more recent policy changes directly prompted by the EU WFD (adoption of the national 2003 Water Law).

At a second stage, this dissertation aims at assessing the realisation of any elements of integrated institutional water regimes in specific Greek basins following a case-study approach. To this aim, two basins, where problem pressure from water degradation and rival use demands indicate an urge for more integrated management, will be closely examined. Furthermore, the aim is to explain why there has been change (if any) or lack of change of the institutional water basin regimes towards a more integrated phase.

Based on the above, the leading research questions of this dissertation are the following:

> "Have there been any changes or attempts to change the institutional water regime in Greece towards a more integrated phase? How can progress (if any) or lack of progress towards a more integrated phase be explained?"

These questions are further specified for the national and for the water basin level of research in chapter 4, following the elaboration of the analytical framework.

Based on the research findings, this dissertation ultimately aims at discussing future prospects of realising integrated institutional water regimes in Greece and at formulating recommendations to assist a regime change in this direction.

Given the EU membership of Greece, the EU WFD requirements for integrated water management are an important part of the policy context of this research. Thus, it is hoped that the research conclusions will also be relevant to the WFD policy targets for Greece. However, it should be kept in mind that the WFD requirements are considered only in as far as they relate to the theory-based definitions of integration and institutions used in this dissertation. Therefore, this dissertation is not explicitly about the implementation of the WFD one-by-one requirements in Greece.

All in all, this dissertation can be considered as a useful assessment of the Greek water management system and institutions to date, to support longer-term efforts for the achievement of integrated water resource management (IWRM). According to van Hofwegen (2001), the desired level of IWRM is a compromise between the existing and the ideal water management situation, the latter being derived partly from theories on IWRM and partly from relevant water policy principles (such as the EU WFD). In reality, the "ideal" IWRM situation does not exist. Local and regional conditions combined with conscious policy choices should determine the most appropriate level of IWRM which should be pursued in each nation and each water basin.

1.5   Structure of dissertation

This first chapter introduced the research problem at hand and formulated the overarching research purpose statement of the dissertation.

Chapters 2 and 3 review literature relevant to the research theme, in preparation for the elaboration of an analytical framework. More specifically, chapter 2 presents some practical examples which explain the (successful or failed) realisation of integrated water management in selected parts of the world. Chapter 3, then, reviews selected institution-based analytical approaches for assessing the achievement of integrated water management. Subsequently, chapter 4 presents the theory-based analytical framework adopted for use in this dissertation and illustrates the research methodology.

Chapters 5-9 are dedicated to the empirical findings of the dissertation, which are largely presented according to the variables and links of the adopted analytical framework. Chapter 5 refers to the development of the Greek institutional water regime on the national level. Chapter 6 describes the selection of two Greek basins as case studies for the water basin level of research. Chapter 7 and chapter 8 deal with the development of the institutional water regimes in the two selected basins (the Vegoritida basin and the Mygdonian basin). Chapter 9 summarises key issues learned on the process of institutional water regime change in the two examined Greek basins and formulates recommendations on how to change these regimes towards more integration.

Chapter 10 draws key cross-chapter conclusions. It first makes closing observations on the analytical framework used. Then, it revisits the main research questions and summarises key recommendations. Finally, chapter 10 addresses in brief the contribution of this dissertation to research, its relevance to policy and potential implications for future research.

2 Practice-based determinants of progress in integrated water management

This chapter provides specific examples of practical progress in integrated water management[4] in a few selected parts of the world. The specific examples presented include developments in the US State of California, in the Federation of Mexico (featuring progress in a large river basin) and a review of progress in developing countries. The examples presented stem from regions with different levels of development and also feature different progress in integrated water management. The assessments of progress towards integrated water management are largely those of the authors whose work is cited.

In general, the purpose of this chapter is not to present an exhaustive, lengthy literature review of the practice of integrated water management. The purpose is rather to present a few cases, which provide practice-based explanations for the success or failure of integrated water management, before looking at more theoretical explanatory approaches in the next chapter. It is argued that, although several variables facilitating or hindering the realisation of integrated water management can be gained from practical case studies, these cannot replace the value of a more systematic theory-based analytical framework.

2.1 Progress in the US State of California

The information presented in the following is based on the analysis of Svendsen (2001) on water basin level management functions in the large interior Central Valley of the US State of California. In general, it is argued that California has achieved progress towards more integrated water resources management, through a combination of effective and well-functioning institutions, coordination mechanisms and the availability of necessary finances. Procedures for managing water at the basin level have solid, even if complex, underpinnings in law and tradition.

California is well endowed with water resources. About half of the water available annually in this State is set aside for in-stream environmental uses. The remaining water is available for agricultural and urban uses. The fact that two-thirds of California's water is in the north, while the bulk of agricultural land and the largest population centres are in the south, led to two massive engineering projects to transport water from north to south. These are the federal Central Valley Project and the State Water Project. Water moving through both these systems must transit the Sacramento-San Joaquin Delta, where it mixes with water in the Delta and contributes to it. The Delta is important environmentally and serves as the nexus of the debate over the future of California water. In fact, environmental values in the water management of California were especially promoted by means of law with the introduction of the Federal framework Endangered Species Act, protecting the salmon of the Sacramento

---

[4] It is noted that, in this chapter, the interpretation of integration is not systematic yet. Instead, integrated water management is interpreted according to the perspective of the authors whose work is here reviewed. Indeed, integration can mean different things to different scholars. To achieve consistency throughout this dissertation, in chapters 3 and 4, a specific interpretation of integration, linked to institutional resource regimes, is presented and adopted before proceeding with the empirical analysis.

River, and of the Central Valley Improvement Act (1992) which reallocated a portion of water from irrigation to the Sacramento-San Joaquin Delta.

As concerns the system of water rights, this is complex for surface water but relatively well-specified in law and through cumulative court decisions. Surface water rights are based on both riparian[5] and appropriative[6] doctrines. On the other hand, groundwater is lightly regulated and is seriously overdrafted by about 12% of its renewable total. In California, sustainable management of groundwater is hindered by the inability to come to consistent interpretations of terms regarding the use of groundwater. Furthermore, there is no permitting process for groundwater exploitation and groundwater is available, in the first instance, to owners of overlying land for reasonable beneficial use on their land. Groundwater users, thus, establish rights simply by use. Rights are correlative with the rights of other owners, meaning that if the water supply is insufficient, the supply must be equitably apportioned. Subject to future water needs on overlying lands, "surplus" groundwater may be appropriated for use on non-overlying lands. Again, no permit is required. This vague and permissive specification of rights to groundwater has two important implications. Firstly, as pressure on the nearly fully allocated surface water continues to build, users turn more to groundwater to make up deficits, leading to a serious and growing problem of over-drafting in many parts of the State. Secondly, groundwater is a magnet for litigation because water users cannot agree over terms such as "surplus", "sufficient", "reasonable", "equitable" and "beneficial". Development of a suitable institutional framework for managing groundwater in the California is urgently needed but proceeding slowly.

Despite problems in groundwater regulation, there is a broad group of actors (public and private), who participate and influence decision-making for basin-level water management through public hearings, media and courts. All major interested parties are well-represented and financed. These parties include municipal water districts, agricultural water districts, public water supply agencies, the State and federal environmental regulators as well as environmental NGOs. Although environmental NGOs entered the picture relatively recently, they participate very actively. Environmental concerns are represented by NGOs, which constantly grew over the past 25 years in number, resources and influence, and are supported by federal and State Acts.

In terms of water planning, a 5-year California Water Plan is produced by the State Department of Water Resources, which holds comprehensive planning and development functions, as well as various regional plans. In the context of this planning process, there is extensive interaction among a variety of stakeholders. Except for the State Department of

---

[5] Riparian rights to surface water are available, under common law, to the owners of property next to streams. Water abstracted under a riparian right cannot be applied to land which is not next to the stream and cannot be transferred to other uses removed from the riparian land.

[6] Appropriative rights to surface water are more flexible and comprise the majority of non-imported surface water rights. Appropriative rights are granted through a permitting process managed by the State of California. Appropriative rights can be for use at points removed from the stream of origin and are subject to transfer and change of purpose.

Water Resources, there is a second important managing State organisation, the State Water Board. The State Water Board performs functions which are managerial, regulatory and quasi-judicial in nature. Since 1967, it consolidated water quantity and quality management. The Board allocates rights to use surface water and settles disputes over rights to water bodies. It also establishes water quality standards and supervises nine Regional Water Quality Control Boards.

It is also important to mention the water-related process named CALFED, which was set up in 1994. This is a consortium of federal and State agencies with management and regulatory responsibilities in the Bay-Delta system of the State. Its mission is to develop a long-term plan to restore ecological health and improve water management for uses of the Bay-Delta system. It is an open forum with participation from the entire range of interests, including environmental NGOs. What makes all parties committed to make CALFED work, is the credible threat that if the CALFED process fails, years of litigation will follow in a far more adversarial process. Many of the actors realised that past reliance on litigation to resolve dispute does not necessarily lead to optimal solutions, thus, there is great willingness to experiment with this alternative model of decision-making and conflict resolution. However, CALFED was conceptualised as an ad-hoc body. New institutional forms with a legal basis are required to legitimise and implement the consensual agreements reached by CALFED.

All in all, progress towards a more integrated model of water management in California and its large Central Valley has been assisted by a number of favourable conditions. In the first place, the water basin management system is endowed with generally adequate financing and human resources. Secondly, there is also a sound legal framework underlying user-based districts, which provide services such as irrigation, drinking, groundwater management and wetland conservation. The user-based districts are self-financing, self-governing and work quite effectively. Thirdly, representation is generally well-developed, with groups having similar interests allied into various associations. Forth, there is availability of technical information and increased transparency in decision-making processes. These changes have been driven by environmental protection laws and a growing public demand for more information on water management processes. Finally, legal authority over water resources is well-established and to a great extent enforceable. The system of water rights is well specified in law (with deficiencies still existing for groundwater) and through cumulative court decisions. An impartial court system (federal and State) also plays an important role in resolving water disputes.

2.2 Progress in Mexico and the Lerma-Chapala basin

The information presented in the following is based on Wester et al. (2001), who discussed the management of a water transition in Mexico and its Lerma-Chapala basin.

Mexico is a highly centralised federation composed of 31 States and a Federal District. In the context of a transition effort from supply to demand management in the Mexican water sector and the reshaping of relationships between water users and the three levels of government (federal, State and municipal), several important developments took place: a National Water

Agency was created in 1989, government irrigation districts were transferred to users in 1991 and a new water law was promulgated in 1992. The 1992 Law defined an integral approach for managing surface and groundwater within river basins, it promoted decentralisation, participation, control over water abstractions and wastewater discharges and full-cost pricing.

The most known developments linked to this water reform have taken place in the Lerma-Chapala basin, which is a large basin of 54,300 km². The basin faces both quantitative problems (overpumping of the aquifer, drying-out of Lake Chapala) as well as water qualitative problems due to serious water pollution. Especially to counteract water depletion in the basin, several institutional innovations occurred under the influence of the national water reform.

In 1989, a River Basin Coordination agreement was signed between the federal government and the five States sharing the basin. Additionally, a Consultative Council was set up to manage this agreement.

In 1991, a treaty was signed between the federal government and the five States sharing the Lerma-Chapala basin for fair surface water allocation. This treaty aimed at maintaining the Lake Chapala level, satisfying irrigation and urban supply needs and safeguarding the environment. It is here clarified that the surface water is defined in the Mexican Constitution as national property and is placed in the trust of the federal government. The federal government has the right to grant surface water use rights as concessions to users.

In 1993, the Consultative Council of the River Basin coordination agreement was turned into a River Basin Council (based on the 1992 federal Water Law), which was the first in the country. The role of the Council is coordination and consensus-building between the federal government, the States, municipal and user representatives. In fact, user representatives entered the River Basin Council later in 1998. Initially, user representatives were nominated by the National Water Agency, but then a more bottom-up approach was to be attempted, whereby user representatives would be directly elected through a User Assembly of the basin.

As regards groundwater, it was also made national property with concessions being granted on a volumetric basis, based on a ruling of the Supreme Court in 1983. In the Lerma-Chapala Basin, the overdraft of aquifers is a serious challenge. In 1993, the River Basin Council signed a coordination agreement to regulate groundwater extraction but the implementation was slow. Even though prohibitions were placed on groundwater throughout the basin (by prohibiting the drilling of new wells), the number of wells increased much. Part of the problem of deficient control is that the River Basin Council does not physically control the wells (as it controls dams for surface water allocation through the National Water Agency). In 1995, aquifer councils started being set up, also under the supervision of the River Basin Council. The aquifer councils aim at stimulating organised participation of aquifer users and establishing mutual agreements to reverse groundwater depletion. By 2001, there were no visible results yet from this institutional attempt.

All in all, it is argued that in the Lerma-Chapala basin, an initially highly centralised water management approach developed gradually into a more decentralised approach, whereby States, municipalities and water users have a larger say, especially thanks to the function of the River Basin Council. The partial improvement of governance was facilitated by the respective reforms promoted by the 1992 federal Water Law as well as the institutional capability in the specific basin to put this Law into practice. An essential component supporting the decision-making process was the availability of professional data and the capacity to process and analyse it in order to propose alternative water management and allocation models in the basin.

However, even more drastic changes are needed to enable sustainable water management (within water basin boundaries). Firstly, further decentralisation of decision-making and control over financial resources to the basin and State levels is needed. In its initial decentralisation effort, the federal government promoted the delegation of water sector responsibilities and programmes to the States, but notably not of financial resources. Secondly, the overdraft of the basin aquifers remains a serious challenge. Thirdly, user representation needs to be ensured in decision-making on the basin level, using a more bottom-up rather than top-down approach. Finally, the fact that poor people, especially farmers, are losing their access to water has to be tackled. The lack of access to water results from the reductions of surface water allocations to irrigation and increased costs for groundwater irrigation. Namely, in the context of institutional innovations in the basin (1991 Treaty for fair surface water allocation mentioned above), a scheme for water re-allocation from the dams to the various uses was put into practice. This scheme is based on hydrological modeling and aims at reallocating water between uses for the preservation of the Lake Chapala level. As a result of increased scarcity, allocation of water to irrigation districts is reduced in order to satisfy urban and environmental demands.

2.3 Progress in developing countries - "Limits to leapfrogging"

In several developing regions of the world, progress is taking or has taken place towards integrating institutions for water resource management especially with the support of international donors. Such developments can be found in Sri Lanka, Zimbabwe and Indonesia to mention but a few (Lenton, 2004).

Especially the experience with integrated water resource management in the pilot Brantas River Basin in Indonesia is considered as positive. In this case, a river basin agency was established by the government in the form of a state-owned company financed through fee contribution. The aim of the company was to address water management in the basin and facilitate the operation and maintenance of river structures. Stakeholders were organised through a water resource committee. Despite the positive developments in this pilot basin so far, problems remain with the issue of deficient enforcement especially in terms of compliance with water pollution standards. The notion of water protection is also still under-developed among farmers mainly due to the lack of awareness (Usman, 2001).

On the other hand, some other attempts at integration in developing countries, such as the attempt of India to transpose the Tennessee Valley Authority model of the US, were a resounding failure. In the case of Sri Lanka, a Water Resources Board was established as early as 1964 to promote integrated water resource planning and river basin development, but it only managed to focus on hydrological investigations and the drilling of wells (Shah et al., 2001).

Here, the main intention is to review some overall lessons learned from the experience of different developing countries with integrated water management. In an international workshop on "*Integrated water management in water-stressed basins in developing countries*"[7] (with a focus on Africa and Asia), it was concluded that the key pre-condition for integrated water basin management is pressure from scarcity of water (in the form of hydrological change, pollution, resource capture or flooding). The impacts of scarcity involve increasing levels of conflict, growing insecurity of supply, complaints and social unrest. Ultimately, pressure for a change in the water management paradigm comes from society in general, downstream users of water, researchers, NGOs and international standards, while those who directly take actions are politicians, industries, local communities and technical managers. A reform process towards more integrated water management begins only when the need for this type of management exists. Subsequently, there must be, first of all, political will and a concrete strategy. When these basic conditions exist, then the legal, institutional and financial constraints must be overcome (Abernethy, 2001).

Table 2.1 summarises success factors and constraints, which influence the establishment of integrated water management in water-stressed basins in developing countries.

Table 2.1   **Factors influencing integrated water management in water-stressed basins of developing countries**

Success factors	Constraints
- Political factors (law, political will, hydraulic bureaucracy committed and convinced) - Social factors (social acceptance, adaptive capacity) - Full stakeholder participation, awareness - Financial factors, human resources - Information factors (data available, transparency)	- Political constraints (water not on the national agenda) - Lack of administrative will and inertia - Legal constraints (fragmenting, contradictory or conflicting laws) - Stakeholders (low awareness, willingness, capacity, power and influence) - Financial constraints, lack of human resources - Lack of information - Strategy (effective strategy to launch the RBM concept is needed)

Source: Summarised from Abernethy (2001).

Regarding the success of efforts to establish integrated water management (especially within water basins) in developing countries, Shah et al. (2001) concluded that the institutional reform which makes best sense should emerge from understanding the realities of the water basins or nations in question. A broad understanding of what has worked elsewhere, including any success stories in developed countries, might offer "a good backdrop to the

---

[7]   At Loskop Dam, South Africa, on 16-21 October 2000.

design of institutional interventions", but it is unrealistic to expect much more. In this context, Shah et al. (2001) argued that the effectiveness of a pattern of institutional development of a river basin is determined by at least four realities: hydro-geological reality, demographic reality, socio-economic reality and the organisation of the water sector. Shah et al. used these four realities to account for material differences between developed and developing countries, in support of the argument that there are "limits to leapfrogging" in the transposition of successful water basin management institutions from the developed to the developing world.

On the issue of the organisation of the water sector, developed countries have evolved in this respect over decades of public intervention. Countries with highly developed water institutions are also those which have evolved industrially. Their water sectors tend to be highly formalised and organised with the bulk of water delivered by service providers in the organised sector. When the water sector is formalised and organised, resource governance becomes feasible and reforms of regulations can be carried out with the confidence that they will (most probably) stick. This is not the case in low-income countries, where a large part of users is served by an informal water sector and service providers are not even registered. In general, it is difficult to implement reforms where the formal water sector is little organised beforehand, since there is very little already in place which can be actually "reformed". For instance, India adopted a water policy in 1987 but nothing changed as a consequence. Especially regarding the issuing of laws or regulations for water property rights, there is a problem of enforcement in societies with a large number of small stakeholders, who operate in the informal sector with few or no links to higher levels of resource governance structures (Shah et al., 2001).

Additionally, in developing countries, local-level, traditionally-based solutions to water resource problems tend to be of greater importance than in more developed countries which work with the model of integrated water management. Iyer (2004) argues that "while both integrated water resources management and basin planning are well-meant terms, they carry within themselves the seeds of centralisation and gigantism. They fail to incorporate adequately the elements of decentralised local community-led planning and of traditional knowledge and wisdom".

Moreover, the problems, which successful institutional models for integrated water management have resolved, are problems with water quality, wetlands, sediment accumulation, navigation and flood protection (mainly in developed countries), which are often not of paramount interest in developing countries. Many of the problems that developing countries are facing have either remained unresolved also in developed countries, such as groundwater depletion, or are rendered irrelevant by their evolutionary process, such as using irrigation to provide food security to poor people (Shah et al., 2001).

It should finally be noted that society's needs for environmental amenity increases with the increase of living standards and the rate of development. Therefore, conditions for environmental regulation (and also for integrated water management which includes ecosystem aspects) tend to be more favourable in developed countries rather than in less developed ones.

Hartje (2002) also made a distinction between the developing world, whose bulk is situated in arid/semi-arid regions,[8] and western developed countries, most of which are situated in temperate regions of the world. Developed countries in temperate regions are dominated by industrial and urban uses, water quality concerns and ecosystem aspects due to the greater strength of the environmental movement. In drier countries and regions, such as many developing countries but also southern Europe, western US and Australia, irrigated agriculture is often the most important use, leading to consumption levels which exceed the available supply. The presence of different water resource problems in the developing and developed world can be related to the adoption of different water management strategies and approaches in efforts for policy change towards more integrated water resource management.

2.4 Conclusions

The presented examples on the practical realisation of integrated water management in different countries and water basins of the world can be used as a useful source of empirical factors which facilitated or hindered the establishment of integrated water management.

The examples of the US State of California and Mexico revealed that integrated water management (within water basin boundaries) is facilitated by effective and well-functioning institutions, institutional innovations accompanied by the capability to implement them, coordination mechanisms, the availability of necessary finances, nationalisation of water resources (of both surface and groundwater), the growth of trust between stakeholders, negotiated decisions, sufficient representation of all interests and the availability of information. On the other hand, constraints are related to factors such as the inability to effectively regulate the over-exploitation of groundwater, the incomplete devolution of power (e.g. financial power) to decentralised levels of government and the lack of adequate participatory processes.

In developing countries, it was argued that relevant success factors could include effective organisations such as river basin agencies, the availability of finances and human resources, the existence of political will and a strategy for implementation, the organisation and participation of stakeholders, the social acceptance of the intended reform, and the availability of information. Constraints to the establishment of integrated water management could involve the lack of awareness, lack of willingness of stakeholders, lack of information, lack of financial and human resources, unbalanced distribution of power and influence, administrative inertia, fragmented legal framework, low need of society for environmental amenity and lack of enforcement potential. Jaspers' conclusion that a major requirement for the implementation of any institutional development towards integrated water management is

---

[8] Regions like India and West Africa are humid for a small part of the year but arid during the rest of the year (Shah et al., 2001).

the presence of sufficient human and institutional capacity at the right time and at the right place (Jaspers, 2003) is especially relevant for developing countries. Namely, reform attempts in developing countries often suffer from the fact that the existing water sector is usually not formalised and not organised enough.

The identification of the above success and failure factors in different contexts gives some first ideas on how to formulate a proposition for assessing and explaining the realisation of any aspects of integrated water management in Greece. The identified factors point, in general terms, to the following: "that the realisation of integrated water management may be dependent on a mixture of administrative and regulatory, financial and human capacity, motivation and awareness determinants". In a similar way, even more success and constraint factors for the establishment of integrated water management could be gained by reviewing a larger sample of relevant practical case studies.

However, this multitude of single determinants cannot be used, as they are, as a solid basis for an analytical research framework. The explanations, which these determinants have offered, may have been valid in the context of the individual studies they stem from. These determinants may also be related in some way to each other but this is still not obvious, i.e. there do not seem to be explicit causal relations between them. As Marty (2001) argues, in such a case "using the determinants would mean to confine analysis to the testing of single independent variables".

From an analytical point of view, a more consistent approach is needed, which could offer a set of variables linked to one another in a more systematic and causal way. In other words, it is necessary to use a theoretical framework with explanatory power. Such a framework should specify a number of core determinants for the change of water management towards more integration, combining dependent and explanatory independent variables.

The need for a theoretical framework with explanatory power is taken up in the next chapters 3 and 4. Firstly, chapter 3 describes a selection of relevant theoretical concepts and approaches and then chapter 4 proposes a differentiated analytical framework, considered as appropriate for answering the questions posed by the present study. The proposed analytical framework is, in fact, able to reflect several of the success and failure factors identified in the practical examples of this chapter.

3 Institutional analytical concepts for integrated water management

This chapter is devoted to analytical and conceptual approaches for the assessment of integration in water management. The literature is, namely, rich in approaches for defining integration and for assessing its realisation in natural resources management, and especially in water management. Many of these approaches are based on or profit from the analysis of institutions, which is thus an explicit focus of this chapter.

3.1 Importance of institutions in water resource analysis

To date, different definitions of institutions have been put forward. The majority of definitions are based on Elinor Ostrom's work who first equated institutions with sets of rules (Ostrom, 1986). Ostrom referred to the term "institution" in the sense of rules, norms and strategies adopted by individuals operating within or across organisations. More generally, humans use shared concepts in repetitive situations organised by institutions (rules, norms and strategies). Marty (2001) regards institutions as a set of rules to guide the behaviour of those who use a resource or impact on it in one way or another. Other scholars have defined institutions in a more detailed manner. For instance, Bandaragoda (2000) defines institutions as a combination of policies and objectives, laws, rules and regulations, organisations, their by-laws and core values, operational plans and procedures, incentive mechanisms, accountability mechanisms, norms, traditions, practices and customs.

In the field of water resource management, the analysis of institutions has long been seen as an essential component of water resources planning and analysis. Already in 1984, Helen Ingram and her colleagues argued that institutional problems in water resources development and management are more prominent, persistent and perplexing than technical, physical or even economic problems (Ingram et al., 1984).

3.2 Institutional concepts in the study of integrated water management

In the context of research on integrated water resource management and water management on a basin level, several scholars focused their investigations on institutional aspects and arrangements (e.g. Mitchell, 1990; van Hofwegen, 2001; Jaspers, 2003; Bandaragoda, 2000). They were, namely, analysing institutions in an effort to answer the question of "how integrated water management and integrated water basin management can be successfully applied".

Bandaragoda (2000) used an institutional framework to study integrated water management in a water basin context, based on laws (legal framework), policies and administration (organisations involved in water management and their internal roles).

Mitchell (1990) suggested an analytical framework for the study of conditions for achieving integration in water management, consisting of six aspects: context, legitimation, functions, structures, processes and mechanisms, culture and attitudes. He stressed that each aspect represents a necessary but not sufficient condition for the achievement of integration. His framework specifically stressed the significance of the human dimension, and reminded us that ultimately people are responsible for implementing any policy or programme favouring integrated water management. He argued that any design of institutional arrangements for the promotion of integrated water management should incorporate some "bottom up" features to counterbalance "top down" directives and initiatives.

Jaspers (2003) also worked with institutional concepts in his study of integrated water management within basins and identified the following as necessary institutional arrangements: stakeholder participation, decentralisation and subsidiarity, management based on hydrological boundaries, organisational set-up for functional river basin and sub-basin authorities, an integrated water planning system based on integrated water basin plans and the introduction of a system of water pricing and cost recovery.

Especially the institutional arrangement of decentralisation (in the sense of management of water resources at the "lowest appropriate level") is often associated with integrated water management (ICWE, 1992; World Bank, 1993; Mody, 2004). According to Blomquist et al. (2005a), decentralisation has two components: a) organising management responsibilities at the river basin scale, which often involves devolution of authority from a central government, and b) involving stakeholders within the basin in decision-making and/or operations concerning water management activities. In their analytical framework for assessing institutional arrangements in eight river basins around the world, Blomquist et al. (2005a) selected several institutional variables hypothesized to be related to the success or failure of decentralisation in efforts at integrated water management and river basin management. The institutional variables were grouped into four categories, which are elaborated in Table 3.1.

Table 3.1 **Institutional variables related to decentralisation efforts for integrated water basin management**

Contextual factors and initial conditions
Level of economic development of the nation
Level of economic development of the river basin
Initial distribution of resources among basin stakeholders
Class, religious or other social/cultural distinctions among basin stakeholders
Local experience with self-governance and service provision
Characteristics of the decentralisation process
Top-down, bottom-up or mutually desired devolution
Extent of central government recognition of local basin governance arrangements
Consistent central government policy commitment to decentralisation and basin management through transitions in central government administration
Central government and basin-level institutional arrangements
Extent of actual devolution of authority and responsibility to the basin level
Financial resources and autonomy at the basin level
Basin management participants' ability to create and modify institutional arrangements tailored to their needs
Extent of other experience at the local or regional level with self-governance and service provision
Distribution (particularly asymmetries) of national-level political influence among basin stakeholders
Characteristics of the water rights system in the country which facilitate or hinder basin management efforts
Adequate time of basin-level institutions for implementation and adaptation of basin management activities
The internal configuration of basin-level institutional arrangements
Presence of basin level governance institutions
Extent of clarity of institutional boundaries and their match with basin boundaries
Recognition of sub-basin communities of interest by the basin-level institutional arrangements
Availability of forum for information sharing and communication
Availability of forum for conflict resolution

Source: Blomquist et al. (2005a).

The above are only but a few of many examples of institutional concepts used in different studies on integrated water management. All in all, previous research on integrated water management shows that the study of institutions is of special significance, since the most critical element of integrated management is the need for coordination among various human efforts to use the available water. Coordination is needed between administrative levels and geographical units as well as between different sectors (Bandaragoda, 2000). Bearing this in mind, institutions are suitable vehicles of research, since they serve as instruments for human cooperation and for reducing uncertainty by establishing a stable structure for human interaction (North, 1990). Aspects of human interaction and their links to institutions will, in fact, be one of the aspects gaining attention in this dissertation, especially in the context of investigations within specific water basins.

3.3 Integrated water management through institutional resource regime theory

A differentiated approach for assessing the achievement of integration in water management was adopted in the context of recent European research from a social science and institutional perspective. Namely, a study on "European water regimes and the notion of sustainable status" (in short Euwareness project, see Kissling-Näf & Kuks (2004a); Bressers & Kuks (2004a)), was based on a framework of institutional resource regime theory initially developed by Knoepfel et al. (2001; 2003).

Knoepfel et al. (2001; 2003) studied institutional regimes, which promote sustainability in the use of natural resources (focusing on soil, forest and water in Switzerland). Their framework of institutional resource regimes was further modified in Kissling-Näf & Kuks (2004a) and Bressers & Kuks (2004a) and was further applied to the evaluation of water regimes. In short, the institutional-resource-regime framework of Knoepfel et al. (2001; 2003) was extended to include an elaborate model of public governance and a set of contextual institutional conditions which, together with external change stimulating factors, influence regime changes. In the following, the development of this framework based on institutional resource regime theory is presented in detail, since it serves as key starting point for the analytical framework adopted for this dissertation in chapter 4.

3.3.1 Defining institutions through the concept of institutional resource regimes

Knoepfel et al. (2001) defined an *institutional resource regime* as an institutional framework which combines property rights for the goods and services provided by a natural resource (= regulative system) with public policies for resource use and protection (= public policy design).

This specific concept of institutional resource regimes aims at integrating knowledge from institutional economics into classical policy analysis (Knoepfel et al., 2001). The resulting theoretical framework for institutional regime analysis, thus, combined property rights theory and institutional rational choice theory with approaches from policy analysis (in particular policy design theory).

Kissling-Näf & Kuks (2004b) made use of this concept of institutional resource regime in order to specify institutional regimes for water resources. In this context, the institutional components of a water resource regime were defined as the public policy design and property rights systems, which have been developed in different ways through the years to manage conflicting water uses and to guide these uses in a sustainable way. Also, experience in OECD countries suggested that water use rivalries can best be resolved through a body of laws that establish a system of water management and clear principles to govern the allocation of rights to use water for different purposes (OECD, 1983).

In this context, public policy design theory concentrates on the effects of policies and property rights theory on the effects of bundles of rights on the management of natural resources. In the case of water resources, the regime component of public policy focuses on the management of water from a public domain (although in interaction with private actors). The regime component of property rights focuses on the accessibility of water in a broader sense, including the private domain, the domain of collective property as well as the domain of "no property" (Bressers & Kuks, 2004a).

The central assumption of this concept of institutional resource regimes is that the dimensions of public policy design and property rights are complementary and both must be considered to achieve sustainable resource management (Kissling-Näf & Kuks, 2004b). Consideration of the distribution of property rights alone is not sufficient for the analysis of the institutional framework for resource use. In most cases, there are several public policies which regulate the use of a resource and which, as a result of insufficient coordination, can result in the degradation of that resource. In fact, one should consider the influences of all other public policies on a specific commodity or the entire resource (Bressers & Kuks, 2004b). All in all, resource regimes are considered as social institutions, in which the public and private domains interact with each other. Especially, the property rights component emphasises that public policies need to take into account the property rights of individual persons or self-regulating groups in society, which can be a restraint but also an opportunity for more sustainable resource management (Kuks, 2004c).

Additionally, in the context of this institutional resource regime theory, institutions are not considered as just a framework within which actions take place, but as the result and an integral part of the political process. Institutions are both the result of past actions and the framework within which new activities take place. Institutions, and hence institutional resource regimes, can change over time and become increasingly differentiated. Therefore, the evolution of the institutional components of the resource regime is at the centre of attention (Bressers & Kuks, 2004b).

Another scholar who worked with the concept of resource regimes earlier on was Young Oran. According to Young, resource regimes are a special case of social institutions or practices that serve to order the actions or patterns of behaviour of those interested in the use of natural resources (Young, 1982). Interestingly, Young's interpretation of resource regimes was also quite focused on the dimension of property rights, but only implicitly and to a minor extent on the dimension of public policy (with main reference made to policy instruments). He argued that, in the substantive core of every resource regime is a structure of rights and rules and that, property rights (ownership, use) constitute the most important category of rights incorporated in resource regimes. Rules are well-defined guides to actions or standards setting forth actions that members of some specified subject groups are expected to perform (or to refrain from performing) under appropriate circumstances. Rules may, among others, be the work of designated legislative bodies or actors possessing sufficient power to impose rules on subjects, who gradually come to acknowledge the authoritativeness of these rules with the passage of time. There can be rules to order relations between holders of various rights and other actors subject to a regime as well as use rules and liability rules. According to Young, except for the substantive core (rights and rules), there is a procedural component to resource regimes. Problems of social choice arise, when there are conflicts of interest among actors interested in a resource. Social choice has much to do with the way allocation of rights is done (e.g. principle of "first come, first served" or issuing permits and licenses by the administration). In this context, also policy instruments pertain to the operation of resource regimes as soon as some administrative apparatus is in place (e.g. changes in bundles of exclusive rights, the promulgation of restrictive regulations or decisions on individual permits) (Young, 1982).

3.3.2 Components of institutional resource regimes

It was mentioned above that the institutional components of resource regimes (including water regimes) are public policy and property rights. These components are presented in more detail in the following, with specific emphasis on water resource regimes.

3.3.2.1 Public policy and public governance

In the Euwareness project on "European water regimes and the notion of a sustainable status" (Kissling-Näf & Kuks, 2004a; Bressers & Kuks, 2004a), both a narrower and a broader concept of public policy were adopted to assess the status and evolution of water resource regimes. The narrow concept considered mainly the evolution of public policy, analysing how and why public policies intervene in the use of water in order to reduce policy problems reflected in use rivalries. The narrow concept focuses, thereby, on the traditional elements of public policy, such as policy aims, instruments, target groups and implementation, and considers policy as it has been adopted, i.e. in terms of policy assumptions and planned policy interventions (Kissling-Näf & Kuks, 2004b). The broader concept deals with public policy in terms of public governance as additional to the traditional elements of public policy, thereby, replacing the initial regime component of public policy design used by Knoepfel et al. (2001).

The interpretation and mode of use of the term "governance" has been numerous in recent literature (also in the field of water resource management). Kooiman (2002) explains that the growth of the governance approach can be due to several factors:

- A growing awareness that governments are not the only crucial actor in addressing major societal issues.

- Traditional and new modes of government-society interactions are needed to tackle these issues.

- Governing arrangements and mechanisms will differ for levels of society and will vary by sector.

- Many governance issues are interdependent and/or become linked.

Kooiman himself defines "social-political" or "interactive" governance as "all those interactive arrangements in which public as well as private actors participate, aimed at solving societal problems, or creating societal opportunities, attending to the institutions within which these governance activities take place and the stimulation of normative debates on the principles underlying all governance activities".

On the European water policy level, the Water Framework Directive also promotes the principle of "interactive" and "participatory" governance for water management and planning.

The interpretation of public governance presented here as a component of institutional resource regimes, is based on a governance model which Bressers & Kuks (2003) developed through a synthesis of various approaches in studies of public administration and policy science, including (among others) the rule-based institutional rational choice approach[9] (Ostrom, 1990; 1999), the advocacy coalition framework[10] (Sabatier & Jenkins-Smith, 1999), the flow model of policy process (Zahariadis, 1999) and actor-centered institutionalism (Scharpf, 1997). From all these theoretical approaches, different aspects were used to enrich the meaning of five proposed structural elements of public governance:

- Levels and scales of governance.
- Actors in the policy network.
- Perspectives and objectives.
- Strategy and instruments.
- Responsibilities and resources for policy implementation.

Table 3.2 presents questions to help understand and operationalise these five elements of the governance model for water regimes, as used in the Euwareness project (see Kuks (2004b); Bressers et al. (2004); Bressers & Kuks (2003)).

---

[9] Institutional rational choice focuses on how institutional rules alter the behaviour of intendedly rational individuals motivated by material self-interest.

[10] The advocacy coalition framework focuses on the interaction of advocacy coalitions – each consisting of actors from a variety of institutions who share a set of policy beliefs – within a policy subsystem. Policy change, therein, is a function of both competition within the policy subsystem and events outside the subsystem.

Table 3.2   The public governance system of a water resource regime

Governance elements	Questions for a national water regime	Questions for a water basin regime
Levels and scales of governance (Key question: Where?)	Who were the policy-makers and how does the policy initiative fit into separate administrative levels? Which ministries at national level are involved? Is the initiative coordinated or fragmented? What discretion does the national policy leave to the lower administrative levels?	Which levels of governance dominate policy and in which relations? What is the relation with the administrative levels of governance? Who decides or influences such issues? Do administrative levels of management fit with the natural scale of water basins? How is the interaction between the various administrative levels arranged?
Actors in the policy network (Key question: Who?)	Does the policy initiative recognise the interests of various non-public actors? Which water users, stakeholders and/or advocacy coalitions participated? Is the policy based on judgements of an expert community? Is there increasing interrelatedness of water policy making with other policy sectors (land use planning, environmental management, nature conservation)?	How open is the policy arena in theory and practice, and to whom? Who is actually involved and with what exactly? What is the structural inclination to cooperate among actors in the networks? What is the position of the general public versus experts versus politicians versus implementers? Is there increasing interrelatedness of water policy making with other policy sectors (land use planning, environmental management, nature conservation)?
Perspectives and objectives (Key question: What?)	Which policy objectives are accepted? What is seen as a problem and how serious is it considered to be? What are seen as the causes of the problem? To what degree is uncertainty accepted?	What is seen as a problem and how serious is it considered to be? What are seen as the causes of the problem? Is the problem seen as a relatively challenging topic or as a management topic without much political salience? To what degree is uncertainty accepted? What relations with other policy fields are recognised as coordination topics? Which policy objectives are accepted?
Strategy and instruments (Key question: How?)	Which instruments belong to the policy strategy? How much flexibility do they provide? Are multiple and indirect routes of action used?	Which instruments belong to the policy strategy? How much flexibility do they provide, e.g. for redistributing use rights? Are multiple and indirect routes of action used? Are changes in the ownership and use rights in the sector anticipated? Do they provide incentives to "learn"?
Responsibilities and resources for implementation (Key question: With what?)	Which organisations including government are responsible for implementing the policy? What level of authority and resources are made available by the policy? How complex is the institutional arrangement for implementation?	Which actors including government are responsible for implementing the policy? What is the repertoire of standard reactions to challenges known to these organisations? What authority and resources are made available to these actors by the policy? With what restrictions? How complex is the institutional arrangement for implementation?

Source: Compiled from Kuks (2004b); Bressers & Kuks (2003); Bressers et al. (2004).

3.3.2.2 *Property rights*

Property rights determine the accessibility of a resource to various users and are subject to a sub-set of certain regulations and rules. According to Libecap (1993), they are the social institutions that define or delimit the range of privileges granted to individuals to specific assets, such as water. Bromley (1991) defined a right as the capacity to call upon the collective to stand behind one's claim to a benefit stream. Thus, rights only have effect when there is some authority system that agrees to defend a rights holder's interest in a particular outcome. In other words, rights can only exist when there is a social mechanism that gives duties and binds individuals to those duties. Rights are not relationships between oneself and an object, but between oneself and others with respect to that object. Bromley (1991) even referred to property rights as policy instruments. The relevant issue, where policy usually comes in, is precisely one of who will get rights and who will have the effective protection of the State to do as they wish.

Kissing-Näf & Kuks (2004b) identified four traditional types of property for natural resources in the literature: no property, common property, State property and private property. Private property rights (ownership and/or use rights) are in the hands of one individual or corporation excluding claiming of these rights by another individual. The case of no property is a classical case of resources, for which access is not formally regulated (but usually, the State protects these by means of public law). Common property can also be described as the groups' private property (Ostrom, 1990; Bromley, 1991; Libecap, 1993). State property means that the State or a public authority has private property rights. The difference with other private property rights is that the State has to use these rights in the public interest (Kuks, 2004c).

If a natural resource lies in the public domain (with or without State property), it does not really matter if there is formal State ownership. The public domain is characterised by the fact that use rights to a resource can only be regulated by the State (Kissling-Näf & Kuks, 2004b).

In the empirical analysis of this dissertation, property rights are discussed with regard to two broad different sources of water rights: water ownership rights and rights to use water.[11] Ownership and use rights can both be in one hand, or they can be separated. Ownership of water resources, especially of groundwater, is often related to the ownership of land,[12] but the right to use water is usually in the public domain.

---

[11] There are also other approaches to distinguish between property rights on natural resources. Schlager & Ostrom (1992), who did much empirical research on inshore fisheries, suggested differentiating between different types of property rights according to the attributes of a resource that the owner of "property" (e.g. water) holds rights to. According to this approach, rights categories to a common-pool resource range from access and withdrawal rights to management rights, exclusion rights (giving power to define who may access resources) and alienation rights (giving the right to sell or lease a resource due to ownership). Depending on the kind of rights one holds, right-holders are distinguished into authorised users (which do not have much to say on the definition of rules and governance structures, thus are not very inclined to sustainable resource use), claimants (who also hold management rights, can make rules on who can have withdrawal rights and how to use the resource and have interest in working governance structures) and proprietors (who also have exclusion rights and can determine who has access rights). These concepts were used in the case of inshore fisheries by Schlager & Ostrom (1992), but they were also proposed as potentially useful in the analysis of other common-pool resources including irrigation systems and groundwater basins.

[12] The ownership of major water systems (coastal waters, deltas, major lakes, major rivers and deep aquifers) is often not related to land ownership (Kuks, 2004c).

The system of property rights exercises important influence on the use of water resources since it determines who, when and in what form has access to the resource. Experience shows that existing water rights are often a main constraint and a source of many problems in the optimization and introduction of integrated water resources management (van Hofwegen, 2001). Property rights to water, which are undefined, unclear or of mixed nature can be a potential source of water conflicts.

According to Elinor Ostrom, property rights matter in water resource management also when they are rooted in customary traditions of water distribution and use. She thereby emphasised the importance of regional and local institutional arrangements, which are often based on long traditions of informal but commonly shared water rights (Ostrom, 1990). In a similar direction, Schlager & Ostrom (1992) emphasised the important distinction between *de facto* property rights, which are rights originating among resource users and are not yet recognised by authorities, and *de jure* property rights which are granted to users by authorities. *De facto* property rights are especially relevant in cases of water uses which are not registered with the authorities, such as water rights based on custom or tradition. *De facto* rights can be less secure than *de jure* rights, if challenged (*de jure* rights have the protection of a court of law, while *de facto* rights most probably don't).

In water use practice, there may be conflict because of different claims on a limited available water reserve or because of externalities, which are external effects accompanying a water use and affecting another use or user in an undesired way. One way to solve conflicts is by the redistribution of the property rights in question. In the case of conflicts due to surface water pollution, polluters (especially point source polluters) are regulated by discharge permits or effluent charges as well as connections to wastewater treatment plants. Thus, this is a type of conflict that is resolved more by means of policy regulation rather than the redistribution of property rights. Indeed, EU legislation has dealt extensively with the problem of water pollution by means of public policy regulation. In the case of conflict over groundwater pollution (e.g. of diffuse agricultural pollution versus drinking water production), not only policy regulation is relevant, but also the redistribution of property rights (e.g. by compensating farmers to move farming practices elsewhere). In cases of conflicts over water scarcity, different users claim water quantities from a finite resource. This is a typical case of conflict between property rights holders, which could be managed through the redistribution of property rights (Kuks, 2004c). Bressers & Kuks (2004c) also concluded that protection of water resources by re-distribution of property rights occurs mostly in cases of water scarcity than in cases of water quality degradation, where property rights are often restricted and give way to public governance in order to improve the sustainability of the resource use. As concerns water scarcity conflicts in the context of EU water policy, the WFD is the first which deals with the problem of scarcity, even if only in a brief way. The WFD refers to the direct regulation of water use rights from a quantitative perspective in Article 11(3e), according to which Member States should (as part of their programme of basic measures for each river basin) ensure balance between abstractions and recharge of groundwater. Member States should establish a regime of control and obligatory authorisations for the abstraction and impoundment of surface and groundwater. These requirements, along with the principles of cost-recovery and water pricing, recognise the quantitative dimension of water resource protection (Blöch, 2001) and provide some basis for balancing quantitative water claims.

3.3.3 Changing institutional resource regimes and defining an integrated phase

3.3.3.1 Classification of institutional resource regimes

Institutions are both the result of past actions and the framework within which new activities take place. Institutions, thus, can change over time. Against this background, Knoepfel et al. (2001) made the heuristic assumption that an institutional resource regime, i.e. the system of property rights and public policies, also becomes gradually differentiated over time.[13] This process of differentiation and change can be observed by using a historic approach, thereby assessing the diachronic development of an institutional resource regime on the basis of two core variables: the extent and the coherence of the resource regime. In short, the extent indicates the scope of resource uses recognised by the regime. The coherence reflects the degree of consistency and coordination within and between public policies and property rights. The change variables of extent and coherence are described in more detail below. The change of the extent and coherence in time (from low to high) determines the transitions of an institutional resource regime through different phases, as expressed by Knoepfel et al. (2003) in the following regime classification: non-existent regime → simple regime → complex regime → integrated regime (see Table 3.3). The classification of Knoepfel et al. emphasises that high extent combined with high coherence lead to an integrated phase of the institutional resource regime. At the integrated phase, the institutional resource regime can take account of varied heterogeneous demands in a coordinated manner. Such a regime could regulate all rival use demands in such a way, that it is possible to maintain sustainably the capacity of the resource in question for the production of all these goods and services.

Table 3.3   Classification scheme of institutional resource regimes

Phase of institutional resource regime	Regime extent	Coherence of regime elements
Non-existent regime	Low	Low
Simple regime	Low	High
Complex regime	High	Low
Integrated regime	High	High

Source: Based on Knoepfel et al. (2001).

Having adopted the classification of resource regimes of Knoepfel et al. (2001), Bressers & Kuks (2004b) used the term "integration" as a label for high extent and high coherence of resource regimes. As the extent of a regime increases, the regime can either move to the direction of more fragmentation (complex regime) or more coordination within and between its elements (integrated regime). A resource regime becomes more complex, when it is

---

[13] Also, Young (1982), who worked earlier on the concept of resource regimes, argued that regimes undergo continuous transformation, in response to their own dynamics as well as to changes in their political, economic and social environment.

characterised by multiple formats in its elements. For instance, in terms of the public governance system, a resource regime becomes more complex when more levels and scales, more actors, more perspectives, more instruments and more implementing organizations become involved in resource management. The most eminent feature of increased complexity is the gradual increase of the extent of a regime, usually accompanied by an increase in relevant property rights. Complexity as such is not wrong. Most of the time, growing complexity is an answer to real needs and developments in natural resource management. However, in the complex phase, the elements of the resource regime remain fragmented, as opposed to an integrated phase where the regime elements are better coordinated with one another.

In the following, the regime change variables of extent and coherence are described with specific reference to water resources, in order to help conceptualise the integrated phase of an institutional water resource regime.

3.3.3.2 Extent as a regime change variable

The extent of a water regime is the scope of uses and users to a water resource, which are effectively regulated by the regime. If a certain use of the water resource, e.g. fishing, is not regulated or considered by any of the regime elements, it does not belong to the extent of the regime. Regimes with insufficient extent are by definition weak as guardians of sustainable water use, since several water uses remain unregulated (Bressers et al., 2004).

Knoepfel et al. (2001) had recommended the use of the term "relative extent", referring to the resource goods and services explicitly regulated by the regime in relation to the actually demanded goods and services. When referring to explicit regulations, one is not interested in regulations which simply exist on a higher level, but in regulations which are also actually applied in the area of interest. Explicit regulations also vary according to the degree of their accuracy. Often they involve new goods and services but remain vague in their descriptions (e.g. requirement to consider the needs of the environment, as opposed to regulations with very precise requirements such as for specific rest water quantities).

A change in the extent usually means that more uses or use functions are incorporated in the regime. A larger extent makes the regime more meaningful for the use of the water resource. However, if the incorporation of additional users or uses takes place by means of new separate property rights and/or public governance aspects, this might lead to a decline of the regime coherence. This is how simple regimes evolve into complex and fragmented ones (Kuks, 2004c). Knoepfel et al. (2003) also argued that a larger regime extent is linked to a more integrated resource regime, only when the extent includes protection policies for relevant ecological goods and services and when also high regime coherence is secured.

3.3.3.3 Coherence as a regime change variable

Coherence is understood as the degree of consistency and coordination within and between the property rights and the public governance systems of a water regime. The dimension of "coherence" is particularly important as an aspect of integration which is additional to an increased regime extent. A water resource regime becomes more integrated on a basin level, when it becomes more coherent within the basin boundaries.

Three forms of coherence are discerned in an institutional resource regime (also for water):

- Internal coherence within the public governance system of the regime.
- Internal coherence within the property rights system of the regime.
- External coherence between the public governance and the property rights systems.

As previously mentioned, in the initial institutional-resource-regime framework of Knoepfel et al. (2001), the public governance system was referred to as public policy system. In that context, the internal coherence of the public policy system was mainly interpreted as coordination of the actors of the main public protection- and use-policies (intra- and inter-policy coordination). According to Knoepfel et al. (2003), incoherence in the public policy system could occur, when resource use limitations as formulated by public policy are unclear, because they are contradicting under the command of different authorities (e.g. permit to build and right to wetland protection conceded in the same location). This type of incoherence is linked to uncoordinated policy products and the lack of horizontal or vertical coordination.

In the modified institutional-resource-regime framework, where the public policy system was replaced by the public governance system, coherence within public governance of an institutional resource regime can be achieved when (Bressers et al., 2004):

- The levels of governance dealing with the same resource are recognized as mutually dependent and influencing each others' effects.
- There is a substantial degree of interaction and coordination between the actors involved in the policy network.
- The various objectives of the different users are analysed in one framework, so that deliberate choices can be made if and when goals are conflicting.
- The policy contains instruments mutually reinforcing each others incentives.
- Responsibilities and resources of various persons or organizations which contribute to the implementation of the policy are consistent and coordinated.

As concerns coherence in the property rights system, more and more use rights are established over time with respect to the same resource (e.g. the same water body), and often perhaps at the cost of other existing use rights. As a result, the property rights system of a resource regime develops towards more complexity and the need for coordination increases. According to Knoepfel et al. (2003), incoherent property rights systems may occur, when property rights regulation does not allow a clear allocation of specific goods and services to specific actors. Especially in the case of water abstractions, several individual abstractions can be allowed in the same place and time, so that ultimately the total abstracted water exceeds the available water volume. Such incoherence in the property rights system may often be the result of use limitations which are unclearly formulated by public policy, as in the case of permissions to abstract water without defining the quantity to be abstracted. The internal coherence of property rights may improve, when the increasing use rights (or ownership rights) are restricted over time, or rights are re-distributed, so as to protect other use rights and reduce rivalries (Kissling-Näf & Kuks, 2004b).

In practice, developments in the internal coherence of the property rights system are often closely interlinked to developments in the public governance system. This is because, as water regimes evolve, property rights are increasingly restricted or re-distributed by means of public governance elements (e.g. via policy instruments or even via non-policy-based conflict resolution processes).

Finally, the external coherence between the public governance and the property rights systems is interpreted, in the first place, as "match" between the actors targeted by the public governance system (e.g. by policy instruments) and the actors which hold relevant property rights (legitimate titleholders). A key question is whether the holders of property rights are also the target group of behavioural policy incentives (Knoepfel et al., 2001). For a more coherent regime, it should be ensured that actors obliged or entitled to use a certain resource by public policies are in agreement with those actors holding titles or rights according to property rights basic regulations. In the second place, the external coherence involves a „match" of the goods and services involved in both systems of public governance and property rights (Bressers et al., 2004).

3.3.3.4 Relevance of integrated institutional water regimes to the EU WFD

In this section, it is shown that high regime extent and high regime coherence, which make up a theory-based integrated institutional water resource regime, also largely relate to the more practical elements of integral water management in the EU WFD (see Table 3.4).

The high extent of integrated institutional water regimes is reflected in the WFD requirement for the combined management of all water resources in a basin considering quality and (partially) quantity aspects as well as the environmental needs of water ecosystems.

The WFD requirements are also relevant to the high coherence dimension of a theory-based integrated water regime, especially in terms of the internal coherence in the public governance system.

Firstly, coherence in the governance levels and scales are reflected in the WFD requirement for coordinated administrative water management arrangements to apply the WFD within each river basin district (RBD). This requirement also promotes the coordination of decisions on a RBD level, thus surpassing administrative boundaries and even national borders in case of international RBDs. The requirement to define a competent authority for the WFD implementation in each RBD is, furthermore, equivalent to a requirement for more coordination of the responsibilities for water policy implementation.

Secondly, coherence in terms of actors in the policy network of the public governance system is promoted by the WFD requirement for public participation in the formulation and revision of the RBD management plans.

Thirdly, increased coherence in the perspectives and objectives in the public governance system is reflected by the WFD requirement for the coordinated and participatory formulation of RBD management plans and by the development of common knowledge through monitoring and the analysis of the status of water systems. Especially, the formulation of a single management plan for each RBD, having the achievement of good water status as common vision, will be catalytic to the integration of different perspectives and objectives of the water stakeholders involved (Aubin & Varone, 2004a). Additional coherence in perspectives and objectives can be achieved by the implicit requirement of the WFD for policy integration between the water policy field and other sectoral policies.

Fourthly, the WFD promotes coherence in terms of strategies and instruments by requiring the adoption of a programme of measures, which is the main WFD instrument for reaching its environmental objectives. The WFD also explicitly considers the whole set of EU water legislation in a single picture and requires better coordination of water legislative instruments, which had so far been considered only in a fragmented way. Additionally, its requirement to get the prices right by full cost-recovery pricing supports further coherence in strategy and instruments.

In fact, the WFD promotes the combination of traditional legal-regulatory instruments (known also as command-and-control approaches) with more procedural steering approaches[14] (Moss, 2003). One of the most obvious command-and-control aspects of the WFD is its requirement for combating water pollution from individual pollutants or groups of pollutants. According to Article 16 of the WFD, cessation or phasing-out of discharges, emissions and losses of priority hazardous substances is foreseen. The respective requirements are to be met mainly through emissions-oriented standards (Moss, 2003). The WFD also goes beyond such classical command-and-control approaches and includes a series of control mechanisms to ensure the achievement of its environmental objectives. Its control mechanisms are not oriented only at compliance with certain standards but mainly at compliance with strict procedural rules. For instance, the WFD sets a demanding timeplan for implementation, with deadlines for the preparation of the analysis of water status, of the RBD management plans, for the implementation of the programme of measures and the achievement of the environmental objectives. To control compliance with this time-plan, the WFD included several reporting requirements to the EC. The WFD is rules-oriented also in other aspects. Several of its articles and annexes include concrete and detailed instructions for the implementation of its requirements. For instance, the content of the RBD management plans and the process for their formulation are defined in a uniform way. If requirements are not met to the full and according to the timeplan, penalties are possible according to Article 23. Another strict rule of the WFD implementation concerns the effect of measures on water status. Contrary to earlier EU directives, compliance with the WFD is not

---

[14] Three steering models are often debated: the regulatory model based on command-and-control instruments, the economic steering model based on financial incentives, and the communicative steering model based on knowledge and opinion transfer.

evaluated only on the basis of the adoption of requirements in national law and the production of relevant programmes and plans, but mainly on the basis of achieving the environmental objectives in each RBD (Moss, 2003). The more procedural steering approaches of the WFD, which complement the above command-and-control aspects, include the administrative coordination requirement within RBDs, the public participation requirement, the requirement for transparency in implementation (access of the EC and the broader public to the water status analysis reports, the management plans and the programmes of measures), the flexibility of the WFD with regard to regional peculiarities (derogations from the environmental objectives and deadlines in exceptional cases, varied environmental objectives and reference conditions according to different water types, definition of heavily modified and artificial water bodies to account for economically necessary hydromorphological changes) as well as the applied principle of cost-efficiency in the selection of measures (Moss, 2003).

As concerns the theory-based integration dimensions of internal coherence within property rights and of external coherence between property rights and public governance, these seem to be only partially reflected in the WFD. According to Bressers & Kuks (2004b), only the WFD requirement for "getting the prices right" (polluter pays principle, full cost-recovery in water pricing, principle of affordability) appeals to an improvement of the external coherence between property rights and public governance in the regime. Other options for the improvement of external coherence (especially, through the redistribution of property rights by attributing new titles to new users at the cost of other users, adopting new titles for the expropriation of ownership, limiting use rights or transferring private property rights to the public domain) are not considered by the WFD. Only the WFD requirement for control and obligatory licensing of water abstractions and impoundments can have some additional impact on the external regime coherence and also on the internal coherence of property rights.

All in all, the incomplete reflection of internal property rights coherence and external regime coherence in the WFD can be in part related to the limited consideration of water quantity issues by the Directive. Indeed, water quantity issues ask for even more fundamental rethink about water management (than the WFD offers), because management of these issues involves an "intrusion" into systems of property rights to water (law-based or traditional) and tighter coordination with other resource rights, especially land use rights which are regulated by spatial planning instruments (see also van der Brugge & Rotmans, 2007).

The above overview shows that the WFD requirements for integral water management are mainly equivalent to the theory-based integration dimensions of high regime extent and of high internal coherence within public governance. The additional theory-based integration dimensions of regime coherence within property rights and between property rights and public governance have been proposed as a worthwhile additional focus of attempts at more integrated water management in policy and in practice. This is especially -but not exclusively- the case, when water quantity problems are at stake (Bressers & Kuks, 2004c).

31

Table 3.4   Relevance of integrated institutional water regimes to the EU WFD

Theory-based elements of integrated water regimes	Elements of the WFD for integrated water management
High extent	Considering water as a resource in its entirety (groundwater, surface water, water quality, water quantity, environmental needs of water ecosystems)
High coherence of public governance	
• Levels and scales	Coordination of administrative arrangements for water management within river basin districts
• Actors in the policy network	Participatory approach (public consultation and information in the formulation of river basin district management plans)
• Perspectives and objectives	Development of river basin district management plans
	Development of knowledge on water resources status (also on the basis of monitoring)
	Coordination of water policy with other sectoral policies
• Strategies and instruments	Programme of measures
	Better coordination of water legislative instruments
	Getting the prices right by full cost-recovery pricing
	Combination of command-and-control instruments with more procedural steering approaches
• Responsibilities and resources for implementation	Definition of a competent authority for the WFD implementation in each RBD
High external coherence between property rights and public governance	Partially reflected in requirement for "getting the prices right" (polluter pays principle, full cost-recovery in water pricing, principle of affordability)
	Potentially linked to WFD requirement for obligatory licensing of abstractions and impoundment
High coherence of property rights	Potentially linked to WFD requirement for obligatory licensing of abstractions and impoundment
	Affected by requirement for "getting the prices right"

3.3.4   Causal factors for change in institutional resource regimes

Firstly, Knoepfel et al. (2001) argued that changes of institutional resource regimes can be set in motion by so-called external change triggers. Such change triggers may be external shocks like disasters (e.g. floods), new risks (e.g. climate change, industrial accidents, loss of biodiversity), socioeconomic and technical changes (e.g. decline of agricultural and industrial development), new political paradigms linked to changes in the public sector (e.g. globalisation and liberalisation trends), resource problem pressure (e.g. decrease of water quality), collective learning processes and the adoption of agreed solutions (e.g. the harmonisation of agricultural policy on a European level) or new policy actor constellations which change existing coalitions (e.g. change of party in government).

In this dissertation, the "external change triggers" of Knoepfel et al. (2001) have been renamed to "external change impetus". "Impetus" is considered a more appropriate term to reflect the kind of external disturbances which can stimulate a regime to change. In specific, external disturbances can be both sudden events (like an industrial accident causing river pollution), in which case the term "trigger" would be as suitable as "impetus", but also gradually increasing disturbance (like gradually increasing water pollution), in which case the term "impetus" is more appropriate than "trigger".

Secondly, Bressers et al. (2004) argued that, while regimes usually evolve towards more complexity under the influence of external change impetus, changes of a regime towards a more integrated phase can also depend on the presence of favourable contextual conditions. These contextual conditions were inspired by research work relevant to the development of negotiated or consensual approaches to environmental policy. Thus, Bressers et al. (2004) added their "favourable contextual conditions" to the initial framework of Knoepfel et al. (2001; 2003), as a new set of factors to explain regime change. This modified framework was then used in the comparative research of Bressers & Kuks (2004b) on institutional water regimes on a basin level.[15] In his comparative work on national institutional water regimes, Kuks (2004c) used an alternative set of contextual conditions (called "institutional context"), which were based on key mechanisms decisive for change in water policy and governance. Similarly, in the analytical framework adopted for use in this dissertation, the contextual conditions influencing regime change are defined separately for the national and the water basin level of research (see chapter 4/4.1.1 and 4/4.1.2).

All in all, the key factors causing and explaining regime change, which have been introduced in this section, are illustrated in Figure 3.1. This figure is the graphical representation of the baseline framework for change in institutional resource regimes according to Bressers et al. (2004), rooted in the initial work of Knoepfel et al. (2001; 2003).

Figure 3.1   Baseline framework for change in institutional resource regimes

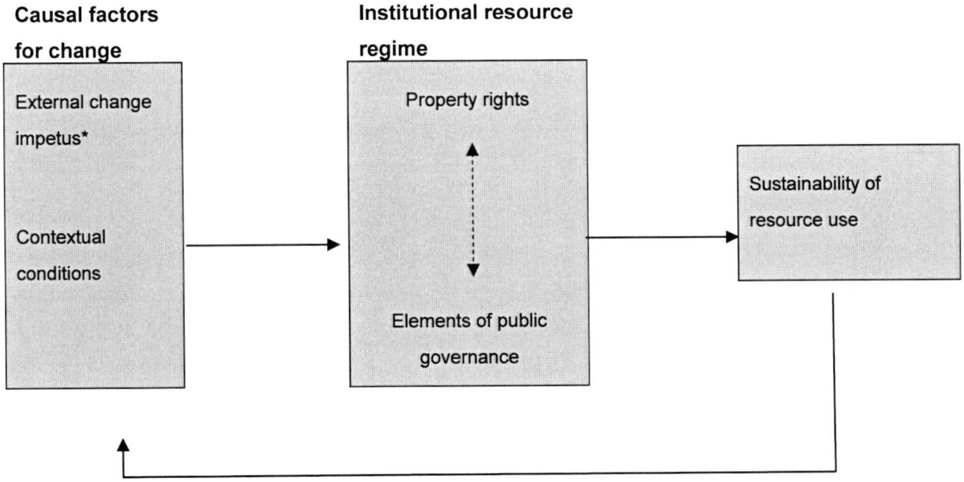

Source: Bressers et al. (2004).

(*): "Impetus" replaced here the term "triggers" or "agents" used by Knoepfel et al. (2001; 2003) and Bressers et al. (2004).

---

[15]   In that comparative research, which was based on information from 24 case studies in 12 European water basins, the presence of less favourable conditions correlated positively with smaller water resource regime changes towards more integration.

3.3.5 Sustainability implications of institutional resource regime changes

Natural resources produce numerous goods and services which are claimed by rival users and this often leads to overuse of the resource. Figure 3.1 shows that institutional resource regime changes may have implications on the sustainability of resource use. Knoepfel et al. (2001) argued that the sustainability of natural resource management significantly depends on the respective institutional resource regime and suggested that more integrated regimes can perform better in terms of sustainability of the resource use.

In the case of water resources, the expectation that more integrated institutional resource regimes perform better in terms of sustainability is part of water management policy ideology on the EU level (especially in the WFD) and in many Member States (Bressers et al., 2004). Based on their comparative empirical research on water regimes in different European basins, Bressers & Kuks (2004c) concluded that, indeed, institutional water regime changes towards a more integrated phase (especially towards a more coordinated public governance system) showed positive correlation with more sustainability implications in water use. They emphasised, however, that this is a probabilistic and not an absolute correlation, leaving the probability explicitly open, that any observed form of integration in the regime may not always lead to more sustainable use but may affect only trivial aspects of water use. Under certain circumstances, a more integrated regime may even lead to more unsustainable water use, for instance, if regime changes are use-driven instead of resource protection-driven.

3.3.6 Studying institutional resource regimes on different levels

As mentioned in section 3.3, Knoepfel et al. (2001; 2003) used the concept of institutional resource regimes to analyse the management regimes of forest, soil and water resources in Switzerland. Their analysis took place on two levels: a) on the national level using an institutional historical approach and b) on a regional level by means of case studies on the use of forest, soil and water resources. Similarly, in the later Euwareness project on "European water regimes and the notion of sustainable status", research took place on water regimes on the national level of different countries and also on the regional level in different water basins. On the national level, the centre of attention was the link between factors causing regime change and the actual changes in the national water regimes (Kissling-Näf & Kuks, 2004a; Kuks, 2004c). Research on the level of water basins could additionally focus on the link between water basin regime changes and their sustainability implications in real-life water use (Bressers & Kuks, 2004a).

4 Research design

The basic elements of any research design are the following: 1) statement of purpose, 2) statement of hypothesis, 3) specification of variables employed for the analysis, 4) explanation on how to operationalise and measure each variable, 5) explanation on how to organise and conduct observations and 6) discussion of how data are analysed (Manheim & Rich, 1995).

The statement of purpose of this dissertation was already presented in chapter 1 (section 1.4). Research hypotheses are formulated in section 4.2, following the elaboration of the adopted analytical framework. The specification and operationalisation of variables to be employed in the analysis is included in the presentation of the analytical framework in sections 4.1.1 and 4.1.2. Finally, section 4.3 deals with methodological issues and, thus, includes an explanation of the organisation and conduct of observations as well as a discussion of how data will be analysed.

4.1 Theory-based analytical framework

As already stated, the main purpose of this dissertation is, in short, to assess whether, to what extent and why Greek water management has experienced any changes towards a more integrated phase to date. To pursue this research purpose, chapter 2 argued for the need to adopt an analytical framework with explanatory power, which would involve causal chains of determinants of integrated water management. After presenting selected analytical approaches for assessing the achievement of integrated water management in chapter 3, this chapter presents an analytical framework to guide data research and to explain the empirical findings of this dissertation.

First of all, institutions are a central element in the proposed framework. Up until now, despite their importance, institutions have not been considered in research literature as critical for the explanation of water problems in Greece. For this reason, the institutions-related analytical framework adopted in the following is especially innovative for the case of Greece.

The adopted analytical framework is based largely on the institutional resource regime framework, which is rooted in the work of Knoepfel et al. (2001; 2003) and was presented in chapter 3/3.3. This framework later merged with work on governance of Bressers & Kuks (2003) and was further elaborated and specified with, among others, contextual conditions for regime change by Bressers et al. (2004) within the European Euwareness project on "European water regimes and the notion of sustainable status" (see baseline framework in Figure 3.1). Furthermore, the contextual interaction theory of Bressers (2004) serves as the point of departure for actor-centered explanatory links which are added to the framework in this dissertation.

Applying the particular analytical framework to Greece will, on the one hand, strengthen the validity of theoretical and empirical conclusions from its previous application to other European countries. On the other hand, it will help to identify the variables whose absence could be responsible for the expected poor record of Greece to date in terms of integrated water management. The specific analytical framework was also preferred for its close conceptual links to the policy requirements of the EU WFD for integral water management. Furthermore, the framework provides an institutional analysis approach, which integrates two narrower analytical approaches from classical property rights theory and from public policy analysis. It is believed that the use of a purely property rights-based or a purely public policy-based approach could overlook certain institutional aspects, which are important for understanding the Greek water management regime.

In the following sections (4.1.1 and 4.1.2), the baseline framework for change in institutional resource regimes (see Figure 3.1) is further modified and adapted for its application to the national level and then to the basin level of research in Greece. The distinction of research and analysis between these two levels helps to reflect on the different variables which can be "at play" for achieving integrated water management in each of these levels. Moreover, this national-regional distinction may also allow useful insight into the outcome of any water management decentralisation attempts.

On the Greek national level, discussions concentrate on key changes in the national institutional water regime. On the water basin level, focus is placed on institutional water regimes developed in the context of real-life water management processes and specific water use rivalries. Additionally, implications of any regime change on the sustainability of water use within the basins are examined.

4.1.1 Analytical framework for the national level of research

The national water regime of Greece is the first to be examined in this dissertation. For the national level of analysis, it is of specific interest to examine whether recent transitions in the water regime were a shift towards a more integrated phase or not. Figure 4.1 illustrates the main components of the analytical framework adopted for the national level.

Figure 4.1   Modified framework for change in national institutional water regimes

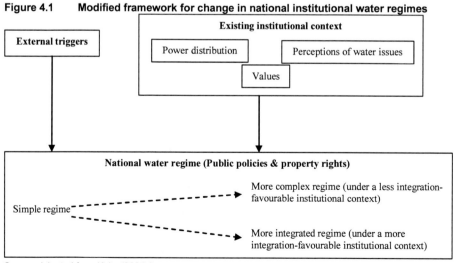

Source: Adapted from Kuks (2004c).

36

The focus of the analytical framework lies on changes of the national water regime, especially from a simple phase towards a more complex or a more integrated phase. Instead of aiming at a complete assessment of the status of the national water regime in certain points in time, the analytical framework concentrates on the assessment of change occurring in the research period covered by this dissertation (mid-19th century until approximately 2005, as far as research on the national water regime is concerned).

4.1.1.1 External change impetus and regime changes

Changes in national water regimes can be set in motion by external change impetus, which may be problem-related (external shocks in the form of calamities or gradual water resource degradation) and institutions-related (e.g. new standards by supranational institutions, federalisation processes or a new level of governance) (Kissling-Näf & Kuks, 2004b). Under the pressure of external impetus, a (simple) national water regime can evolve towards more complexity or more integration. Table 4.1 provides indicators which operationalise the regime change variables of extent and coherence for the national level of analysis.

Table 4.1   National water regime change variables and selected indicators

CHANGE VARIABLES	INDICATORS FOR THE INTEGRATED PHASE OF A NATIONAL WATER REGIME
Extent	**For high extent**
	• All uses are considered by the regime elements, including ecological values
Coherence within public policies	**For high coherence**
Levels of governance	• Administrative levels of management fit with the natural scale of water basins (authority decentralisation on the level of water basins) • Coordination within each administrative level and between different levels
Actors in the policy network	• Participatory arrangements for all actors with an interest in water resources, especially non-public actors • Open policy networks instead of closed policy communities
Perspectives and objectives	• Development of national water visions and river basin management plans incorporating multiple stakeholder perspectives • Recognition of relations between water policy and other sectoral policy fields as coordination topics • Policy objectives (in official visions and visions of stakeholders) recognise new water uses and administer justice to rivalries between different water uses
Strategy and instruments	• Adoption of integrated water legislation • Adoption of flexible, procedural instruments (e.g. planning)
Responsibilities and resources for policy implementation	• Policy implementation is equipped with sufficient and coordinated resources (financial, time, human, information, authority) • Clear institutional arrangement for implementation of policy and measures (as opposed to a fragmented arrangement of competence)
Coherence within property rights	**For high coherence**
	• Coordination, instead of contradiction, of rights of different users and owners, when complex bundles of rights develop (for competing water uses)
Coherence between public policies and property rights	**For high coherence**
	• Property rights holders recognised as target group of policy interventions, e.g. via restrictive regulations, establishment of permits for abstraction or pollution, restriction of private rights

Source: Based on Kuks (2004c).

The regime change variables and their indicators in Table 4.1 are used in the later empirical analysis of this dissertation to conclude on the extent to which the Greek national water regime has moved towards a more integrated phase or not. To reach such a conclusion, certain assumptions are made regarding the evaluation of the individual regime change variables:

- A more complete regime extent leads either to a more complex or more integrated regime phase, depending on the progress achieved in the coherence of the other regime elements.

- The three forms of coherence normally have equal weight in the total assessment of regime change. However, if the institutional regime becomes over time more public governance-oriented, a decreased coherence in the property rights system is made less important. This should be reflected as differentiated weight of the one form of coherence over the other, in the overall assessment of the regime change.

- Within the variable of coherence of public governance, the five sub-variables have equal weight in the assessment of the total coherence of public governance. The more sub-variables show increased coherence, the more the total coherence of public governance increases.

- Each of the public governance sub-variables involves more than one indicator. The more of the indicators are met, the more the coherence of the respective public governance sub-variable increases.

These analytical assumptions are also valid for the set of regime change variables operationalised for the water basin level of analysis (see section 4.1.2, Table 4.5).

4.1.1.2 Institutional context (conditions)

The change effect of external impetus on the national water regime is also determined by the institutional context, which prevails at the time of an attempt to initiate change. In earlier comparative research on European national water regimes, Kuks (2004c) proposed a set of institutional context conditions as important determinants for the change of national water regimes. These institutional conditions are here further complemented with selected institutional variables of Blomquist et al. (2005a) for the success of decentralisation as a process supportive of integrated water management.

Based on the principle that institutional change is path dependent, Kuks (2004c) argued that more integrated national institutional water regimes can be better and faster adopted, the better the required changes fit into the existing institutional structure of a given national water regime. Indeed, it is well known in literature that institutional change is path dependent in the sense that "where you end up is strongly influenced by where you started from". Once institutions are installed, institutional change will, at any rate, be costly. Institutions are hard to reform or abolish, even if the circumstances that brought them about no longer persist (Scharpf, 1997). The path dependency phenomenon of institutions often proves to be an insurmountable barrier to institutional reform (Moss, 2003). Especially, the introduction of institutional change from so-called "peak regimes" on a multitude of subordinate levels (such as the EU regime of the WFD on the Member States level) can be highly problematic, as described in literature (e.g. Ostrom, 1990). The intended effect of "peak regimes" often fails due to the lack of knowledge of the special characteristics and modes of operation of existing institutions (Moss, 2003).

As concerns the institutional requirements of the EU WFD, Kuks (2004c) assumed the following on the ability of Member States to adapt to them (adaptive potential):

- A river basin approach will be easier adopted, if a decentralised water management already exists with administrative scales based on river basin boundaries.
- A participatory structure will be easier adopted, if some form of user participation in water management already exists.
- Integral water visions for river basins will be easier developed, if a planning structure for water management already exists.
- The streamlining of legislation will be easier, if a country has already adopted integral water legislation.
- Full cost recovery pricing will be easier adopted, if a country is already accustomed to water taxes based on the polluter pays and full cost recovery principles in water services.
- Coordination of implementation will be more effective, if a country is characterised by co-governance between different authorities (e.g. central and decentralised authorities) in water management.

Kuks (2004c) combined the above (WFD-specific) assumptions with other broader causal mechanisms related to policy change in the literature (e.g. see Jänicke & Weidner, 1997), in order to define a more complete set of institutional conditions, which determine the change effect of external impetus on national water regimes. Based on a review of relevant literature, Kuks (2004c) concluded that several theories identify intellectual-based and power-based causal mechanisms as decisive for policy change. Kuks (2004c), thus, decided to make use of the following three causal mechanisms, which determine stability and change in governance systems (e.g. see Bressers & Kuks, 2003), to group conditions which affect changes in national water regimes:

- Set of dominant values in the water sector of a nation;
- Perceptions of water resources and water issues in a nation; and
- Power distribution between administrative levels and authorities as well as between public and non-public actors.

Specifically, Kuks (2004c) supported that external impetus will more easily change a national water regime towards integration, when the dominant set of values, the dominant perceptions of water issues and the power distribution in the water sector of a nation are favourable to an integrated water regime. According to Kuks (2004c), if these favourable conditions are not fulfilled, there is a low adaptive potential of the country for water regime changes towards more integration and external impetus is needed that could bring about radical change. If the favourable conditions are largely fulfilled, there is a higher adaptive potential of the country for regime changes towards more integration and it is sufficient to have impetus that brings about incremental change.

Table 4.2 operationalises the separate conditions of such an institutional context, mainly based on the indicators used by Kuks (2004c). Kuks' indicators are therein slightly extended and re-formulated to better reflect the importance of resources committed to reform and decentralisation-supportive conditions, based on the work of Blomquist et al. (2005a) (see a summary of their work in chapter 3/section 3.2). The resulting institutional context, thus, involves a mixture of conditions prevailing in the national water sector as well as in the overall socio-political and socio-economic context of a nation.

Table 4.2   Institutional context favourable to integration in national water regimes

INSTITUTIONAL CONTEXT	INDICATORS FAVOURABLE TO INTEGRATED WATER RESOURCE REGIME
Power distribution (within administration, between public and non-public actors)	• Tradition of co-governance between central and regional/local authorities • Mutually desired authority devolution (instead of top-down or bottom up) * • Central government with sufficient resources committed to integration and decentralization * • Regional/local authorities with sufficient resources and experience in self-governance * • Tradition of public participation and debate • Strong position of green NGOs • Support of water sector by a strong environmental policy sector
Set of values	• Cooperative policy style with participatory values (openness of water policy to rival interests) • Strong environmental awareness in society • Willingness to keep water in the public domain and restrict individual autonomy on water access rights • Adherence to polluter pays and full cost recovery principles (principles based on public values)
Perceptions of water issues	• Common, instead of isolated, understanding of water resource problems (favouring proactive instead of curative resource protection) • Developed water planning and a supportive learning system (national statistics, science and research) • Availability and dissemination of knowledge on increased risks from water degradation • Openness of policy networks, scientific community (importance of epistemic communities) and media to new problems, issues and paradigms for water protection • Ability to adapt existing institutions (and their expertise) to new water functions

Source: Based on Kuks (2004) and Blomquist et al. (2005a).

*: Indicators added to the initial list of Kuks (2004c), on the basis of decentralisation-supportive conditions used by Blomquist et al. (2005a).

For national regime integration to develop, it is not necessary for all above conditions to be favourable in the same time. The absence of some favourable conditions can be outweighed by the presence of others.

Upon closer inspection, some of the conditions of the institutional context (Table 4.2) seem to be similar to elements of the national water regime itself (Table 4.1), e.g. the tradition of public participation (in the institutional context) with the coordination of actors in the public governance system (as a regime element). However, this should not be mistaken for a duplication of variables in the analytical framework. The similarity of variables in fact can be traced back to the concept of institutional path dependency in the explanatory approach adopted. The change-determining institutional context explicitly includes preconditions (e.g. tradition of public participation and debate), which -when fulfilled- support the development of relevant regime aspects in a certain direction (e.g. the coordination of actors in the public governance system of the water regime).

4.1.2 Analytical framework for the water basin level of research

At a second stage, the baseline analytical framework for change in institutional water regimes has been modified to be applied to the institutional water regimes in specific Greek basins. Similarly to the national level of analysis, the framework does not aim at a complete assessment of the status of a water basin regime in certain points in time. Instead, it concentrates on the assessment of regime change through time (from (t) to (t+1)), occurring in the research period covered by this dissertation (from the 1970s-1980s till approximately 2005, as far as the water basin level of research is concerned).

As described in the following, two important modifications have been made to the baseline framework for its application to the water basin level. In specific, based on insights gained during the application of the framework to the national level and during the screening of candidate case studies for the water basin level, two things became obvious. Firstly, in real-life water management processes in specific basins, there seemed to be often lack of progress both in terms of making attempts at regime changes towards more integration as well as in terms of subsequently implementing any attempted changes (like new policies and measures). In general, the implementation of Greek policies for the environment (including water) is often characterised as greatly 'symbolic' rather than effective (Skourtos et al., 2000). Secondly, it seemed important to pay attention to the interactions of humans within real-life water management processes in order to explain progress (if any) or lack of progress in regime change and implementation. To deal with the first issue, the baseline framework is modified in the following into a phased process framework. To deal with the second issue, the framework is combined with an actor-centred explanatory approach.

4.1.2.1 Modifying the analytical framework into a phased process

The modification of the baseline analytical framework into a phased-process approach was influenced by the methodological concept of the policy cycle, which consists of several phases taking the dynamics of policy-making into account (see Figure 4.2).

Figure 4.2   Phases of the policy cycle

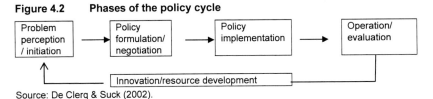

Source: De Clerq & Suck (2002).

The analysis of institutional water basin regimes in this dissertation is, thus, treated as a phased process and is separated into two main phases. Phase (I) consists of the process of making changes or attempts to change the water basin regime including policy formulation. Phase (II) consists of the process of implementing new incentives (mainly policies and measures) resulting from regime change in Phase (I) (see Figure 4.4 for a graphical illustration of the analytical framework proposed for the water basin level).

As mentioned above, such a phased process approach was judged as particularly useful for the Greek basins, to allow for more in-depth explanations of bottlenecks encountered in the making of attempts for regime change towards more integration but also in implementing any changes attempted. The process of implementing policies and measures is also especially relevant to the WFD obligations of Greece, whereby the achievement of the WFD objectives is not only related to the adoption of measures in the river basin management plans but also to the implementation of measures and their actual effect on water status. Also Young (1982), who worked with the concept of resource regimes earlier on, emphasised the importance of the implementation of resource regimes, saying that it is important to consider compliance mechanisms as a principal component when discussing the effectiveness of regimes. Implementation was also mentioned as an important aspect for the realisation of regime change in the theoretical elaboration of Bressers et al. (2004), but this aspect was not treated in detail in their subsequent empirical case study research in the Euwareness project.

4.1.2.2 Modifying the analytical framework into an actor-centered process

The second important modification to the baseline framework concerns the explicit emphasis placed on the characteristics and the interactions of actors[16] involved in the regime change and implementation processes of the basins in question (see the proposed modified analytical framework in Figure 4.4). According to De Clercq & Suck (2002), interaction[17] between actors plays a major role in all public policy-related research, since the need to solve collective problems leads to collective interaction between different stakeholders. Scharpf (1997) points out that actors respond differently to external threats, constraints and opportunities, because they may differ in their intrinsic perceptions and preferences but also because their perceptions and preferences are very much shaped by the specific institutional setting within which they interact. The institutional setting specifically exercises important influence on actor orientations and capabilities, actor constellations and modes of interaction. The main elements of Scharpf's approach (interaction-oriented, actor-centered institutionalism) are shown in Figure 4.3. Combined with problem-oriented policy analysis, the main aim of interaction-oriented policy research within Scharpf's framework is to explain

---

[16] Actors can be individuals and composite actors. Individuals often act in the name of and in the interest of another person, a larger group, or an organisation, and therefore, aggregates of individuals are treated as composite actors (Scharpf, 1997).

[17] A simple definition of interaction is given by Reading (1977): "Interaction is stimulating and responding of persons to one another".

past policy choices and to produce systematic knowledge that may be useful for developing politically feasible policy recommendations or for designing institutions that will generally favour the formation and implementation of public-interest-oriented policy.

Figure 4.3   Interaction-oriented policy research based on Scharpf (1997)

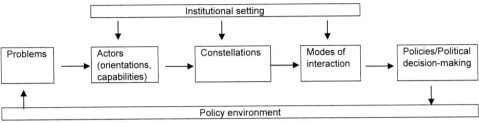

Source: Scharpf (1997).

Although Scharpf's approach is a widely accepted explanatory model for the course and outcome of policy processes, this dissertation makes use of an alternative actor-centered explanatory model based on the contextual interaction theory of Bressers (2004). The advantage of the contextual interaction theory is that, it was developed especially for explaining implementation processes, which are of great importance for the evaluation and interpretation of developments in the institutional water regimes and the water status of Greek basins. In this theory, the main factors that have direct influence on the course and results of interaction processes are the actor characteristics of motivation, information and resources (see Table 4.3). Actors' motivation, information and resources have proven to be exceptionally useful variables for explaining the dynamics of interaction processes in the literature. According to Bressers (2004), any other variables (e.g. general institutional variables), which form the context of actor interaction processes, have an indirect influence on the process outcome via their influence on these three core actor characteristics. Also in Scharpf's actor-centered institutionalism, the institutional setting influences the policy process by influencing actors, their constellations and interactions.

Table 4.3   Actor-characteristics in interaction processes

Actor characteristics	Description
Motivation *to want*	Motivation relates to the aspired positions of the involved actors or actor coalitions relative to other actors. It is a central question to what degree the application of the policy or instrument in question is perceived as contributing to the goals and interests of the actors involved
Information *to know*	It is a central question what information actors have access to and whether they are able to process the information necessary for the (inter)actions they want to pursue. Information may be distinguished into process-wise information (information on positions and circumstances of the other actors) and content-wise information (technical aspects of knowledge)
Resources *to can*	It is a central question whether the actors involved have resources and power to engage in the (inter)actions they want to pursue. Types of resources can be distinguished into economic, political, legal and cognitive resources or even more specific types of resources such as skilled people, physical goods, money, authority

Source: Based on Bressers (2004); Dinica & Bressers (2004); De Clercq & Suck (2002).

43

With some adaptations to its theoretical interaction assumptions, the contextual interaction theory is made suitable for explaining implementation but also regime change processes in this dissertation. In section 4.1.2.3 and 4.1.2.4, more details are given on the way actor-centered interaction is used to refine the explanation of the process of regime change and of the process of implementing regime incentives. In fact, both these processes are treated as conversion processes, whereby certain inputs (e.g. water regime at (t)) are transformed into certain outputs (e.g. water regime at (t + 1)) through activities and interactions of actors.

In total, the addition of an actor-centred approach was considered appropriate for the water basin level of analysis, where real-life water management processes are at stake. This explicit emphasis on the human dimension can help provide more adequate explanations for the course and outcome of the water management processes in question.

4.1.2.3 Process (I): "Making changes in the regime"

External change impetus and regime changes

According to Figure 4.4, the input and output of process (I) are different phases of the water basin regime in time (at (t) and at (t+1)). Changes of the water basin regime from (t) to (t+1) are set in motion by external change impetus. The main types of potential external change impetus for regimes on a water basin level are summarised in Table 4.4.

Table 4.4   Types of external impetus for change in water basin regimes

Type of impetus	Examples
European Union (or supranational) impetus	Policies, infringement procedures
National policy impetus	Adoption of new statutes, new regulations, constitutional rule changes, court decisions
Problem pressure	External shocks (e.g. floods, extreme climatic events), risks e.g. through industrial accidents, or gradual water resource degradation
Technological and socioeconomic changes	New available technologies, decline of agricultural and industrial development
New political paradigms	Globalisation and liberalisation trends (e.g. in energy or water sector) linked to changes in the public sector
Collective learning processes and the adoption of agreed solutions	Harmonisation of the agricultural policy on a European level
New policy actor constellations which change existing coalitions	Change of party in government, rise of green parties

Source: Based on Knoepfel et al. (2001); Bressers et al. (2004).

As a result of external change impetus, interactions of relevant actors can produce an initial change in the resource regime (Bressers et al., 2004). The regime elements will normally evolve towards more complexity. A change towards more coherence (and more integration) will occur only when relevant actors acknowledge that coherence is necessary to prevent further deterioration of the resource. Coherence typically stems from special external impetus that demands some form of coherence, such as an EU directive that demands multilevel cooperation in water resource management planning (e.g. the EU WFD). This means that,

44

unlike an increase in regime complexity, developments of a water basin regime in the direction of more coherence need some sort of deliberate attempt by motivated actors (Bressers et al., 2004).

Table 4.5 operationalises the regime change variables (extent and coherence) for a shift towards a more integrated water basin regime.

Table 4.5   Water basin regime change variables and selected indicators

CHANGE VARIABLES	INDICATORS FOR THE INTEGRATED PHASE OF A WATER BASIN REGIME
Extent	**Indicators for a complete extent**
	All water uses and users are considered by the regime elements (including the recognition of nature as use/user)
Coherence within public governance	**Indicators for high coherence**
Levels of governance	• Administrative levels of management fit with the natural scale of water basins • (Actor) coordination and cooperation within each administrative level and between different administrative levels (relevant to the river basin management process)
Actors in the policy network	• Participatory arrangements for actors, especially non-public actors, with an interest in water resources (formal arrangements such as participatory planning and/or informal consultations) • Increasing interaction, consensus and cooperation among actors from different (policy) networks • Establishment and involvement of local associations
Perspectives and objectives	• Development of a water basin management plan or a water basin vision (preferably official or informal initiative), incorporating multiple stakeholder perspectives • Recognition of relations between water policy and other sectoral policy fields as coordination topics • Policy objectives (in official visions and visions of stakeholders) recognise new water uses and administer justice to rivalries between different water uses • Changes in the dominant water management paradigm, e.g. from hydraulic interventions to "softer" management principles
Strategy and instruments	• Incorporation of flexible instruments in policy; increasing preference for indirect procedural instruments (e.g. participatory planning procedures) instead of command-and-control tools • Availability of water management tools to redistribute use rights and reduce water use conflicts
Responsibilities and resources for implementation	• Policy implementation is sufficiently coordinated and equipped with resources (authority, finances, time, human, information) • Clear, non-fragmented institutional arrangement for implementation
Coherence within property rights	**Indicators for high coherence**
	• Coordination, instead of contradiction, of use and ownership rights when competing uses evolve
Coherence between public governance and property rights	**Indicators for high external coherence**

CHANGE VARIABLES	INDICATORS FOR THE INTEGRATED PHASE OF A WATER BASIN REGIME
	• Changes in policy and public governance are reflected in changes of the property rights (e.g. via restrictive regulations such as restrictions on wells, restriction of private rights, establishment of permits for abstractions or pollution) • Match between actors targeted by policies and actors with use and ownership rights

Source: Indicators are based on the empirical summarised findings in Bressers & Kuks (eds) (2004a).

Influence of contextual conditions and actor characteristics on regime changes

The first level of explanatory factors that influence the regime change process includes a set of contextual conditions, which are relevant to selected institutional interfaces and characteristics of the actor network in place. The set of contextual conditions, which is operationalised in Table 4.6, is based on the conditions defined by Bressers et al. (2004) as favourable for the success of attempts to change water basin regimes towards more integration.[18] The development of these conditions was inspired by the state-of-the-art literature on conditions favourable for the rise of negotiated agreements in environmental policy, which also involve interaction processes in which multiple parties come together (see De Clercq & Suck, 2002). Thus, Bressers et al. (2004) empirically tested whether attempts to integrate water basin regimes need the same sort of favourable conditions like other forms of coordinated collective action. For regime integration to develop, it is not necessary for all conditions to be favourable in the same time. The absence of some favourable conditions can be outweighed by the presence of others.

Table 4.6   Contextual conditions favourable to integration in water basin regimes

CONDITIONS	FAVOURABLE INDICATORS TO INTEGRATION (SELECTION)
Characteristics of the actor network	
Tradition of cooperation	Existing longer tradition of thinking in terms of cooperation and consensus-seeking in water management or such thinking was built during the process early enough to influence later stages of the process; Existing structures to allow coordination
	Mutual respect and trust of actors
Joint problem	Common understanding that non-integrated water management harms sustainability and that this sooner or later will have to be stopped
	Convergent problem perception and shared perception of risk
	Shared sense of responsibility concerning the causes of the water resource problems
	Knowledge in the form of reports and statements from respected sources on water resource deterioration, made available to all actors
Joint chances	Notion of possible joint gains and common interests from a more coherent water regime ('win-win-situations')
Credible threat of dominant actor	Credible threat of a strong and determined actor (e.g. government) to impose an alternative arrangement in his interest, if the other involved actors cannot reach a workable solution/cooperation (imbalance of power)
Institutional interfaces (formal and informal)	

---

[18]   These conditions extended the initial framework of Knoepfel et al. (2001; 2003) for the analysis of institutional resource regimes.

CONDITIONS	FAVOURABLE INDICATORS TO INTEGRATION (SELECTION)
Institutional position of brokers	Actors, independent or within the administration, who act as brokers or mediators
Institutional representation of rival interests	Small number of relevant actors or strong representative organisations that make it possible for large groups to act as a uniform actor in the policy network
Alert mass media	Free and alert mass media to induce awareness and public debate on the various aspects of the problem
Alert public (*)	Alert public to press authorities (public considers water problems urgent)
Legal leeway for more integrative approaches	Water or environment laws (including EU directives) that enable (and not prohibit) attempts to increase coherence in water resource management
Protection of negotiated compromises	Legal or practical possibilities to protect negotiated compromises from continuous litigation

Source: Based on Bressers et al. (2004) and the empirical summarised findings in Bressers & Kuks (eds) (2004a).

(*): Added by the author

Although the above set of contextual conditions appears, at first sight, quite different from the conditions proposed as "institutional context" for the national level (see Table 4.2), it is in fact possible to relate several aspects of these two sets of conditions. The "tradition of cooperation" could be linked to the national "values" context (especially, the condition of "cooperative policy style") and the national "power" context (especially, the tradition of co-governance between administrative levels). The "joint problem" and "joint chances" could be linked to the national "values" and "perceptions" context. The "credible alternative threat" can be linked to the national "power" context. Finally, "institutional interfaces" can be linked both to the national "power" and the national "perceptions" context.

The second level of explanatory factors which influence the regime change process includes the actor characteristics of motivation, information and power. As already mentioned, to initiate change in one or more of the water basin regime elements towards coherence (and integration), actors need to be accordingly motivated and to acknowledge that regime coherence is necessary to protect water resources. The aspect of actor motivation has been recognised as important also in other approaches used to investigate institutional changes towards more integrated water management. For instance, in the comparative study of Blomquist et al. (2005a) on institutional changes towards more decentralised water management on the river basin level, the motivation of involved stakeholders was an important variable for the interpretation of successful or failed institutional changes. Changes of water basin regimes also depend on whether those involved have sufficient information on the regime itself and on potential regime alternatives. Information of the actors can include information on the institutional structure, existing and potential alternative rules, information on potential costs and benefits of any alternatives considered (e.g. the anticipated costs and benefits of introducing new rules). The third actor characteristic influencing the regime change process is the distribution of resources between those involved. Actors directly involved in the regime change process need to have the necessary resources. Moreover, a balanced distribution of power and resources should be created. Necessary resources can include skills of leaders in proposing regime changes, available resources to invest in changing the status quo, capacity to influence and get the support of external key actors such as key national-level policy players.

Box 4.1 proposes in a systematic way the types of actor interactions, which can influence regime change attempts towards more integration, given different constellations of actor motivation, information and resources. The formulation of these assumptions on actor interactions was based on a series of assumptions made by Bressers (2004) in the context of the contextual interaction theory on the implementation of policy instruments (see Annex I).

Box 4.1   Theory-based assumptions on actor interaction in regime change processes

• *Interaction assumption I-1:* Attempts for regime change towards more integration develop through actor interaction, when at least one actor (or group of actors) is positively motivated towards integration.
• *Interaction assumption I-2:* If all actors are neutral in their motivation, there will be no interaction resulting in deliberate actor attempts towards a more integrated regime.
• *Interaction assumption I-3:* If all actors are negative or one actor (or actors' group) is negative and the other is neutral, there will be again no interaction towards a more integrated regime. Two negative actors (or actors' groups) may even act together to prevent any regime change towards more integration (lack of interaction as a form of obstructive cooperation).
• *Interaction assumption I-4:* If one actor or actors' group is positively motivated (the others being either neutral or also positive) but has insufficient information, there will be initially a learning process before an integration attempt can take place.
• *Interaction assumption I-5:* If the motivated actor has sufficient information (the other actors being either neutral or also positive), there will be cooperation for an attempt to introduce more integrative elements in the water basin regime.
• *Interaction assumption I-6:* If one actor or actors' group is positively motivated with sufficient information but the other actor or actors' group is negatively motivated, the outcome of interaction depends on the balance of power between actors. If the motivated actor is dominant, there will be (forced) cooperation for an integration attempt.
• *Interaction assumption I-7:* If the unmotivated actor is dominant, integration attempts will be obstructed (continuation of argument from interaction assumption I-6).
• *Interaction assumption I-8:* If actors are equally powerful, there will be opposition (in the form of negotiation or conflict) leading to intermediate likelihood for an integration attempt (continuation of argument from interaction assumption I-6).
• *Interaction assumption I-9:* If the motivated actor has insufficient information and is also faced with an unmotivated actor, a regime change attempt towards more integration will not take place at least until after the motivated actor becomes more knowledgeable (learning period) (continuation of argument from interaction assumption I-6).

As already mentioned, all variables which form the context of an interaction process have an indirect influence on the process outcome by influencing actor motivation, information and resources (Bressers, 2004). Thus, the contextual conditions described in Table 4.6 above form part of the context for the motivation, information and resources of the actors involved in the regime change process:

- The tradition of cooperation influences mainly the motivation of actors.

- The joint problem perception influences the motivation of actors and their openness to learn information from each other, while they place new facts in a similar frame of reference (e.g. exchange of information between sectoral actors).

- The joint opportunities also influence the motivation of actors and their openness to learn information from each other.

- The credible threat influences mainly the motivation of the actors involved.
- The institutional position of brokers facilitates a smoother information flow between negotiating actors. Trust in brokers may also increase the motivation of actors to participate in a certain process. Furthermore, brokers can bring balance into a process where actors with unequal power are involved.
- The institutional representation of rival interests influences the negotiation capacity of actors and therefore their resources.
- Alert media influence the information level and partly awaken motivation.
- An alert public influences mainly the motivation and increases the resources of actors who depend on social support.
- Legal leeway for more integration is mainly part of the resources that actors have in their hands to work in the direction of regime change.
- Institutional possibilities to protect negotiation may influence the motivation of actors, who may feel more secure to take part in the process. Institutional protection of negotiation may also influence the balance of power and resources between actors.

4.1.2.4 Process (II): "Implementing new incentives from regime change"

According to Figure 4.4, the input to the implementation process (II) consists of new incentives provided by changes in the water basin regime. Such incentives usually are adopted policies and measures. Especially incentives, which are designed to improve water use from a sustainability perspective, are here of interest. In general, regime changes towards a more integrated phase are expected to contribute more towards sustainability than changes towards a more complex phase. Additionally, incentives from a more integrated regime are expected to be more coordinated with one another in their aim to achieve more sustainable water use. Even in case of regime change to a more complex phase, certain sustainability-relevant incentives may be adopted (even if these are initially restricted to simple regulations for water protection). In this process, it is of interest to investigate whether there is a positive water sustainability effect related to the implementation of such new incentives. The output of process (II) consists of the sustainability implications in the use of water resources. If new incentives adopted during regime change are not implemented or implemented inadequately, even elements of the regime which are in principle coherent will be insufficient to halt unsustainable water use (Bressers et al., 2004).

The implementation process of measures and policies is also analysed as an interaction process, whose course and outcome can be explained by the motivation, information and resources of the actors involved. The implementation process involves interactions mainly between the responsible government officials (implementers of policies and measures or usually pro-actors) and the members of the target group(s) (usually contra-actors).

Influence of actor characteristics on the implementation process

For implementation, motivation is needed both from the side of the target group and of the implementers. Implementers have values and interests of their own, which may not coincide with the activities involved in the implementation of a certain policy or measure. 'Symbolic policy' is a well-known phenomenon in many contexts – that is, policy that is not taken seriously by implementers and that is not supported by a serious commitment of resources (Bressers, 2004).

The successful application of policies and measures also depends on whether those involved have sufficient information. The first issue is whether the implementers of a certain policy actually know who the crucial target groups are and how well prepared the information available to the responsible implementers is. On the other hand, if the target group stands to gain from the application of the instrument – for instance, in the case of subsidies – then the information available to the members of the target group will also help to increase the likelihood of application (Bressers, 2004).

The distribution of resources and power between implementers and the target group also influences implementation. It is important to clarify who is empowered to apply the instrument and how far this power goes. The formal power might rest exclusively with the implementers, but in some cases (e.g. in subsidies) the instrument can only be applied at the request of the members of the target group. The target group then enjoys an extremely strong position if it is not in favour of the application of the instrument. Other forms of power may derive from formal sources (e.g. the opportunity to appeal against compliance) and informal sources (e.g. being dependent on another party for the achievement of other objectives). In most interaction processes, informal sources of power may be highly important, and in many cases balance the – often more formal – powers of the implementers (Bressers, 2004).

Box 4.2 summarises in a systematic way the types of actor interactions, which can influence the outcome of implementation, given different combinations of motivation, information and resources of the implementers and the target group. This set of assumptions stems from the assumptions formulated by Bressers (2004) for implementation processes of policy instruments. Separate assumptions are formulated on the likelihood of the initial implementation of policies and measures and separate assumptions are formulated on the degree of adequate implementation (after the implementation process has taken off).

Box 4.2   Theory-based assumptions on actor interaction in implementation processes

Assumptions on the likelihood of initial implementation of policies or measures
• _Interaction assumption II-1:_ For any actor interaction to evolve to initiate implementation, it is necessary that the implementation of a policy or measure contributes positively to the motivation of at least one actor (the implementer or the target group).
• _Interaction assumption II-2:_ If both actors are neutral in their motivation, there will be no interaction, leading to a small likelihood of implementation of the policy or measure.
• _Interaction assumption II-3:_ If both actors are negative or one is negative and the other is neutral, there will be again no interaction to initiate implementation.
• _Interaction assumption II-4:_ If one actor is positively motivated (the other being either neutral or also positive) but has insufficient information to implement the policy or measure, there will be initially a joint learning process.

- *Interaction assumption II-5:* If the motivated actor has sufficient information (the other actor being either neutral or also positive), there will be cooperation leading to great likelihood for the initial implementation of the policy or measure.
- *Interaction assumption II-6:* If one actor is positively motivated with sufficient information but the other actor is negatively motivated, the outcome of interaction depends on the balance of power between actors. If the motivated actor is dominant, there will be (forced) cooperation leading to great likelihood for implementation.
- *Interaction assumption II-7:* If the unmotivated actor is dominant, there will be obstruction leading to small likelihood of implementation (continuation of argument from assumption II-6).
- *Interaction assumption II-8:* If actors are equally powerful, there will be opposition (in the form of negotiation or conflict) leading to an intermediate likelihood for implementation (continuation of argument from assumption II-6).
- *Interaction assumption II-9:* If the motivated actor has insufficient information and is also faced with an unmotivated actor, there will be initially no interaction but the positive actor will try to learn on its own to initiate implementation in the future.

Assumptions on the degree of adequate implementation of policies or measures
- *Interaction assumption II-10:* If both actors are neutral in their motivation to adequately implement the policy or measure, there will be symbolic interaction and a small degree of adequate implementation.
- *Interaction assumption II-11:* If both actors are negative or one is negative and the other is neutral in their motivation, adequate implementation of the policy or measure will be obstructed.
- *Interaction assumption II-12:* If at least one actor is positively motivated but has insufficient information to apply the policy or measure, there will be initially symbolic interaction but also learning by the positive actor. In case the implementer is positive and the target group is also positive or neutral, a process of joint learning will quickly emerge.
- *Interaction assumption II-13:* If at least one actor is positively motivated and has sufficient information (the other actor being positive or neutral), there will be constructive cooperation for a great degree of adequate implementation of the policy or measure.
- *Interaction assumption II-14:* If one actor is positively motivated with sufficient information but the other actor is negatively motivated, the outcome of interaction depends on the balance of power between actors. If the motivated actor is dominant, there will be (forced) cooperation for a great degree of adequate implementation.
- *Interaction assumption II-15:* If the unmotivated actor is dominant, there will be negotiation leading only to an intermediate degree of adequate implementation (continuation of argument of assumption II-14).
- *Interaction assumption II-16:* If actors are equally powerful, there will be negotiation or conflict leading to a relatively high degree of adequate implementation (continuation of argument of assumption II-14).
- *Interaction assumption II-17:* If the motivated actor has insufficient information and is also faced with an unmotivated actor, there will be initially only symbolic implementation but the positive actor will try to learn on its own to improve implementation in the future.

Source: Based on Bressers (2004).

Changes in the water regime, especially in several of the public governance elements, have an indirect influence on the outcome of the implementation process by influencing the three core actor characteristics of motivation, information and resources. Thus, public governance plays a similar role in the implementation process as the contextual conditions in the regime change process by forming part of the context for the motivation, information and resources of the interacting actors. According to Bressers et al. (2004), with more coherence in public governance, the goals of implementers and target groups can be expected to be less likely in discord. A more coherent regime can also be assumed to contribute to a lower degree of experienced uncertainty, an increase in information exchange and a lower degree of distrust.

Certain public governance elements may influence the motivation, information and resources of actors as follows:[19]

- The "perspectives and objectives" affect the motivation and the information of actors. More coherent perspectives specifically influence actors' openness to learn information from each other while they place new facts in a similar frame of reference.

- The "strategies and instruments" influence motivation, since actors can dislike or like a certain strategy. They also influence actor information, since different instruments have different information requirements, and resources, since different instruments provide different sources of power among actors.

- The "resources and responsibilities for implementation" influence motivation giving institutionalised responsibilities to certain actors, which they can take on board or resist themselves against. They also influence information giving actors a certain amount of information gathering and processing capacity. Finally, they influence resources distribution among actors, by influencing the access of actors to the resources made available for policy implementation. Actors, however, can also have additional resources from other sources, other than policy.

Sustainability implications

Having considered the factors that influence the process of implementation, attention turns here to the output of this process, namely the actual water use sustainability implications of measures and policies adopted during regime change. This dissertation does not aim at an extended assessment of resource use sustainability on the basis of sustainability indicators (cf. Knoepfel et al. (2003), who evaluated resource use sustainability on the basis of indicators for ecological, economic and social sustainability). Instead, the purpose is to make some observations on the most obvious implications of adopted measures and policies for sustainability, focusing on changes in the environmental status of water resources in the time period of investigation (see approach of Bressers et al., 2004). In this context, attention is paid both to the status and trends in water quality and water quantity of surface and groundwater. Additionally, account is taken of any important consequences on social and economic development from the environmental implications of measures and policies.

Figure 4.4 illustrates graphically the change dynamics and all causal links of the analytical framework proposed for the water basin level of analysis.

---

[19] "Levels of governance" and "actors in the policy networks" do not influence the motivation, information and resources of actors but determine the composition of actors in the process.

Figure 4.4   **Modified framework for change in institutional water basin regimes**

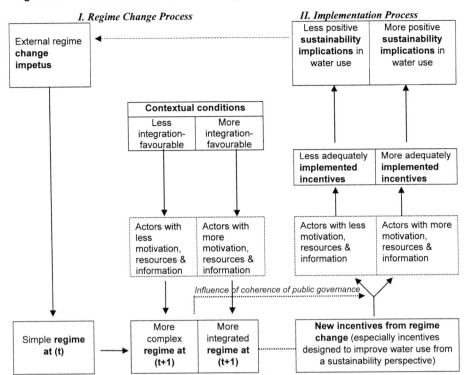

Source: Modified framework based on Bressers et al. (2004) and Bressers (2004).

4.1.3 Conclusion

In this chapter, an analytical framework was elaborated for use in this dissertation. The framework was based largely on a baseline institutional-resource-regime framework, which was rooted in the work of Knoepfel et al. (2001; 2003), later merged with the work on governance by Bressers & Kuks (2003) and was further elaborated and specified with, among others, contextual conditions for regime change by Bressers et al. (2004).

This baseline framework was first applied (in a slightly modified version) to the Greek national institutional water regime (see results in chapter 5). As described in section 4.1.1, for the national level of analysis, the baseline framework was combined with the "institutional context" conditions proposed by Kuks (2004c) for national water regimes and was further complemented with institutional variables from the work of Blomquist et al. (2005a) on decentralisation for integrated water management.

Subsequently, the baseline framework was further modified for application to institutional water regimes in specific Greek basins (see empirical results in chapters 7 and 8). Section 4.1.2 described how the baseline framework was modified into an actor-centred phased-process approach for this water basin level of research and analysis. The contextual interaction theory of Bressers (2004) served as the point of departure for actor-centered explanatory links, which were added to the framework.

Looking back at the conclusions of chapter 2, several single determinants were identified from selected practical case studies on the success or failure of integrated water management. In that context, the need was expressed for a theory-based analytical framework, which could combine single determinants in a more systematic way in order to explain the extent of achieving integrated water management. Indeed, several variables and causal links of the analytical framework presented in this chapter seem to adequately reflect several of the single determinants which were identified back in chapter 2. Both success factors (e.g. "representation of all interests") and failure factors (e.g. "lack of willingness of stakeholders", "lack of information" or "lack of resources") can be reflected in the explanatory contextual conditions and the characteristics of interacting actors, which influence change of institutional water regimes towards a more integrated phase.

4.2 Hypotheses and questions

Based on the adopted analytical framework and the leading research questions posed at the start of this dissertation, this section formulates hypotheses to be examined empirically on the national level and the water basin level of research and analysis.

4.2.1 National level

Hypothesis 1

The less integration-favourable the institutional context of the Greek national water sector, the more likely that attempts at a more integrated national institutional water regime are unsuccessful.

Research questions:

- Can changes or attempts to change the national water regime towards a more integrated phase be identified? Did the extent of uses regulated by the national water regime and the coherence of the regime elements (property rights and public policy) change over the investigation period? If yes, how did they change?
- To what extent was the existing institutional context (power distribution, values and perceptions) favourable to integration attempts? Can the institutional context explain progress (if any) or lack of progress towards a more integrated phase?

4.2.2 Water basin level

Hypothesis 2

The less favourable the contextual conditions and the characteristics of interacting actors for integration attempts, the more likely that Greek water basin regimes fail to reach a more integrated phase.

Research questions:

- In Greek basins where water degradation and rivalries indicate an urge for more integrated water management, were there any changes or attempts to change the water regime towards a more integrated phase? Did the extent of uses regulated by the water basin regime and the coherence of the regime elements (property rights and public governance) change since the start of the water rivalries under investigation? If yes, how did they change?

- To what extent were the contextual conditions in place favourable for integration?
- To what extent were the motivations of the actors involved favourable for changes towards a more integrated water regime? To what extent did the actors possess of adequate information and resources to contribute to changes towards a more integrated water regime?
- Can the existing contextual conditions and the motivation, information and resources of interacting actors explain progress (if any) or lack of progress of the regimes in terms of integration?

Hypothesis 3

The less new incentives provided from regime change and the less adequate the implementation of any new incentives, the more likely that there are less positive sustainability implications of regime change in water use.

Research questions:

- To what extent were new incentives (measures and policies) adopted during regime change adequately implemented? To what extent were the motivations of the interacting actors favourable to the adequate implementation of the adopted policies and measures? To what extent did the actors possess of adequate information and resources for the adequate implementation of the adopted policies and measures?
- To what extent were there positive implications of any regime change on the sustainability of water use in the basin?
- Can the positive (or negative) implications on sustainability be explained by the adequate (or inadequate) implementation of regime incentives?

4.2.3 Interchange of national and water basin level

On the empirical level, it could also be of interest to investigate links between regime developments on the water basin level and on the national level of analysis. In this case, questions of interest may include:

- Did any observed changes in the Greek national institutional water regime lead to transformations in the examined institutional water basin regimes, and vice versa?
- Are there any other water regime dynamics in place on the basin level, which cannot be explained by the development of national water regime determinants?

In this respect, also Knoepfel et al. (2003) had supported the possibility for regional institutional resource regime developments to take place before their counterpart developments on the national level. It could even be possible for regional developments to be very different from what the national policy defines and requires.

4.3 Empirical focus and methodological issues

The empirical research aiming at assessing "whether, to what extent and why integrated institutional water regimes are realised or not in Greece to date" was carried out separately on the national and on the water basin level. On the national level, the recent evolution of the Greek national water regime was investigated by tracking its historical development. Regime investigations on the water basin level were carried out in two selected case study areas.

As the following sections indicate, the overall methodological approach used was of qualitative nature, which is a commonplace approach for explanatory studies like this dissertation. According to McNabb (2002), in studies of qualitative explanatory type, the issue is often *why* some phenomenon occurred or did not occur (in this case, development to a more integrated water regime phase) and *how* cause-and-effect relationships between two or more variables can be interpreted.

> "Qualitative research is not looking for principles that are true all the time and in all conditions, like laws of physics; rather, the goal is the understanding of specific circumstances, how and why things actually happen in a complex world. The underlying assumption is that if you cannot understand something in the specific first, you cannot understand it in the general later." (Rubin & Rubin, 1995)

4.3.1 Research methodology for the national level

The methodological approach used on the national level involved mainly a comprehensive review of existing documentary material (laws and regulations, ministry reports, minutes of government meetings, parliament records, reports of NGOs, academic publications and press articles) in search for data on the main variables of the institutional-resource-regime analytical framework, i.e. on the development of property rights and public policies in the national institutional water regime, external impetus for regime change and the institutional context influencing change in the national water regime. Effort was made to locate and analyse documents produced (public and internal) on the formulation of important water management laws, especially the 1987 and the 2003 Water Laws.

The findings of the documentary analysis were further complemented and verified through informal discussions with selected national water actors (see Annex II for a list of people contacted). During these communications, open discussions were made on key water policy changes related mainly to the 1987 and to the 2003 Water Laws. Some of the issues concerning the national institutional water regime were also taken up in the semi-structured interviews with water officials from Regional and Prefectural authorities (see Annex III) in the two case study basins of this dissertation.

To ensure that the analytical framework was correctly interpreted and applied to the Greek national water regime, results from its previous application to the national water regimes of other European countries were also used as reference, when evaluating change in the Greek regime and when making outlook recommendations (for details see chapter 5/5.7).

4.3.2 Research methodology for the water basin level

4.3.2.1 Case studies as a research strategy

For the water basin level of analysis, case studies were chosen as a research strategy. Case studies are preferred over other social science research strategies, when "how" and/or "why" questions (explanatory questions) are being asked about a contemporary set of events, over which the investigator has little or no control (see Yin, 1994). This setting seemed to match the questions being asked in this dissertation on "whether and how" Greek institutional water regimes have changed to date and "why" they have changed (if at all) or not changed towards a more integrated phase.

4.3.2.2 Research design of case studies

Two case studies on two Greek water basins were carried out (see chapters 7 and 8). A two-case design is the simplest form of a multiple-case research design. According to Yin (1994), the analytic benefits from having two (or more) cases may be substantial. Analytic (common) conclusions independently arising from two cases will be more powerful and will largely expand the external generalisability of the findings compared to those from a single case alone.

The research design of case studies includes the following elements (Yin, 1994):

- A study's questions.
- Its hypotheses.
- Its unit(s) of analysis.
- The logic which links data to the propositions.
- The criteria for interpreting the findings.

Questions and hypothesis

The questions and hypotheses of this dissertation for the water basin level of research were already presented in section 4.2.

Unit of analysis

The unit of analysis of the case studies is here defined on the basis of a spatial and topical dimension. The spatial boundaries of the case studies should follow the boundaries of distinct water basins. With regard to the topical dimension, a set of criteria were defined for choosing case studies. These criteria were used in the screening of candidate case studies described in chapter 6:

- Firstly, the water resource problem structure should be such, so that it is more likely to find attempts (if any) towards more integrated water management. In this context, two important aspects were considered: the presence of water use rivalries and the distribution of externalities from the key water problems present.

Cases were preferred, where the water management process is related to clear water use rivalries due to water resource deterioration. In such cases, it is more likely to find attempts towards more integrated water management, since it (sooner or later) should become obvious that without more integration, opportunities to solve the rivalries in a more sustainable manner could be missed.

It was also assumed that the distribution of externalities from water problems in the candidate cases may affect chances to find attempts at more regime integration. This assumption was influenced by Dombrowsky's (2005) hypothesis on the effect that the distribution of externalities may have, on the cooperation potential over transboundary water problems. According to this hypothesis, in case of transboundary water problems with reciprocal externalities (e.g. water abstraction from a shared lake or aquifer), cooperation is in the interest of the parties involved and potential for cooperation is better. In case of water problems with unidirectional externalities, such as upstream abstraction with downstream effects, the advantaged upstream party does not necessarily have an interest to cooperate and potential for cooperation is worse (see also Table 4.7). A similar effect can be expected in case of water problems of non-transnational character, where parties with similar stream benefits from certain water use (homogeneous parties) can be identified. Thus, it was assumed that in national Greek basins with reciprocal externalities, cooperation potential is likely to be higher than in basins with predominantly unidirectional externalities. A higher cooperation potential is also considered to offer higher chances to find regime integration attempts, considering the importance of actor coordination and cooperation in a more integrated phase of water regimes.

- Secondly, case studies should involve water use rivalries between economic and also ecological uses of water.

- Thirdly, one or more water use rivalries should be the subject of interactions between actors involved in the water management process.

Table 4.7   Distribution of externalities for different types of water problems

Type of externality	Unidirectional	Reciprocal
Negative	E.g. upstream water abstraction; upstream water pollution	E.g. water abstraction from a shared aquifer or lake; wastewater discharge into a shared lake
Positive	E.g. upstream wastewater treatment; upstream water storage or flood control	E.g. wastewater treatment at a shared lake or contiguous river; water storage or flood control at a contiguous river

Source: Dombrowsky (2005).

Linking data to the hypotheses

The way of linking the data collected to the hypotheses is related to the mode of data analysis presented in the next section on the case study working steps.

Criteria for interpreting the findings

The criteria for interpreting the findings of case study research were already illustrated in section 4.1.2, which elaborated the variables of the theory-based analytical framework for the water basin level.

Additionally, in order to increase the external validity of the case study findings from the Greek water basins, these findings were checked in a comparative manner against the findings of case studies on the institutional basin regimes of other countries, which had been carried out previously using a similar analytical framework and methodological set-up. The advantage of using case studies following a similar framework is obvious, since in order to compare different cases, one needs to be sure that these cases were studied by applying similar methods and techniques (Dente et al., 1998). Although reference to case studies on other basins shall not go into depth, the case study causal assumptions of this dissertation will be more strongly supported, if other cases provide contrasting or similar results for predictable analytical reasons.

4.3.2.3 Case study working steps

The individual case study working steps included the screening and choosing of case studies, the preparation of data collection, field work and the analysis of case study data.

Screening and selecting case studies

In order to select two case studies, which are suitable for the purposes of this dissertation, seven different candidate water basins in Greece were screened according to the set of criteria for the case study unit of analysis mentioned above. The screening and selection of two case studies is described in chapter 6.

Preparation for data collection

The two main sources of case study data included documentary evidence ('desk research' approach) and interviews with actors involved in the water management process in the two case study areas of interest.

In the preparation phase for field data collection, it was important to identify which types of documentary evidence to look for during field work, to identify which actors to interview and to prepare the interview process.

Firstly, in terms of documentary evidence, the following potential sources of information were identified: policy and legislation documents from different administrative levels, minutes of formal and informal meetings related to the water basin issues, archives of written correspondence between the actors involved, pamphlets by actors, technical and scientific studies on water management issues of the basin, archives of the national and regional press for relevant articles, conference proceedings, professional and academic journals.

Secondly, the preparation of interviews started with the identification of relevant actors to be interviewed in both case study areas, following Rubin & Rubin's (1995) guidelines for interviewee selection. A first list of potential interviewees was prepared through the evaluation of the first collected documentary evidence on the water management processes in question. Interviewees were selected from the following main actor groups: relevant authorities on different administrative levels, water user interest groups, other non-governmental groups (including e-NGOs), academic experts and technical consultants. The list of potential interviewees was further extended and modified during field work, since actors interviewed often proposed other important actors which should be interviewed but were missed out in the initial interviewees' list of the author.

The type of interviews carried out is mentioned in the literature as focused semi-structured (Rubin & Rubin, 1995) or as in-person semi-structured interviewing (O'Sullivan & Rassel, 1999). This type of interviews follows a list of questions prepared by the interviewer whose content and order can be slightly modified from interview to interview. To this aim, a semi-structured questionnaire was prepared whose general outline is given in Table 4.8. This questionnaire was adjusted and further specified according to the content of each case study and according to each individual's knowledge on specific subjects of the questionnaire (an effort was made to limit the main topics dealt with in each interview). In general, the qualitative interview type chosen fits the explanatory research approach of this dissertation.

Table 4.8   Questionnaire for semi-structured interviews

Actor information

1. What are the aims and actions of your organization in water resource management in the water basin? How are your activities financed?

2. Do you have access to the water management decision-making? If yes, how?

3. What resources do you use in the water management process (financial, political, legal, knowledge)?

Problem definition and water use rivalries

4. To your opinion, are there water use rivalries in the basin? If yes, which and what is the issue at stake?

Chronology of water management process

5. Questions on the specific events of the water management process of each case study and according to the competence of each interviewee

    a. What action was taken to deal with the water use rivalry(s) (action may include formal and informal decisions, studies and management plans, legal action and protests, technical measures etc)?

    b. Why was this specific action (mention which) preferred? Which actors were activated to promote it?

    c. Was action linked to any participatory process? If yes, who participated?

    d. Was the specific action in agreement with the perspectives of multiple actors (your perspectives?)?

    e. What was the outcome of action taken? Were any relevant decisions and measures implemented?

f. Did any of the proposed solutions lead to conflict of interests and new rivalries? Did negotiations take place?

Property rights to water

6. Questions on property rights to water according to the competence of each interviewee

   a. Are rights to use water (for different water uses) confirmed by legal regulations?

   b. Is there illegal use of water? What is the role of older rights not yet treated by recent regulations?

   c. Discuss the regulations and restrictive measures imposed on water use by authorities

   d. Discuss the implementation problems of regulations and restrictive measures

   e. Has any action been taken to establish the right of the environment and nature to water?

Relationships with other actors

7. How do you view other actors' actions and behaviour in the water management process?

8. Is your cooperation and coordination with other water actors satisfactory? How has quality of communication with other actors developed over time?

9. In your opinion, which actors have become more sensibilised to more sustainable water use? What has contributed to this?

10. Do different actors perceive joint opportunities from the resolution of water rivalries in the direction of more integrated water management?

11. Is there overlap of water responsibilities and competences in the water basin?

Note: The original questionnaire was drafted in Modern Greek.

The preparation of data collection was made more systematic by developing a case study protocol (according to Yin, 1994) for use in each case study. Each case study protocol included the following information to guide the field work:

- The case study research questions and hypotheses.
- A brief overview of the analytical framework for the water basin level.
- An overview of the techniques to be used in the analysis of the case study data.
- Names of sites to be visited and a schedule for data collection and interviews.
- A list of the potential interviewees and their contact details.
- A list of documentary sources which should be consulted in the field.
- The questionnaire of Table 4.8 appropriately extended with details for the water management process of the specific case study area.

Field work

Empirical research on the basin level was carried out during field visits in the two case study areas. The first case study area (the Vegoritida basin) was visited in January-March 2005 and in November-December 2005. The second case study area (the Mygdonian basin) was visited in March-April 2005 and in July 2006.

Most documentary evidence was collected during field visits, since only limited information could be collected a priori from on-line sources and via email communications. To this aim, the investigator organised access to the file archives of different authorities, academic institutions and non-governmental groups, which had information available on the water management process of each basin. The availability of relevant documents was also included as a standard question during or after personal interviews with actors.

Additionally, a series of in-person interviews were carried out in each case study area. Lists of the actors interviewed are presented in Annex III. Interviews were carried out mainly with actors directly involved in the water management process of each case study. However, also a few more neutral, knowledgeable observers or informants of the process were interviewed, who were willing to explain and give details on several aspects and events of the process. The investigator made an effort to interview such observers or informants as early as possible in the field work period. This way, the investigator could gain factual, descriptive information which could be omitted from later interviews with actors holding more specific, critical information and having less time available. In general, the arrangement and execution of interviews followed relevant guidance provided in the literature by Rubin & Rubin (1995) and O'Sullivan & Rassel (1999).

At the start of each interview, it was emphasised that interview material would be used exclusively for the research purposes of this dissertation. To protect to some extent the identity of interviewees, use of interview material in the dissertation text refers to the organisations of the interviewees but not to the names of the individual persons interviewed. Finally, permission was asked to record the interview and/or take notes. In case the interviewees felt uncomfortable with the tape recorder, the investigator held a written protocol by taking thorough notes throughout the interview. The interview started with a brief presentation of the subject and aims of this dissertation. In the following, the questionnaire presented in Table 4.8 was used, although often the content and order of questions were modified to suit the knowledge and experience of the specific interviewee. As soon as possible after each interview, interview transcripts were produced (in Greek) on the basis of the tape recording or on the basis of the written notes taken during non-recorded interviews.

Analysis of case study data

The general framework for the analysis of the case study data is based on the three phases of reconstructing decision-making processes proposed by Dente et al. (1998), including:

- The construction of the chronology of the decision-making process (replaced in this dissertation by the chronology of events in the water basin management process).
- The analysis of the actors involved.
- The analysis of the patterns of interaction and the definition of success factors (replaced in this dissertation by the analysis of the variables and causal links of the analytical framework).

These three phases also determined the main structure of each case study in chapters 7 and 8. The first part of each case study includes a general description of the water basin, problem pressures and the reconstruction of the main events of the water basin management process. The second part includes an actor analysis and the systematic assessment of the key analytical variables and their causal relationships.

To construct the chronology of important events, information was used from the analysis of the press, formal and informal documents as well as interviews, especially those interviews with a more descriptive, factual character (initial interviews with observers or informants).

Secondly, the different governance levels and main actors in the water management process were analysed. In fact, an important result from the construction of the chronology was a list of the main actors involved in the process. The following step was to construct monographs for the most important actors participating in the water management process. Although several documents were useful in this respect, the most important sources for actor monographs were the actor interviews. Box 4.3 presents the main issues to be considered in the construction of an actor monograph. These issues were largely integrated in the interview questionnaire presented in Table 4.8.

Box 4.3   Issues for an actor monograph

Role/competency:	What is the role of the actor in the water management process? What main actions and events define their role?
Problem perception:	What is the problem from the point of view of the actor? Every actor defines the problem in reference to its own knowledge, resources and objectives. What is the stake of the process (what is here going to be decided about?)
Objectives:	What objectives (motivations) does the actor have in this process?
Resources:	What resources does the actor have available and which he uses in the process (financial, political, legal, cognitive resources)?
Perception of others:	How does the actor view other actors' behaviour?

Source : Based on Dente et al. (1998).

The phase of overall interpretation was equivalent to the analysis of data to systematically assess the variables of the analytical framework, i.e. the regime components of property rights and public governance, the influence of external change impetus on the regime, the contextual conditions influencing regime changes and the actor characteristics influencing the regime change and implementation processes.

The systematic assessment of the analytical variables was carried out by identifying repeating patterns and common themes in the document analysis and the interview material and gathering those in extended table data matrixes. In fact, data analysis involving the production of document and interview summaries started early in the research process once the first data was collected.

Finally, the assessment of the individual analytical variables was followed by the analysis of the causal links between variables in the hypotheses. Each hypothesis formulated in section 4.2 was checked using information from various written documents as well as actor replies to individual questions of the interview questionnaire (see Table 4.9 for links between hypotheses, research questions and the interview questionnaire).

Table 4.9   Linking case study hypotheses to research and interview questions

Hypotheses	Research questions	Related interview questions
The less favourable the contextual conditions and the characteristics of interacting actors for integration attempts, the more likely that Greek water basin regimes fail to reach a more integrated phase (Hypothesis 2)	Did the extent of uses regulated by the water basin regime and the coherence of the regime elements (property rights, public governance and policy) change since the start of the water rivalries under investigation? If yes, how did they change? Do changes justify a shift of the water basin regime to an integrated phase?	Questions 1, 2, 5, 6, 7, 8, 9, 11
	To what extent were the motivations of the actors involved favourable to a more integrated management?	Questions 1, 5, 6, 7, 9
	To what extent did the actors possess of adequate information and resources to contribute towards more integrated management?	Questions 2, 3, 5, 6, 11
	To what extent were conditions favourable to integration present?	
	• Was there a tradition of cooperation among the actors involved?	Question 5, 8, 11
	• Did the actors in the process assess that there were 'joint chances' stemming from more integrated management?	Question 10
		Questions 4, 5, 7, 9
	• Did the actors in the process assess that there was a 'joint problem' stemming from the lack of integrated management?	Questions 5, 7
	• Was there a credible alternative threat provided as a solution by a dominant actor?	Questions 5, 6
	• Were there well-functioning institutional interfaces to support a change towards integrated management?	
The less new incentives provided from regime change and the less adequate the implementation of any new incentives, the more likely that there are less positive sustainability implications of regime change on water use (Hypothesis 3)	To what extent had the implementation of measures and policies adopted positive implications on the sustainability of water use in the basin?	Questions 5, 6d, 6e
	To what extent were the motivations of the actors involved favourable to the successful implementation of policies and measures adopted?	Questions 1, 5, 6, 7, 9
	To what extent did the actors possess of adequate information and resources for the successful implementation of policies and measures adopted?	Questions 2, 3, 5, 6, 11

5 National institutional water regime in Greece

5.1 Introduction

This chapter uses the analytical framework for change in national institutional water regimes (see chapter 4/4.1.1) to characterize and explain the development of the Greek national water regime. The main concern is whether, to what extent and why the Greek national regime has so far changed or has not changed towards a more integrated phase.

In short, it is shown that although attempts at a more integrated national water regime have taken place since the 1980s, the national regime still remains quite complex and fragmented. It is checked whether failure to change towards a more integrated regime can be explained by the influence of an unfavourable institutional context which prevailed during the integration attempts. This institutional context is examined in terms of power-based, value-based and perceptions-based conditions, which are considered decisive for change in national water policy and governance in the direction of more integration.

The empirical research on the Greek national water regime followed the methodological approach described in chapter 4/4.3.1. To set the characterisation of the Greek national water regime into context, this chapter starts off with a description of key administrative levels in Greece (also related to water management) followed by a brief overview of the status and uses of Greek water resources (sections 5.2 and 5.3). Thereafter, the chapter describes the two key regime components, i.e. the system of water property rights and of public governance (sections 5.4 and 5.5), which have been developed in Greece to deal with the main water management problems of the country. Particular emphasis is placed on the first Water Management Law of 1987, which was the first national effort to shift towards a somewhat more integrated institutional water regime. More recent policy changes were induced by the EU WFD and led to the adoption of a new national Water Law in 2003. Such recent developments are presented only up to 2006, when investigations for this dissertation were completed. In sections 5.6.1 and 5.6.2, this chapter discusses the diachronic development of the Greek national water regime, on the basis of the regime change variables of extent and coherence. Section 5.6.3 is then dedicated to the external change impetus and the institutional context, which influenced the outcome of attempts to change the national water regime to a more integrated phase. The final section 5.7 summarises key conclusions on the Greek national water regime and makes recommendations for changing towards more integration, also on the basis of a comparative discussion of the national water regime of Greece and of six other European countries.

5.2 Key administrative levels

Greece can be indisputably categorized as a unitary state, which operates largely on the basis of a centralized bureaucracy. The centre has until now conferred a minimum level of powers to its periphery (Kapsi, 2000). The tradition of concentration of the Greek politico-administrative system hindered in the past vertical coordination in several policy areas and the set up of a powerful network of decentralised institutions such as regional environmental services (Spanou, 1995).

Recently, there have been certain reform initiatives of the State to increase decentralization, largely driven by evolutions on a European level. Local self-administration at the level of communities and municipalities was established very early in Greece, but until recently there was no intermediary level of an elected administration on a regional level. In the meantime, on a European level, a progressive strengthening of regionalism was taking place, especially after the Maastricht Treaty in 1992 established among others a Committee of the Regions. Therefore, for the sake of better programming and development, it was imperative that an intermediary level of administration that would ensure the interests of the regions were established in Greece (Kapsi, 2000).

By Law 2218/1994, the Greek territory was administratively organized into 13 Regions that correspond to the target-regions of the European structural funds (see Figure 5.1). The Regions among others play an important role in the planning and implementation of European programmes (including the management of European regional funds through their Regional Operational Programmes). The Regions, however, do not constitute yet a third level of regional self-administration. They are administered by non-elected officials that are directly appointed by the central government and represent the State (Kapsi, 2000).

Figure 5.1   The 13 administrative Regions of Greece

Source: MEPPW (2006).

The same law (2218/1994) established a 2nd level of local government and self-administration at the level of Prefectures. Several Prefectures are part of each Region. Prefectures existed also prior to 1994 but they were not forms of self-administration. In 1995, the first directly elected Prefects assumed office following Prefectural elections. One of the most important competencies conferred to the Prefects was the management of Prefectural funds, a large part of which came from the European Union, and were destined to cover the cost of certain types of public constructions. Since 1995, however, the institution of Prefectures was seriously put to the test due to phenomena of clientelism, abuse of power by newly elected Prefects and instances of corruption (Kapsi, 2000). All in all, this whole decentralisation process can be seen as an improvement, if only because little had previously existed in terms of regional governance and vertical coordination. Nevertheless, the reform process has been criticized to have resulted in a conventional, and questionably operational, form of regional organization with restricted competencies (Kapsi, 2000).

The 1st level of local government (municipalities and communities) also went through a fundamental reform in the mid-90s, whereby local authorities were merged into larger units of self-administration to overcome fragmentation. Beforehand, too many small local authorities lacked adequate political representation, had insufficient capacity to provide desired services to the community, and had limited participation in the procedures of local and regional development. The reform attempted to reorganise the country into economically viable units of local government. The aim was to define the powers of local authorities, new geographical boundaries that merged local authorities and improved economies of scale, new financing arrangements to allow for the delivery of services including water supply and sewerage, adequate staffing for those services, and specific monitoring and enforcement mechanisms.

In the field of water management, responsibilities are distributed to the national, regional and local tiers of administration, with the greatest share of decision-making and policy-making still enjoyed by the national level. The most important national ministries with responsibility over water resources are:

- The Ministry of Agriculture, which plays a key role in the use of water in irrigation.
- The Ministry of Development, which plays a role in the management of water for industrial and energy use. Until 2003, the Ministry of Development was also the national supervising ministry for water resource issues according to the national 1987 Water Law.
- The Ministry of the Environment, Physical Planning and Public Works (in short, Environment Ministry), which has national coordination competence in water management since 2003, after the transposition of the EU WFD via the national 2003 Water Law.
- The Ministry of the Interior, Public Administration and Decentralisation (in short, Ministry of Interior), which is responsible for municipal water supply, with the exception of the cities of Athens and Thessaloniki which lie in the competence of the Environment Ministry.

On the regional level, the Regions are responsible for the approval of planning and financing of projects involving water supply and irrigation (Assimacopoulos et al., 2002). In the context of the 1987 Water Law, also Regional Water Management Departments were set up in each Region, with competence to implement the specific law on a water district level (see map of water districts in Figure 5.3). These Departments were re-structured into Regional Water Directorates by the 2003 Water Law (see section 5.5.4 for details). Other Directorates of the Regions also have competence in issues related to water management, e.g. land reclamation works and authorisation of activities with an impact on water.

Prefectures are responsible for the more operational management of water resources on the user level. In this context, they have responsibility for issuing effluent disposal permits. Until 2005, they also had partial responsibility for the issuing of permits for water abstractions and for the construction of water works. This responsibility was concentrated to the Regions through the 2003 Water Law. The Prefects also have the power to issue regulatory decisions for water resources within their territory. Such Prefectural decisions may deal with the quality of water resources and define conditions for the disposal of wastewater and effluents into water. They may also deal with the quantity of water resources, thereby setting measures for the protection of surface and groundwater quantity.

The 1st level of local government (municipalities and communities) has traditionally been responsible for the maintenance of local infrastructure, including water supply, sewage networks and treatment plants, or for the creation of public companies responsible for the provision of water and sanitary services (Pridham et al., 1995).

As far as the sharing of responsibilities on a horizontal level is concerned, there is considerable fragmentation along sectoral lines in terms of water protection and management. Such fragmentation was also observed in terms of general environmental protection and reflects a more general tradition of intense compartmentalization within the public administration of Greece (Pridham, 1994). On the national level, coordination among ministries dealing with similar or related issues tends to be minimal. In the water management field, ministerial coordination formally existed in the form of a joint ministerial water committee (set up by the 1987 Water Law). In practice, however, ministerial cooperation was subject to the vagaries of political interest. The lack of intersectoral coordination on the national level (and consequently on the regional level) has significantly hindered progress towards more integrated water management given that several Ministries co-managed and often competed over water.

In brief, the characteristics of the politico-administrative set-up of Greece, which are of interest for the further discussions in this dissertation, can be summarized as follows: Greece is a centralized country with recent tendency to decentralization via a State reform process. Water management tasks are divided between national, regional and local levels of administration with the national level playing so far a rather dominant role. Apart from still weak vertical coordination between central and decentralised levels, there is also considerable horizontal fragmentation between different sectors related to water management.

5.3 Water resources and uses

Distribution of water resources in space and time

Greece is a small country (132,000 km²), mountainous to its largest extent. It has a long coastal line (15,021 km) and a significant number of islands. As a result of its geomorphology, the country is segmented into relatively numerous small hydrological basins (MDEV, 2003) with small rivers, torrential streams and small lakes (see river basins in Figure 5.2). Thus, relatively numerous small water bodies have been formed rather than few large water bodies with limited spatial distribution. There are 765 recorded streams, of which only 45 are classified as perennial. Four rivers flow from the northern neighbouring countries into Greece and one crosses the border in the opposite direction into Albania. There are also several lakes (almost 60 in number according to Assimacopoulos et al. (2002)).

Figure 5.2   The river basins of Greece

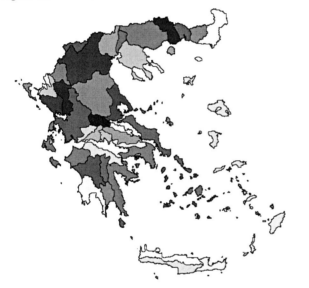

Source: Assimacopoulos et al. (2002).

In 1987, the water landscape of Greece was divided into 14 water districts for administrative management purposes by the 1987 Water Law. Water districts included groups of water basins with as far as possible similar hydrological-hydrogeological conditions. The 14 water districts were maintained in the initial implementation phase of the EU WFD and the national 2003 Water Law, as the 14 River Basin Districts of the country (see Figure 5.3).

Figure 5.3   The 14 River Basin Districts of Greece

Source: MEPPW (2006). Note: See names of numbered River Basin Districts (previously called water districts) in Table 5.1 below.

Contrary to many central and northern European countries, the most significant water problem of Greece is quantitative. Moreover, the available water resources and water demand in Greece are not evenly distributed in space and time (MDEV, 2003). In terms of distribution over time, there is high concentration of rainfall in winter. In southern Greece, 80-90% of rainfall falls in winter, while the summer rainfall reaches maximum 20% of the annual distribution only in northern Greece (MDEV, 2003). Water demand increases in May-September, reaching its peak in July, when water availability is at its lowest due to the decrease in precipitation. The summer peak is, on the one hand, due to the heat that increases drinking water consumption, also linked to summer tourism. On the other hand, agriculture, which is the largest water consumer in Greece, consumes water mainly in the dry summer period due to the current irrigation practices and cultivated crop types (Assimacopoulos et al., 2002; MDEV, 2003). As regards water distribution in space, the following are the main relevant issues as summarised by MDEV (2003):

- The highest water demand concentrates mainly on the coastal zone and in the large plains. Additionally, the two large urban centres of Athens and Thessaloniki are both water deficient areas.

- The maximum precipitation is recorded in the western part of the country, which however has a relatively low rate of development and therefore low demand in relation to its available water resources. On the contrary, eastern Greece, including the area of Athens, the Aegean islands and the plains of Thessaly, has less water resources available but it is characterized by high demand. To a large extent, shortage in eastern Greece is dealt with via water transfers, which have been carried

70

out or are planned across the country (mainly serving the water supply of large urban centres and of large agricultural plains). The islands face intense water problems but of small scale, mainly related to the seasonal high water demand for tourism. Table 5.1 provides information on water supply and demand for each of the 14 water districts.

Table 5.1   Water supply and demand in the 14 water districts (in July 2003)

Water Districts	Supply (hm³)	Demand (hm³)	Water Balance
01 Western Peloponnesus	73	55	Surplus
02 North Peloponnesus	122	104	Surplus
03 Eastern Peloponnesus	56	67	Deficit
04 Western Sterea Ellada	415	82	Surplus
05 Epirus	193	33	Surplus
06 Attica	56	54	Slight surplus (1)
07 Eastern Sterea Ellada	128	187	Deficit (2)
08 Thessaly	210	335	Deficit
09 Western Macedonia	159	136	Slight surplus
10 Central Macedonia	137	130	Surplus
11 Eastern Macedonia	354	132	Surplus
12 Thrace	424	253	Surplus
13 Crete	130	133	Slight deficit (3)
14 Aegean Islands	7	25	Deficit
Country total	2464	1726	

(1) Water resources are mainly transferred from neighbouring water districts.

(2) The irrigated areas, according to the National Statistical Service of Greece, appear over-estimated. Therefore, although the water district has almost sufficient water resources, it appears as largely deficient.

(3) Demand is covered mainly through springs and wells (data as of 2003).

Source: MDEV (2003).

Finally, it is important to note that northern Greece depends largely on surface waters of transboundary nature. Approximately 13 km³/y of water entering Greece (30% of total average annual water resources) originates in the northern neighbouring countries, where water is both abstracted and used for effluent disposal (Assimacopoulos et al., 2002).

Water demands of different uses

Water demand is unequally distributed among the main water uses in Greece. Total water demand equals 8,243 hm³/y. 83% is used for irrigation, 13% for drinking water supply, 3% for industry and energy and 1% for animal farming (MDEV, 2003).

As concerns irrigation, it is interesting to note that 32% of the total agricultural land and 60% of the plains are irrigated. Except for being the main water consumer, irrigation faces serious problems of water losses since its methods have not been yet 'modernised' or adapted to current requirements and standards (MDEV, 2003). In regions where surface water is not sufficient or not easily exploited, groundwater is used for irrigation through collective or private wells. Several groundwater aquifers already face serious problems of quantitative and qualitative degradation due to their over-exploitation for the purpose of irrigation. Over-abstraction leads to the lowering of the groundwater tables, which in its turn results in increased pumping costs and in some cases to land subsidence. In coastal aquifers, over-abstraction has led to their salinisation through sea water intrusion (Dosi & Tonin, 2001).

In general, in Greece, only 60% of water consumed annually is surface water, while the average surface water consumption in the EU reaches 70-90%. Groundwater abstraction covers 40% of annual water demand, which is double the EU average (Gorny, 2001). According to Margat (1999), the use of groundwater abstracted in Greece per sector is: 37% for public water supplies, 58% for agriculture and 5% for industry (data of 1990). The use of groundwater is of great importance for covering local needs (MDEV, 2003).

The extensive drilling of private wells to use groundwater is an important issue in the Greek context. Evidence suggests that a large part of the water used for irrigation is obtained through private wells, often without a permit. This results in lack of exact measurements of the full volume of water used for irrigation, since only the volumes provided by the public irrigation networks are known (Assimacopoulos et al., 2002).

As concerns water consumption for drinking, the water supply needs of the permanent population are 200 litres/day/person and of the seasonal tourist population 300 litres/day/person. Because of tourism, demand is particularly high in coastal areas and islands. Drinking water needs are covered either by groundwater and springs or by surface water. Water supply through groundwater is preferred due its lower cost. Actually, the domestic sector in Greece is quite dependent on groundwater supplies since aquifers provide 64% of the water directly consumed by households (Margat, 1999). Except for the large urban centres, which are served through large-scale water transfers, in most cases needs are met using local water resources (MDEV, 2003).

As concerns water demands in the industrial sector, the industrial units in Greece have been developed mainly within or around the urban areas. Therefore, in terms of water supply and effluent discharges, they are served by the infrastructure of these areas. Industries which are located in remote areas are obliged to arrange for their own water supply, sewerage and effluent treatment (MDEV, 2003). It needs to be mentioned however that relevant sufficient regulations and control mechanisms have not been developed yet.

Among the non-consumptive uses of water, hydropower is quite important in Greece. Because of its topography, the country has significant water potential, whose greatest part is concentrated on the large rivers of western and northern Greece[20]. There are fifteen main hydropower plants of the Public Power Corporation and thirteen additional small ones. According to the MDEV (2003), only one third of the economically available water potential has so far been used or is under development for hydropower.

Water quality issues

Greece has in general water of good quality. Nevertheless, human activities combined with lack of planning and of control over long periods of time, have affected the quality of both surface and groundwater.

---

[20]   Rivers Acheloos, Arachthos, Aoos, Aliakmonas and Nestos.

Domestic effluents and agrochemicals are the major sources of surface and groundwater pollution in many parts of the country. Although progress has been made in terms of urban pollution, still more effort is needed for the treatment of wastewater from small agglomerations. As concerns the treatment of industrial effluents, there is no significant progress made so far (MDEV, 2003). Although industry is not highly developed, it pollutes water with high local intensity due to its concentration near urban centres and the uncontrolled dispersion of small-scale but polluting industrial activities. As concerns agriculture, the widespread overuse of fertilisers and pesticides (most of them not adapted to the Greek climatological conditions) results in high rates of run-off to surface water with disastrous biodiversity impacts and in the contamination of groundwater through leaching (Kouvelis et al., 1995).

As concerns surface water, mainly the water quality of large lakes and the quality of small streams located close to inhabited areas is largely degraded (MDEV, 2003). Water quality problems in rivers mainly concern transboundary pollution and are concentrated on a few large transboundary rivers, whereby the disposal of raw domestic and industrial effluents in the upstream countries has had profound effects on the quality of water entering Greece. Other than that, water quality is satisfactory in Greek rivers at least as far as hazardous substances are concerned. This is due to the relatively limited industry, which produces other types of pollution (not toxic) (MDEV, 2003). The quality of Greek rivers is also dependent on the time of the year. Water levels decrease significantly in the summer, thus allowing higher concentrations of pollutants (Assimacopoulos et al., 2002).

As concerns groundwater, overabstraction and saline intrusion exacerbates the deterioration of water quality and increases water shortage, especially in coastal areas and on the Aegean islands (MDEV, 2003; Assimacopoulos et al., 2002).

5.4 Property rights and regulation of water uses

5.4.1 Evolution of water property rights

Rights with respect to the quantitative use of water

In the independent Greek State of the 19th century (previously occupied by the Ottoman Turks from the 15th century until 1821), water ownership and use rights were often unclear. This was partly due to the frequent lack of clarity over land ownership titles as explained in the following. During the Ottoman occupation, most land belonged to the Ottoman authorities, but reports also exist on acts of sale of private land. Water use rights were also often sold along with or without private land among private persons (Kallis, 2003). Indeed, in Islamic theory, although water cannot be sold, the rights of water use can. Private land could, thus, be sold without water rights and vice versa, meaning that water rights were not necessarily annexed to land (Forni, 2003). The independent Greek State assumed public ownership of all land previously belonging to the Ottoman authorities, which covered the largest part of the State. As concerns lands privately owned under the Ottomans, it became difficult to control their ownership titles due to the lack of a land register in the new Greek State and the destruction of most Ottoman records (Kallis, 2003). As a result, there was frequently no clear picture of the actual private land ownership titles. Moreover, trespassing and "privatizing" public land at will occurred very often, leading to court conflicts between the State and private persons. In this time period (19th century and beginning of 20th century), land owners (whether they were lawful or not) also implicitly claimed the right to use water on and below their land. There is also evidence that water rights could be bought off, usually for the purpose of water supply.[21] In general, however, the State often faced difficulties to proceed with several hydraulic works, due to conflicts between State and individuals over unclear or established private land ownership. Finally, in 1920, a national law introduced the possibility for the obligatory expropriation of private land by the State in order to facilitate the construction of public hydraulic works (earlier in 1911, there was a general Constitution modification allowing the State to expropriate private property against compensation for the purpose of works of "public benefit").

In 1940, a Greek Civil Code was also introduced which defined "things in common use" including therein waters of free and perpetual flow (rivers), the banks of navigable rivers, large lakes and their shores (article 967).[22] Things in common use can belong to the State, towns, communities, or private persons when this is authorised by statute. The owner of a thing in common use may exercise all prerogatives of ownership that are compatible with common use (Yiannopoulos, 1988). In the case of water resources, article 969 defined that any person could use waters of common use and for cases of conflict among entitled users, an order of use priority was set as follows: 1) most important use for the common benefit, 2) use promoting social economy at the most, 3) oldest use, 4) use for an enterprise connected to a certain location, 5) use for the benefit of the riparian. Groundwater was not included in "things in common use". The Civil Code (article 954) defined that groundwater (and springs), regardless the amount of water, are inseparable elements of the overlying land and belong to the land owner.[23] The land owner can use groundwater at will, as long as he does not discontinue or significantly reduce water already used by any local village for its needs (article 1027).[24] Still, it was often the case in the past that water became object of fierce

---

[21] An example is a legal act of 1936 (nr.226/1936) on the approval of funds for buying off water use rights for the water supply of the settlement of Agios Stefanos.

[22] The substance of the Greek Civil Code Book on Property (articles 947-1345) was derived from the Romanist tradition, indigenous Greek variations developed in the 19th century, and from the modern, at the time, civil codes. The influence of the French, German and Swiss Civil Codes is here particularly noticeable. Prior to the promulgation of the Greek Civil Code, civil laws of the Byzantine Emperors remained in force in the independent Greek State of the 19th century (Dacoronia, 2003). The Byzantine-Roman Law which preceded the Civil Code defined the air, sea, shores and running waters as "common to all" (Karakostas, 2000).

[23] The Greek Civil Code distinguished between movable (personal property) and immovable things (real property). Water under the surface (groundwater) is a component part of real property, insusceptible to separate real rights.

[24] In most European countries, groundwater is traditionally considered as "closed water", and as such is part of land ownership rights together with rainwater, ponds dug by the landowners, and springs generating minor flows and used within the property. However, countries with stronger customary traditions have tended to limit the abstraction of groundwater to uses non-detrimental to neighbours, while countries which have increasingly relied on Roman Law tradition initially placed no restrictions on the use (and abuse) of groundwater by landowners, be it eventually detrimental to his neighbours. In all countries, however, when

commercial exploitation especially on islands, where certain privileged land owners sold their water to other island inhabitants.[25]

In 1948, Law 608 (on the administration and management of water for irrigation) introduced a definition for "public waters", including all flowing or non-flowing surface and groundwater, except for those defined as "private". This Law defined "private waters" as follows: springs on or groundwater underneath private land, groundwater underneath private land abstracted to the surface, public waters abstracted for private use based on titles or concessions, rain water, standing water and small lakes on private land.

As concerns water abstractions in general, until the 1980s there was restricted State control, mainly by the Ministry of Agriculture. Law 608/1948 defined that the use of public waters for irrigation by any private or public entity could be conceded only by the Ministry of Agriculture, while for other water uses also the decision of the Minister of Public Works was necessary. Private use of public water was limited to the use for which the right was conceded and to certain water quantity up to the real needs of the beneficiary. The Service for Land Reclamation of the Ministry of Agriculture, set up in 1959 with central and regional branches, had a Water Administration Department with competence for public water use concessions and expropriation of existing private rights. Concerning groundwater, a license had to be acquired from the services of the Ministry of Agriculture to drill wells for irrigation (Law 1988/1952). For wells serving other uses (e.g. water supply and industry), also other ministries were involved in their planning and licensing. All in all, at that stage, only the right to drill a well was linked to an official permission, but not the right to abstract groundwater. Since 1958, however, the possibility was given to Prefectures (by Law 3881/1958 on land reclamation works) to impose restrictive measures on abstractions for the protection of groundwater, including restrictions on the drilling of new wells.

In 1987, the first national Water Law (Ministry of Development, L. 1739/87) characterized water as "natural good" and "means for meeting the needs of society". The Law introduced a permit system for the use of surface and groundwater, by private and public users. Exceptions were personal and domestic water uses, where no water supply network existed to cover these needs and consumption was less than 3m³/day (these exceptions were not reinstated by the new Water Law later in 2003). The permit system, in essence, brought abstractions for all important water uses into the public domain. For groundwater, this meant that the right to use it was detached from private land ownership and transferred to public law. The right to use water was no longer a private right, but in order to exercise it, an administrative permit was needed.[26] The intention was to "rationalise", in general, the system of water ownership and use rights, which was until then a prohibitive factor to sustainable

---

public water supplies rely on groundwater, public control on water and on the overlying land has gradually developed (Barraqué, 1998).

[25] Parliament plenary records on the draft 1987 Water Law, 14.10.87, p.288, Member of Parliament S. Korakas.

[26] Parliament records on the draft 2003 Water Law, 29.10.2003, p.352, R. Zisi (Deputy Minister for the Environment, Spatial Planning and Public Works).

water management and to the fair allocation of water resources. In practice, until 1987, any public or private person or entity used water resources in such quantity, quality and use to serve its own needs (Ministry of Industry, Energy and Technology 1988).

The 1987 Water Law also introduced a permit system for the construction of water works by entities outside the public sector. The construction of water works by the public sector was to be regulated separately in the context of national water resource development programmes (which were, however, not drafted, as illustrated later in this chapter). A Common Ministerial Decision (43504 of 12/2005), complementing the more recent 2003 Water Law, finally obliged also public sector entities to apply for water works' permits.

The details for the operation of the system of permits, introduced by the 1987 Water Law, were set in a Presidential Decree of 1989 for water use permits (256/1989) and a Ministerial Decision of 1989 for water works permits (5813/17.5.1989).

Various permit-issuing authorities were defined, depending on the water use. For agricultural and industrial use, permits were issued by the Prefectures, for water supply, by the Unions of Communities and Municipalities, and for multiple water use and energy production, by the Regions. Additionally, all permits had to be approved and registered by the Regions, before being issued. The 2003 Water Law later concentrated permit-issuing authority to the Regions, aiming at better coordination of the permit-issuing process.[27]

Permits are issued for a limited time period (a period of maximum 10 years) for the following water uses: water supply, agricultural use (including animal farming), industrial use, energy production and recreation (including bathing). Water use permits should define the water quantity to be abstracted (maximum und minimum), the appropriate water quality, the water source, the mode of abstraction, the duration of the permit, the time of the year in which water can be used and any other terms for the use of water. Especially permits for private wells are issued only for locations on one's private land or lawfully rented land. Permits to use surface water are not related to land ownership or rental.

The Presidential Decree 256/1989 on water use permits defined that a permit should be issued, bearing in mind the quantities of water available in a specific location and for a specific use, any water use restrictions valid in the specific Prefecture and the relevant water development programme. It was also defined that use permits should be issued for water quantities not exceeding the highest limit of permit holders' real needs, leaving surplus water at the disposal of the State for possible allocation to other uses. It was very controversial, though, already during parliament discussions on the draft 1987 Water Law, how to define "the highest limit of real needs". In an effort to clarify "real needs", the Ministry of Development issued Common Ministerial Decisions (CMDs) in 1989 and 1991, which defined lowest and highest limits of water quantities for rational water use in irrigation, water supply

---

[27] A Common Ministerial Decision (CMD) defining a revised system for water permits, according to the 2003 Water Law, was issued in 12/2005. For the preparation of this CMD, also written comments from the existing Regional Water Management Departments were invited.

and tourism. In the case of irrigation wells, permits define the maximum discharge of abstracted water, depending on the kind of crop and irrigated surface. Since 2002,[28] permits also need to determine the maximum annual water consumption. The usefulness of this requirement, however, is questionable, considering that no water meters are installed nation-wide.

An additional requirement in the permit procedure for the construction of wells is the execution of an environmental impact assessment study (EIA) (required by Law 3010/2002). There is however critique, that this requirement only adds bureaucratic burden to authorities and unnecessary costs to water users. Instead of an EIA for each single well, there is need for river basin management plans, which should define limits to water consumption and environmental degradation (Greek Committee for Hydrogeology, 2003) in each basin and, if necessary, in specific parts of the basin.

The Civil Code and the 1987 Water Law also included some clauses for the resolution of differences between water users (due to over-abstraction or other reason). In specific, the 1987 Water Law referred private differences on water ownership, use and disposal rights to political courts. The Civil Code (articles 57 and 59) defined that, in case of impairment of the public benefit character of water or of a water use right, the affected user (regardless whether he has ownership rights) is entitled to request the discontinuation of impairment at present and in the future as well as compensation. According to article 914 of the Civil Code, whoever causes illegally and on purpose damage on common use or public waters has the obligation for compensation (which is monetary).

All in all, throughout the 1990s, the water permit system could be gradually established in the administration nation-wide but water overexploitation, unfortunately, remained commonplace. Especially groundwater underneath private land is still used at will by land owners (Delithanasi, 2004). The unsustainable groundwater consumption behaviour of private users could be encouraged by the maintained private character of groundwater ownership (groundwater remains an inseparable element of private land in the Civil Code). Given the importance of groundwater as a resource in Greece, Karakostas (2000) proposed to define it by law as a "thing in common use" similarly to running surface waters. Then, ownership rights to groundwater should only be allocated by law when this does not harm the "common use" of groundwater.

There are also additional factors related to the regulation of property rights, which contribute to the unsustainable use of groundwater and of water in general. A first factor is the omission of current public law to deal with those abstraction rights, which were already established before the regulations of 1989 requiring water use and works permits on the basis of the 1987 Water Law. A Presidential Decree, which was expected to define a permit procedure for these pre-1989 abstraction rights, was never issued. Thus, to date, permits are issued

---

[28] CMD 25535/3281 of 20-11-2002 on the approval of environmental conditions by the Secretary General of the Region for works and activities listed in subcategory 2 of category A of the CMD 15393/2332/2002 "Listing of public and private works according to categories".

only for new water uses, while pre-1989 abstraction rights based on law or custom are maintained without being subject to the permit system.[29] Secondly, there is inadequate control of illegal water use and water works. According to expert estimations, an extra 30% of illegal wells should be added to each official figure for legal wells (Delithanasi, 2004). Additionally, the requirement to issue permits for the use of water by public entities (like municipalities) was often ignored in practice. Thirdly, there is inadequate inspection of the specific terms defined in permits. Relevant compliance-check responsibility was until recently in the hands of the 1st level local self-administration and rural police forces (assisted by the various permit-issuing authorities: Prefectures, Regions and Unions of Communities and Municipalities). This competence allocation was considered ineffective, since most local self-administrations did not have organised technical services for this task (Papalimnaiou, 1994). In practice, water users could abstract uncontrolled large amounts of water, which were often tolerated by the local authorities (MDEV, 1996). The 2003 Water Law concentrated control responsibility of compliance with issued permits to the Regions, whose effectiveness remains to be seen.

Limiting the right "to pollute water": Regulation of water pollution

A system of permits for the disposal and treatment of wastewater and effluents into surface water, groundwater and onto the ground was introduced as early as 1965 by means of the Sanitary Regulation (E1b/221/25-1-1965, amended in 1971 and 1974). The Sanitary Regulation emphasises the intended use of waters receiving effluents in order to set requirements for treatment and disposal (Tsagarlis, 1998). This regulation is still in force today and it is the basic instrument of authorities (mainly the Prefectures) for the evaluation of polluters' applications for effluent disposal permits. In some Prefectures, also a one-off administrative fee is charged for the issuing of these permits. Other than that, no effluent charging system exists in Greece and sewage is discharged into the wastewater system for a nominal charge (Assimacopoulos et al., 2002).

The Sanitary Regulation has been specified in more detail through Prefectural decisions which set specific terms for the disposal of effluents and wastewater into the receiving waters on a Prefectural level. Such Prefectural decisions define the intended use of important waters receiving effluents as well as maximum limits for the concentration of certain pollutants in the disposed effluents and wastewater.

Tsagarlis (1998) noted that the Sanitary Regulation did not succeed by its own to impose the construction of effluent treatment units on industries. Law 1650/1986 on Environmental Protection and its complementary Common Ministerial Decisions (CMDs) gave even higher

---

[29] Decisions on the interpretation of legality of pre-1989 rights had to be taken often on a case-by-case basis to be able to interpret ambivalent cases. For instance, the Prefecture Council of Athens decided in 1996 that legal are those pre-1989 water works which were carried out according to the valid regulations at the time of their construction, operated for the past 5 years and are operational at the moment. In case of applications to replace a pre-1989 abstraction water work, the year of construction and start of operation needed to be certified by the rural police or any other public authority (cited in the proceedings of the Council of State 3952/2001, E, p. 89-92, Journal of Environment and Law 1/2003, in Greek).

priority to the environment and, thus, gave the authorities more power to impose environmental conditions on industries. Thus, most industries were forced to install systems of effluent treatment, mainly due to the implementation of the Environment Law 1650/1986 rather than of the earlier 1965 Sanitary Regulation. Also the CMD 5673/400 of 1997, which transposed the EU Urban Wastewater Treatment Directive (UWWTD), is of equal importance in promoting wastewater treatment by industries and human settlements.

5.4.2 Regulation of the main water uses

The following briefly describes the core national regulation for five water uses, which have been here selected as the most significant ones: irrigation, industry, hydropower generation, living environment and urban use.

Irrigation

Legislation on irrigation dates back to the 1940s followed by legislation on drainage, land reclamation works and land reclamation organisations starting in 1958. On the level of the Prefectures, the Divisions for Land Reclamation, which were earlier regional branches of the Ministry of Agriculture, have competence relevant to water management in agriculture.

Water management in collective irrigation networks is in the hands of obligatory cooperatives of farmers, so-called General Organisations for Land Reclamation, which consist of two or more Local Organisations for Land Reclamation using common water resources. Land Reclamation Organisations are legal entities of public law (Bartzoudis, 1993) administratively supervised by the Divisions for Land Reclamation of the Prefectures. Nowadays, the Local Land Reclamation Organisations are self-managed by an elected general assembly of community representatives and self-financed through member contributions (Mantziou, 2003).

As already mentioned, the quantitative use of water in agriculture (irrigation mainly) is regulated by the system of water permits introduced by the 1987 Water Law.

Industry

The quantitative use of water in industry is also regulated by the system of water permits introduced by the 1987 Water Law. Earlier than 1987, industries had to get a license for drilling a well from the Ministry of Agriculture or its regional services. If the industry's operation was linked to the use of public (surface) waters, the use right could be conceded by the Ministries of Agriculture and of Public Works (Law 608/1948).

The discharge of industrial effluents into surface water or groundwater is regulated, as mentioned above, by the 1965 Sanitary Regulation, the 1997 CMD transposing the UWWTD as well as regionally-specific Prefectural Decisions.

Hydropower generation

The exploitation of the force of running waters for the production of energy was first regulated with legislation in the 1920s. In 1950, the Public Power Corporation (PPC) of Greece was founded as a public utility company (Law 1468/50), in essence enjoying the monopoly of energy production and distribution in Greece.

According to its founding law (Law 1468/50), the PPC could use and manage running waters, rivers and lakes to such an extent as it judged necessary in order to accomplish its purpose for hydropower generation (exclusive right of water use). The PPC could thus define itself the water quantity it needed for its own purposes. The 1987 Water Law (Art.15) took away this privilege of the PPC to use water unlimited, at the expense of other uses. From then on, the PPC had to propose the water quantities it needed, to be considered in the national water resource development programmes. As far as the reservoirs of the PPC were concerned, water quantities could be allocated also to irrigation purposes via ministerial decisions. The Joint Ministerial Water Committee, set up by the 1987 Water Law, was, among others, responsible for apportioning the stored reservoir water between power production and irrigation, each year before the irrigation season (and also defined respective compensations for the loss of energy).

On 1st January 2001, pursuant to new national legislation of 1999 and 2000, the PPC was converted to a Societé Anonymé. This signalled the start of deregulation of the electrical power in Greece in the context of the integration of the domestic electrical power market in the European Union.[30] The PPC as a Societé Anonymé owns the land of its lignite mines, electricity generation stations and the dams at its hydropower stations. Its ownership of hydropower stations includes the reservoirs formed (PPC S.A., 2005).

The conversion of the PPC into a profit-driven company may also result into changes in its relationship to water uses other than energy production. For instance, concerning the PPC relationship to irrigation, as long as the PPC behaved as a public utility company, there was little room for conflict between hydropower production and irrigation. Namely, the PPC was willing to bear the cost of irrigation (and domestic water supply) as imposed by the State. The amounts apportioned to irrigation were not negotiable but they were fixed according to the PPC obligations to secure the necessary electricity production (PPC, pers.comm., 2004). In the future, there is bound to be an effect of the new status of the PPC on the sharing of waters of its reservoirs with other uses such as irrigation and water supply.

Environment

The 1987 Water Law was the first law to institutionalise the "protection and preservation of the aquatic ecosystem" as a use of water (Article 11). The 1987 Water Law was also the first law requiring the definition of minimum constant vital flows for rivers and of minimum levels for lakes, as a necessary precondition for the existence of the respective aquatic ecosystems. Article 11(6) even defined that the use of water for ecosystem protection is the first priority water use (even prior to drinking water supply) in case of water resources which are under protection by other laws or international conventions.

The more recent 2003 Water Law recognises more explicitly the environmental dimension of water management and emphasises the protection and the preservation of the environment

---

[30] Information from the website of the Public Power Corporation. Accessed online on 30/06/04: http://www.dei.gr.

pursuant to the provisions of the EU Water Framework Directive. However, although the 2003 Law mentions that "water demands are satisfied according to the capacity of water reserves, considering the need to preserve ecosystems and the balance needed between withdrawal and recharge of groundwater", it does not make any reference to the definition of minimum constant vital flows and levels.

So far, minimum constant flows or levels have not been studied adequately nation-wide (MDEV, 2003) and they have not been defined for individual Greek water basins. In some cases, minimum flows were set according to Environmental Impact Assessment studies, which were nonetheless not complied with most of the time in practice. The Ministry of Development only recently assigned the definition of minimum flows and levels for major rivers and lakes, as a specific task within the first water district management plans being produced by consultants (research duration: 2003-2006).

In general, it seems that so far the main environmental considerations in practical water resources management in Greece have mostly been driven by "non-water" EU environmental policies. These included the nature conservation directives (Birds and Habitats Directives) and the Directive on Environmental Impact Assessment which inspired parts of Law 1650/86 on Environmental Protection. The latter imposed, among others, restrictions to projects and activities with adverse impacts on water protection. International legislation has also been very influential, the main one being the Ramsar agreement on the protection of internationally important wetlands.

Urban use

The construction, maintenance and operation of water supply and sewerage systems are under the competence of Local Organisations of Administration and of Municipal Enterprises for Water Supply and Sewerage (DEYA). DEYA are municipal and/or communal enterprises of common benefit and were founded as legal entities of private law. They are the main managers of water resources for drinking and waste water all over Greece. Running costs of DEYA are covered through municipal taxes for water supply and sewerage. The foundation and operation of municipal water supply enterprises, for any municipality with a population over 10,000 people, is regulated by Law 1069/1980. Municipalities' responsibilities for water supply, sewerage services and refuse collection were transferred to the enterprises, which also assumed the ownership of all networks and infrastructure related to those duties, as well as the responsibility for the provision, financing and monitoring the quality of services (Assimacopoulos et al., 2002).

The quantitative use of water for drinking was regulated by the system of permits introduced by the 1987 Water Law. The 1987 Water Law also defined drinking water supply as the use of highest priority.

The adoption of EU environmental regulation on drinking water quality and its transposition into Greek law (the Drinking Water Directive 80/778/EEC was transposed in 1986) was also an important turning point in drinking water regulation in Greece.

As concerns the discharge of urban wastewater into surface water or groundwater, this is regulated by the 1965 Sanitary Regulation, the 1997 CMD transposing the UWWTD as well as regionally-specific Prefectural Decisions.

5.5 Public water policies

A chronological account of the main Greek water policies from the mid-19th century until 2006 is presented in Table 5.2.[31] The list of legal acts is not exhaustive but summarises the most important legal and policy developments with regard to water use. More detailed emphasis is given to recent legislation after the 1970s. Earlier periods are treated in a less detailed way. Section 5.5.1 discusses the development of water policy prior to 1987. Section 5.5.2 then concentrates on the first Water Management Law of 1987, which was the main attempt towards a more integrated national water regime until recently. Upon careful inspection, several of the principles of the 1987 Water Law were in agreement with some of the main requirements of the EU WFD. Section 5.5.3 attempts to briefly reflect on the relevance of the 1987 Water Law to the EU WFD objectives. Finally in section 5.5.4, recent policy changes due to the WFD (reflected in the recently adopted 2003 Water Law) are presented.

Table 5.2   Chronology of Greek main water policies

Year	Policy milestones	Main relevant water uses	Main actors
1852	First law (ΣΗ'/1852) on the drainage of the Voiotikos plain	This was the first of a series of legal acts on local land reclamation and hydraulic works aiming at: Flood protection as well as drainage of marshes to protect public health and to gain land for farming	
1867	Law ΣΚΘ on lake drainage		
1915	Law 550/1915 on the construction of hydraulic works (and law 2853/1922)	Drainage, river regulation/flood protection, irrigation	
1920	Law 2040/1920 on obligatory expropriation of land for the construction of hydraulic works (earlier in 1911, there was a Constitution modification allowing the State to expropriate private property against compensation)	Facilitation of hydraulic works overcoming unclear or established land property and accompanying water rights	
1922	Royal decree on the issuing of water supply permits and permits for technical works	Water supply	
1923	Decree on the exploitation of the force of running waters	Energy production	
1932	Law 5501/1932 on protective hydraulic measures of emergency		
1938	Sanitary rule with articles on water	Sanitation	

---

[31] In the Byzantine period of Greece before the Ottoman occupation in the 15th century, Greek customs and the remnants of ancient Greek law influenced the codification of the Laws of the Byzantine Empire. Shortly after the Greek War of Independence in 1821, the 1st Greek Constitution was adopted which largely copied the Constitution of the French revolution. The Laws of the Byzantine Emperors were then reintroduced and remained in effect (with many amendments via subsequent legislation) until the introduction of the various codes (Dacoronia, 2003). In the period 1833-1862 under the ruling of King Otto (of Bavaria), the first laws/codes were introduced and a court judiciary system was set up. Royal power ended with the 7-year dictatorship regime of 1967-1974. In 1975, the Constitution of democratic Greece was adopted.

Year	Policy milestones	Main relevant water uses	Main actors
	supply, sewage, effluents, irrigation, bathing area, standing water		
1940	Civil Code (in force since 1946)		
1943	Law 481/1943 on management and administration of waters used for irrigation (complemented with further acts in 1948, 1949, 1952, 1957)	Irrigation * Irrigation rules for specific regions existed since the end of 19th c.	Ministry of Agriculture (remains main water manager until mid-1980s)
1950	Public Power Corporation (PPC) founded (Law 1468/50)		PPC
1952	Law 1988/1952 on wells		
1958	Decree 3881/1958 on land reclamation works and Royal Decree on land reclamation organisations	Legislation on irrigation and land reclamation dominates the water regime until the 1980s	
1965	Sanitary Regulation on the discharge of urban and industrial effluents Introduces permits for the discharge of wastewater and effluents issued by the Prefectures	Sanitation	Ministry of Health
1968	Decree (801/1968) on compensation after expropriation of water for the supply of agglomerations	Water property expropriated for water supply	
1968	Decree on the quality of drinking water (disinfection rules since 1957/58)	Drinking water quality	
1969	Acts to limit polluting activities (cattle feed, industry etc) around water supply sources (especially for the capital)	Regulation of conflict on water quality (water supply vs. production)	
1970	Code 420/1970 amended by Law 1740/87 defines that in case of danger of disturbance of the aquatic ecosystem, the Ministry of Agriculture can impose limitations to fisheries in rivers and lakes	Fisheries	
1972	Directorate for natural resources set up at the Ministry of Coordination (later Economy)		Ministry of Economy (role for short time)
1974	Ratification of the 1971 RAMSAR convention on the protection of wetlands of international importance	Environment/wetlands	
1975	New Constitution (environmental protection as an obligation)	Environment	
1980	Law 1069/1980 on water supply and sewerage under the competence of Local Organisations of Administration (OTA) and of Municipal Enterprises for Water Supply and Sewerage (DEYA)	Water supply/sewerage	Local actors (OTA and DEYA)
1980	Ministry of Environment set up		
1981	EU membership		
1981	Presidential Decree 658/81 for the protection of fish in lakes and rivers	Environment	
1983	Directorate for Water Potential and Natural Resources transferred to the Ministry of Development		Ministry of Development
1983	Sanitary regulation for the protection of drinking water for the capital		
1985	Common ministerial decision for the transposition of EU Directive 79/409 for the protection of wild birds	Environment Indirect implication for water management in areas used by wild birds	EU
1986	Law 1650/86 on the Protection of the Environment	Water quality, environment, EIA	Environment Ministry, EU

Year	Policy milestones	Main relevant water uses	Main actors
1986	Common ministerial decisions transpose EU water quality legislation (EU directives on bathing water, drinking water, fish and shellfish water)	Water quality (bathing, drinking, fish and shellfish water) – sanitation improvement	Ministry of Health, Ministry of Merchant Marine, Environment Ministry, EU
1987	Law 1739/87 on Management of Water Resources - Introduced but hardly effectuated - Quality/ quantity apart (no integrated legislation)	All water uses relevant to water quantity Environment (minimum constant flow and level) Vague reference to quality (quality should be assured according to Law 1650/86)	Ministry of Development
1988	Common ministerial decisions for the transposition of the EU directives on groundwater and surface water protection from dangerous substances	Groundwater quality protection, surface water quality protection	EU
1989	Legal acts on water use permits as well as permits for the construction of water works	Water permits	
1988/89	Legal acts for the establishment of Water Management Directorates in 6 Water Districts		Regional Water Management Departments
1990	Common ministerial decision implementing the EIA requirement of the Environment Law 1650/1986	EIA for wells and any activity affecting water resources (industry, energy production, mining etc) etc.	Environment Ministry
1997	Transposition of the EU directives on water quality: the Nitrates Directive and the Urban Waste Water Treatment Directive	Water quality protection consideration of diffuse pollution sanitation improvement	EU, Environment Ministry, Ministry of Agriculture
1998	Common ministerial decision for the transposition of the EU Directive 92/43 on the conservation of natural habitats, flora and fauna	Environment Indirect implication for water management in areas of habitat conservation	EU
1999	Partial privatization of the water supply company of Athens	Water supply partial privatization	Private/public actors EU influence
2000	Deadline of the UWWTD not met for effluent treatment in all agglomerations with more than 15000 inhabitants		
2001	Partial privatization of the water supply company of Thessaloniki (transformed into a Societé Anonymé in 1998)	Water supply partial privatization	Private/public actors EU influence
2001	Public Power Corporation converted to a Societé Anonymé, pursuant to new legislation of 1999 and 2000.	Commencement of the deregulation of the electrical power in Greece	Private/public actors EU influence
2003	New National Water Law 3199/2003 – WFD transposition	All water uses Environment	Environment Ministry, EU
2005	Common Ministerial Decisions on the organisation of a Central Water Agency at the Environment Ministry (CMD 49139), on the structure of Regional Water Directorates (CMD 47630) and on permits for water use and water works (CMD 43504)	All water uses	Environment Ministry

Note: Much of the information on the legal water acts in the table is based on an overview of Greek water legislation by Aggelakis (1993).

5.5.1 Complexity and lack of coordination until mid-1980s

The development of modern Greek water legislation started in 1852 with legal acts on land reclamation, flood protection and hydraulic works to protect public health (e.g. through the drainage of marshes) and to gain farmland. Until the mid-1960s, legislation was dominated by irrigation and land reclamation acts and the Ministry of Agriculture was the main actor in water management (Bodiguel et al., 1996). The narrow scope of water policies was enriched with only few additional acts on hydropower production and on water services for a growing urban population (water supply and sanitation for specific locations). In the late 1930s, sanitary rules began to apply and sewer construction took place in the two main urban centres of Athens and Thessaloniki.

From the mid-1960s onwards, the complexity of the water policy system gradually increased with the entrance of new actors and the regulation of new uses. In 1965, the national Sanitary Regulation on urban and industrial effluents emphasized 'public health' and the role of the Ministry of Health. The Sanitary Regulation was the basis for a series of Prefectural decisions on the types of uses allowed in certain waters throughout the country, the conditions of urban wastewater and industrial effluent disposal, as well as on the qualitative parameters of water. Nonetheless, efforts to treat wastewater were, until the early 1980s, very limited. The OECD noted that "in 1983, enforceable national effluent standards did not exist yet in Greece". The Greek authorities acknowledged that "the enforcement of measures for the control of industrial wastes was not strict in most cases" (OECD, 1983). On the other hand, domestic water supply was significantly extended thanks to heavy investments until the 1980s. In 1961, 38% of the population was connected to public water supply; by 1982 this had reached 90% (OECD, 1983). In 1980, a national law was also issued to support the creation of municipal and communal enterprises for water supply and sewerage.[32]

As concerns environmental considerations, these were still very low in water policy as well as in other sectoral policies. Despite a new Constitution (in 1975) proclaiming that 'the protection of the environment constitutes an obligation of the State' (Art. 24), Greece during the 1970s was still an associate EU member and no limits were placed upon the exploitation of natural resources in order to promote economic growth (Kousis, 1994). Only few developments, such as the ratification of the international Ramsar treaty on the protection of wetlands (1974) and the establishment of the Ministry of the Environment (1980), indicate first attempts to conserve the environment.

In 1981, Greece became a member of the European Union and, since then, the EU has exercised decisive influence in shaping Greek environmental policy. Upon its EU accession, Greece faced considerable problems to adjust to the administrative requirements attached to the implementation of EU environmental policy and also to the internal philosophy of this policy area. The lack of prior experience with pro-active environmental policies was a fundamental handicap, and integrated environmental planning was an unfamiliar concept to Greek domestic bureaucracy (Koutalakis, 2006). Until the 1980s, Greek environmental policies only consisted of primarily reactionary, legalistic and command-and-control driven

---

[32] Later figures by Aggelakis & Diamantopoulos (1995) indicate that the number of households being connected to a water supply network increased from 600,000 in 1961 to 3,300,000 in 1991 (corresponding at that time to 98% of all households in urban and semi-rural areas).

regulations responding to occasionally alarming environmental problems (Weale et al., 2000). Additionally, despite pressure from EU environmental policy, on the road towards fulfilling the convergence criteria of the European Monetary Union, growing concern over Greece's ability to keep pace with European economic and monetary targets still tended to reduce the attention paid to the environment (Pridham & Konstandakopoulos, 1997).

In 1986, a national framework Law on the Protection of the Environment (Environment Ministry, L. 1650), among others, required water quality monitoring and introduced environmental impact assessment for water-related projects. The Environment Ministry was defined as the competent authority for environmental protection including water quality issues. The competence of the Environment Ministry over the issue of water quality was to be shared with other ministries, mainly that of Health and of Merchant Marine. Additionally, important EU water quality legislation was transposed into Greek law in the 1980s and 1990s (EU directives on bathing, drinking, fish and shellfish water, nitrate pollution and urban waste water treatment).

All in all, due to the addition of new acts, especially for water quality regulation to protect human health and the environment from the mid-1960s till the mid-1980s, water policy developed into a fragmented mosaic of often overlapping regulations on specific aspects of water management and/or specific geographical areas (MDEV, 2003; OECD, 1983).[33] Early attempts to deal with the increasing institutional complexity in the water sector were discussions to coordinate water management and administration within an Interministerial Water Resources Committee in 1976-1982 (under the leadership of the Ministry of Coordination, which later became the Ministry of Economy). In 1977, a Directorate for Water Potential and Natural Resources was set up, which was attached in 1983 to the Ministry of Development (MDEV, 2003). In the early 1980s, the Greek government announced plans for a framework water law to coordinate water legislation and institutional arrangements (OECD, 1983).

5.5.2 First attempt for partial integration: The 1987 Water Law

The 1987 framework Water Law (Ministry of Development, L. 1739) was the first to define coherent water management principles following an intersectoral approach. Till then, no holistic legal framework for water resource management existed in Greece, while '[...] dealing with [water] problems was done empirically and according to custom'[34]. Until the later adoption of the 2003 Water Law which transposed the EU Water Framework Directive, the 1987 Water Law was the most significant law in the field of water resources management in Greece. The 1987 Law focused on water quantity (but not quality) management and

---

[33] According to MDEV (2003), from 1900 until the mid-1980s, approximately 300 legal acts relevant to water were passed whose main characteristics were the following: Attempt to put forward the position of the institutions which issued them; Piecemeal account of sectoral problems; Lack of dealing with modern water problems; Lack of coordinated and systematic programmes to acquire and evaluate field data, which are necessary for the implementation of regulations; Lack of organisations to monitor and adapt their implementation; Lack of links to development goals of sectors of production and regions in the country.

[34] Parliament plenary records, 12.10.1987, p.164, Member of Parliament P. Zakolikos.

indicated the State's intention to coordinate national water policy and to institutionalize water management bodies on a regional level.

The 1987 Water Law was not short of critique. Although it had the dual aim of increasing production and, in the same time, lessening water rivalries and contributing to water and environmental conservation, it was often criticized for being too development-oriented and focusing on water supply increase instead of demand management. Moreover, the separation of water quantity and water quality issues according to two different laws (the 1987 Water Law and the 1986 Environment Law respectively) was still a significant disadvantage of the national water policy system at this stage. The 1987 Water Law was indeed often criticised for not dealing adequately with the issue of water pollution to achieve coherent water management.

After a brief description of the law-making process based on the information available at present, the following paragraphs describe selectively the most important principles introduced by the 1987 Water Law as well as their practical implementation nation-wide. In fact, as described in the following, the 1987 Water Law was only partially activated and implemented. Some background factors influencing implementation are here discussed but a more systematic analysis of the institutional context conditions which hindered the 1987 Water Law is made in section 5.6.3.2.

The law-making process in brief

Before the 1987 Water Law was issued, Greece was facing two kinds of water problems. Firstly, there was imbalance between physical water offer and demand. There were increasingly rivalries between different water users and for the first time it was recognised on the policy level that Greece was facing scarcity in water resources. Secondly, administrative structures and policy were insufficient to provide for coordinated and unified management in this field.

To provide solutions to these problems, the Minister of Development and his Directorate for Natural Resources and Water Potential, took the initiative to formulate a draft bill on a framework Water Management Law. In other words, the initiative for the 1987 Water Law was ministerial (top-down approach). Relevant draft water bills existed already since 1981 at the Ministry of Coordination during the government term of the conservative party of New Democracy. The socialist party (Pasok) came into power in 1981 and was also in power when the draft water bill was finalized in 1987 by the Ministry of Development.

The final bill was prepared by the Ministry of Development considering contributions from other ministries, such as the Ministry of Agriculture. However, the Ministry of Development was criticised for not consulting with other relevant stakeholders, scientific and professional collective groups before bringing the draft bill to Parliament. During discussions at Parliament, certain stakeholders (e.g. the Greek Committee of Hydrogeologists) even requested to postpone the bill discussions and give enough time to relevant stakeholders and organizations to comment on the draft bill. Also the Technical Chamber of Greece, which is recognized as the technical advisor of the government, complained for not being involved in the process. Amidst complaints for the lack of dialogue, the Minister of Development admitted that the Ministry of Development did not take the initiative to contact stakeholders individually but emphasised that the preparation of the bill was announced in the press six months before bringing it to Parliament. Moreover, the Ministry of Development had specifically addressed the Technical Chamber in written at the start of the bill-drafting process. Therefore, stakeholders were also blamed by the Ministry for not expressing their interest and comments themselves earlier on in the process (Parliament records, 12-19/10/1987).

Despite disagreements over the lack of transparency, the draft bill was adopted with parliamentary majority in October 1987.

Administrative organisation for water management

The 1987 Water Law defined competent ministries for each water use and competent institutions for water research (see Table 5.3). It thereby aimed at clearer competence allocation, largely overcoming the overlap of responsibilities of the previous water policy phase. The Law, however, did not eliminate competence fragmentation among multiple ministries. According to the Law, the Ministry of Development maintained its competence for water use in industry and energy production, and the Ministry of Agriculture in agriculture. The Ministry of Interior[35] was defined as competent for the supervision of municipal water supply and sewerage, the Environment Ministry for water protection, the Ministry of Health for drinking water quality and the Ministry of Foreign Affairs for transboundary water issues. A key innovation in competence allocation was the definition of the Directorate for Water Potential and Natural Resources at the Ministry of Development as national authority for water management coordination. This Directorate was also defined responsible for the elaboration of national water policy and national water development programmes, water allocation to different uses, the organisation of regional water management authorities, the coordination of research and the establishment of national monitoring and a central hydrological database to collect and register information for water development programmes.[36]

---

[35] The Ministry of Interior is also responsible for issues of personnel in the public sector, for the organisation and operation of public services and for transferring competence from central to regional authorities. Thus, this Ministry is also relevant to water management in as far as the decentralisation of competence to the regional and local administration is concerned. The Ministry of Interior defines the structure of the administrative Regions and has, in general, competence to enact changes in their overall structure. It also participates in the allocation of funds to the Regions together with other co-responsible ministries, especially the Ministry of National Economy.

[36] Information from the website of the Ministry of Development. Accessed online on 11/12/2003: http://www.ypan.gr/fysikoi_poroi/emne_yd.htm.

Table 5.3   Allocation of water competence by the 1987 Water Law

Authorities and other bodies	Areas of competence
Administration	
Ministries	
Ministry of Foreign Affairs	International water issues
Ministry of the Interior, Public Administration and Decentralisation	Supervision of municipal water supply and sewerage, except for the cities of Athens and Thessaloniki (competence of Ministry of Environment)
Ministry of Development	Allocation of water to use sectors, industry, energy, research-technology, commerce
Ministry of the Environment, Physical Planning and Public Works	Water protection, study and construction of large works for water supply, sewerage and irrigation
Ministry of Agriculture	Irrigation, forestry, husbandry, fisheries, land reclamation organisations
Ministry of Economy	Participates in water programming
Ministry of Health and Welfare	Drinking water quality
Ministry of Culture	Use of water in sports
Ministry of Transport and Communication	Use of water for transport
National Tourism Organisation	Use of water for recreation and spas
Regional and local authorities	
Regional authorities	Water management on the water district level, small hydropower works, water quality, approval of planning and financing for water supply and irrigation works
Prefectures	Permits for water abstraction, restricting measures, construction of works
Municipal authorities (OTA/organisations for local administration) - Municipal water supply and sewage corporations, Water supply and sewage corporations of Athens and Thessaloniki	Water supply, sewerage
Institutes, research centres, enterprises	
National Meteorological Service	Meteorological observations
Public Power Corporation (PPC)	Energy production, supervised by the Ministry of Development
National Institute of Geology and Mineral Exploitation	Hydrogeological research, thermometallic waters, supervised by the Ministry of Development
Enterprises for water supply and sewerage	Water supply, sewerage
National Centre for Maritime Research	Water resource research
National Institute for Agricultural Research	Agricultural research

Greece was also divided into 14 water districts (see Figure 5.2 in section 5.3) for the purpose of monitoring and managing water resources regionally. The Law required setting up one Regional Water Management Department (RWMD) in each district, as a regional service of the Ministry of Development, to adjust national water policy to regional conditions, collect and assess data for calculating water balances for each river basin, determine the water district supply-demand balance, develop a mid-term water district development programme and carry out or assign any studies necessary to fulfill its role. RWMDs were also given competence to issue water permits (for energy production and multiple water uses) as well as to approve and register permits issued by other authorities.

The 1987 Water Law also established advisory bodies on a national and regional level. A Joint Interministerial Water Committee (of the Ministry of Economy, of Interior, of Agriculture, of Environment and of Development) was set up to give its expert opinion on legal acts for the full activation of the Law as well as the formulation of national water policy and national water development programmes. On a regional level, the Law required setting up Regional

Water Committees (one in each district) to consult over issues such as the water district development programmes. Their participants included representatives of the relevant Regions and Prefectures, farmers' unions, local self-administration and the Technical Chamber of Greece (RWMD official, pers. comm., 2005).

Unfortunately, the practical reality of policy implementation did not quite reflect these planned innovations in water administration.

Although the Central Water Directorate at the Ministry of Development existed prior to the 1987 Water Law, it was not further equipped with sufficient personnel to fulfill its role according to the Law. In fact, over time, its personnel, instead of increasing as planned,[37] was reduced from 17 to 6 employees (MDEV, pers. comm., 2003). The limited human resources of the Directorate were one of the reasons why many legal acts needed for the full activation of the 1987 Water Law were delayed or not issued at all. In specific, ca. 40 complementary Presidential Decrees and 33 Common Ministerial Decisions needed to be drafted and issued. This is not unusual in the Greek law-making system, where the operation of framework laws is supported by numerous complementary legal acts.[38] An additional reason for delays in the full activation of the 1987 Water Law by the Ministry of Development was that other co-responsible ministries were often unwilling to sign certain complementary acts (MDEV, pers. comm., 2004). This was related to the general unwillingness of certain sectoral public actors to accept the water coordination competence of the Ministry of Development, loosing thereby part of their own power and control over water. Signs for this unwillingness were already made obvious during discussions on the draft Law at Parliament (Parliament records, 12-19/10/1987). During Law implementation, the main Ministries, which co-managed (and competed over) water, continued acting in a sphere of mistrust in order to safeguard sectoral interests.

As regards the Regional Water Management Departments (RWMDs), it was initially planned to set them up gradually, while, until their establishment, the Central Water Directorate of the Ministry of Development would carry out their tasks (MDEV, 1988). Until 1994-5, only 6 of them could be set up. In 1997, RWMDs for all water districts were set up (as departments of the Regions' Directorates for Planning and Development), when they were transferred from the jurisdiction of the Ministry of Development to the 13 administrative Regions of the country (by L.2503/1997). This occurred in the context of the national decentralization reform taking

---

[37]   A Report of the General Accounting Office of the Ministry of Economy, which accompanied the draft bill of the 1987 Water Law at Parliament, defined a number of 37 members of staff for the Central Water Directorate (MDEV, pers. comm., 2003).

[38]   In the Greek law-making system, a law is adopted by Parliament on the basis of a draft bill, which is prepared by the respectively competent ministry and then sent for comments to the co-responsible ministries and interested stakeholders. The operation of adopted laws is supported by numerous Presidential Decrees (PDs), Common Ministerial Decisions (CMDs) and other legal tools. Draft PDs and CMDs are sent for comments to co-responsible ministries and may be sent for comments to interested stakeholders. Draft PDs are also checked from a legal perspective by the Supreme Court of Greece before being signed by the President of Democracy. CMDs are only signed by the co-responsible Ministers, therefore, the procedure for their adoption is considered less bureaucratic than that of PDs.

place in the mid-1990s (see section 5.2), which affected most governmental sectors (MDEV, pers. comm., 2004). Each of the 13 in total established RWMDs was given administrative competence for a specific water district (in some cases, for two) (Presidential Decree 60/1998). Most RWMDs had different territories of competence than the Regions, because the 14 water districts did not coincide in all cases with the boundaries of the 13 Regions.

To date, despite their establishment nation-wide, the RWMDs have had only a marginal role in water management. Their competence for water planning and programming never became operational; in practice, their main role was bureaucratic and restricted to their involvement in the water permit procedure and the formulation of restrictive regulations on surface and groundwater quantitative use (prohibitions, restrictions on the distance between wells and other regulations) to be issued by the Prefectures. The latter regulations, which aimed at water conservation and protection, also set out the specific terms for approving water abstractions and issuing water permits in each Prefecture (RWMD official, pers. comm., 2005).

Moreover, RWMDs lacked adequate personnel resources both in number and qualifications of staff (Delithanasi, 2004). Most RWMDs were equipped with only 1-3 employees, despite plans for ca.10-15 employees in each RWMD (MDEV, pers. comm., 2003).[39] Some RWMDs did not even employ hydrogeologists, to carry out tasks such as the calculation of water supply-demand balances. In many RWMDs, the few employees present were often responsible also for issues other than the management of water resources, such as management of minerals and renewable energy (Delithanasi, 2004). Additionally, the RWMDs were unable to collect the necessary data for carrying out their water management tasks (e.g. hydrological and water consumption data). The main data in their hands were restricted to the registers of permits for water use and works issued in their territory of competence. As far as financial resources are concerned, the RWMDs were placed under the administrative and financial mandate of the General Secretaries of the Regions. Therefore, resources allocated to them depended on the financial capacity and thematic priorities of the Regional administrations (MDEV, pers. comm, 2003). The RWMDs had no income sources of their own (e.g. through water charges), and neither did they manage funds for projects on water supply, wastewater treatment or irrigation. Funds for such water projects are usually managed either by the central government (national funds) or by the Managing Authorities of the Regions (Regional Operational Programmes, which, in their turn, stem from the EU community support framework).

As concerns the Regional Water Committees, these were set up with delay, considering that the first ministerial decisions for their establishment were issued in 2000. Only a few of the

---

[39] A report of the General Accounting Office of the Ministry of Economy, which accompanied the draft bill of the 1987 Water Law at Parliament, defined to have 10-15 employees in each RWMD. The urgent demand of the RWMDs for personnel unfortunately coincided with a period when new recruitment in the public sector, in general, was for a long time "on hold". Thus, the RWMDs were not equipped with sufficient personnel resources (MDEV, pers.comm., 2003).

established Committees were activated and actually met up some occasions (e.g. in the Region of Central Macedonia and in the Region of Crete). Even so, no substantial water management issues were brought up for discussion. The limited activation of these Committees was related to the absence of water district development programmes to consult upon (see below) and to the limited human resources of the RWMDs which were responsible for providing organisational and secretarial support to the Committees (MDEV, pers. comm., 2003). The Joint Interministerial Water Committee also convened rarely and only a few important decisions were based on its opinion (e.g. decision to share reservoir water of the Public Power Corporation with irrigation users) (MDEV, 2003). Overall, there was a lack of willingness of the participating ministries to bring up substantial issues for joint discussion in this Committee (MDEV, pers. comm., 2003).

Water resources planning and water pricing

The 1987 Water Law introduced the concept of intersectoral water planning through its requirement for national and water district development programmes to protect and develop water and to support development policy. These programmes should be coordinated with national programmes for social and economic development. The basis of the water programmes, as main planning instruments, would be the balance of water supply and demand considering the present status of water resources and plans for future development. In order to gain necessary data for the water programmes, the 1987 Water Law required the establishment of a central hydrological database at the Ministry of Development. Furthermore, the Law introduced the concept of water pricing. The price of water should be set by the Ministry of Development and the co-responsible Minister depending on the water use and following the opinion of the Joint Interministerial Water Committee. In the following, the implementation of these planning and pricing requirements is briefly reviewed.

Shortly after adopting the 1987 Water Law, the Ministry of Development carried out two pilot water management studies in western Greece, one on a river basin and one on a water district level, which calculated water balances and produced management modelling tools for the specific water systems. These pilot studies were a preparatory exercise to support the drafting of water district development programmes nation-wide (MDEV, pers. comm., 2004). Throughout the 1990s, however, there was no further progress in this respect, except for the Water District of Crete whose RWMD - on own initiative and engagement – secured funds for a study on the integrated management of Crete's water resources as a basis for a water district management plan.[40] There was also failure in the timely creation of a National Hydrological Database at the Ministry of Development. Finally, in 2003 and still under the legal framework of the 1987 Water Law, the Ministry of Development hired consultants to develop water resource management systems and tools for all water districts. The hiring of consultants was considered necessary due to the lack of adequate personnel in the RWMDs

---

[40] Also a study on water resource management of the Cyclades islands was carried out by the Prefecture of Cyclades as well as of the River Viotikos Kifissos by the Environment Ministry.

- both in number and qualifications – to draft such studies. Consultants should – in cooperation with the RWMDs - collect and analyse data to determine water balances, record current and future water needs, record the present state of the water environment, define environmental terms for managing water and develop management systems for planning water works. It is now considered important to harmonize the water management systems and tools developed for the Ministry of Development with the new national water policy framework (2003 Water Law of the Environment Ministry) and to use these results also for the EU WFD reporting requirements.

In 2003, the Ministry of Development also issued a National Water Master Plan, which attempted to assess the quantitative and qualitative status of water resources nation-wide and to propose solutions for the main water problems. Its ultimate aim was the approximation of the required national water development programme (MDEV, 2003).[41] However, except for assessing the water balance and supply-demand balance of the 14 water districts, the National Master Plan did not provide complete information on different sectoral water works and did not contribute to the coordination of water use, development, protection and research. Therefore, despite attempts, the required national water development programme to harmonize water sectoral policies and coordinate sectoral water works could not be formulated by the Ministry of Development.

Due to the non-activation of the water programming procedure, there continued to be little coordination in the construction of water works by different users and public bodies all over the country. In this context, each public body with a stake in water continued its own planning and activities on the use of water on the national, regional and local level (MDEV, pers. comm., 2004). This situation is also related to the unwillingness of certain public bodies to accept the principles of the 1987 Water Law and the coordination competence of the Ministry of Development (Aggelakis & Diamantopoulos, 1995).

As far as water pricing is concerned, no effective pricing policy has been defined so far. Especially, water used in agriculture is not priced at all (Karageorgou, 2003).

Summary

In brief, the 1987 Water Law was the first law defining principles for more intersectoral and coherent water management in Greece and it formed the general legal framework for every water use. The 1987 Water Law also constituted an initiative for the decentralisation of water management responsibilities from the central government to the river basin level. It required (top-down) the set up of Regional Water Management Departments (RWMDs) with administrative competence on water district level. The reform attempt of the 1987 Water Law was accompanied by a broader decentralisation reform in Greece promoting transfer of authority from the centre to the Regions. In fact, the RWMDs were at first under the

---

[41] The first phase of a National Water Master Plan was completed in 1994 by the Ministries of Environment, of Development and academic experts. The Master Plan was then finalised in 2003 by the Ministry of Development only, without coordination with the Environment Ministry (Papaioannou, 1998).

jurisdiction of the central government (branches of the Ministry of Development), while since 1997 they are part of the Regional administration.

On a practical level, some new water management structures could be set up nationally and regionally in the process of implementing the 1987 Water Law. However, most of these structures in practice sub-functioned. The procedural impact of the Law, in terms of water management and planning, was even weaker, since no management plans were produced and coordination mechanisms were not supported. The result of the weak implementation of the 1987 Water Law was that water management continued in a piecemeal and opportunistic manner throughout the 1990s. Management remained re-active and focused on short-term responses to water challenges which are driven by crises (e.g. acute shortages confronted with supply augmentation options) (Assimacopoulos et al., 2002).

5.5.3 Intersection of the 1987 Water Law and the WFD

Before dealing with the impact of the EU WFD on Greek water policy (in the next section), it is interesting to emphasise in brief that also the 1987 Water Law reflected some requirements of the WFD for integrated water management.

From an administrative point of view, the definition of 14 water districts by the 1987 Water Law served a similar purpose as the river basin districts of the WFD, i.e. as administrative units for water management following natural boundaries. Also, the RWMDs of the 1987 Water Law reflected the WFD principle of water management coordination on a river basin district level. Moreover, the Regional Water Committees and the Joint Interministerial Water Committee were in the spirit of the public participation requirements of the WFD.

From a planning perspective, the water resource development programmes required for each water district by the 1987 Water Law had similarities (at least as a concept) to the river basin district management plans required by the WFD. Additionally, the concept of the cost and pricing of water for each use (Art. 10 of 1987 Water Law) reflected the WFD requirement for water pricing.

Finally, the 1987 Water Law placed importance on the development of knowledge to achieve more integrated water management, much like the WFD does. In specific, the National Hydrological Database, which was required by the 1987 Law to secure the collection and registration of information on water resources, was in the direction of the information requirements of the WFD. Unlike the WFD, however, the 1987 Water Law did not explicitly require a regular monitoring programme of the quantitative and qualitative status of surface water, groundwater and protected areas.

5.5.4 Recent integration attempts under the influence of the EU WFD

In 12/2003, Greece transposed the EU WFD by adopting a new Law on the Protection and Management of Water (Environment Ministry, L.3199). Due for the most part to the WFD emphasis on the environment and water quality, the Inner Cabinet and Prime Minister decided to transfer national water coordination competence from the Ministry of Development to the Environment Ministry. This transfer of competence, however, did not take place free of critique. There were indeed claims by the Ministries of Development, of Environment as well as of Agriculture over the main competence on water. The essence of these conflicts lied in the unwillingness of each of these ministries to "lose" water as an important part of their competence (Papanidis, 2002). To some, it seemed logical for the Ministry of Agriculture to be national water competent authority, since agriculture is the main consumer of water in Greece. It was contra-argued, however, that the Ministry of Agriculture had not been able in the past to develop the capacity to face up to such holistic responsibility (Zisi, 2002). On the other hand, the Ministry of Development had a strong advantage by already being the competent authority for water management under the 1987 Water Law and also being competent ministry for natural resources such as mines, energy and raw material. However, even during the parliament discussions of the draft bill of the 1987 Water Law, there were arguments that the Ministry of Development might not be the right national water competent authority because of its jurisdiction over a limited group of water users (industry and energy) and because of its limited budget (Parliament records, 12/01/1987).

After the transfer of national water competence from the Ministry of Development to the Environment Ministry had been decided, a Working Group of Environment Ministry representatives, the National Centre for Environment and Sustainable Development as well as academic and legal experts worked on a new water bill to transpose the WFD. The first draft was then revised internally by the Environment Ministry. Before being sent to Parliament, comments were invited from other ministries, different experts (public scientific institutions, water-related corporations and research centers, technical professional chambers) and NGOs (MEPPW, pers. comm, 2005).

In terms of its principles, the new adopted 2003 Water Law emphasizes the environmental dimension of water resources and explicitly requires water pricing and cost recovery. The Law is also important in bringing together the management of water quantity and quality both in policy and in the administrative structure. In terms of planning, it requires the development of 6-year river basin district management plans for water protection and management (according to the WFD). These plans should include programmes of measures, monitoring and specific measures to control groundwater pollution.

In administrative terms, the 2003 Water Law required setting up several administrative and advisory bodies, with similarities and differences to those set up by the 1987 Water Law.

On a national level, the Law required setting up a Ministerial National Water Committee (Ministry of Environment, of Economy, of Interior, of Development, of Health and of Agriculture). This Committee should determine the overall policy for water protection and management, monitor policy implementation, approve national water programmes and define river basins, river basin districts and their competent authorities.

A National Water Council, headed by the Environment Ministry, should also be set up to assess and give its expert opinion on water policy and law implementation. Apart from the six ministries involved in the National Water Committee, participants of the Council also include main parliamentary parties, the association of Prefectural and municipal councils, the association of municipal water and sewerage utilities, the Greek association of agricultural producers, the Greek association of industrialists, the Public Power Corporation, the central trade union association, the Technical Chamber, the Geotechnical Chamber, the Institute of Geology and Mineral Exploitation, the National Centre of Seawater Research, the Greek Centre of Biotopes and Wetlands, the National Centre of Natural Sciences, two environmental NGOs, the Consumers Chamber, the National Foundation of Agricultural Research and representatives of the National Committee for Desertification.

The Law also required the establishment of a Central Water Agency at the Environment Ministry as the principal national competent authority for water protection and management. The Agency should draw up national water programmes and prepare annual reports on water status, law implementation and EU compliance. The Central Water Agency was established in 12/2005.

On a regional level, the Law required setting up one Regional Water Directorate (RWD) in each administrative Region with competence for the protection and management of all river basins within the Region's boundaries. The Law defined a river basin according to the EU WFD, as "the area of land from which all surface run-off flows through a sequence of streams, rivers and, possibly, lakes into the sea at a single river mouth, estuary or delta". In fact, the 2003 Water Law became the target of critique for linking the management of river basins to the administrative borders of the Regions contrary to the WFD principle of management within natural river basin boundaries (Haintarlis, 2003). In case of river basins crossing the boundaries of two or more Regions, the RWDs' competence should be exercised in common, but the National Water Committee can also determine a single competent RWD.

Each RWD should develop and implement a management plan for all river basins of its competence, programmes of measures, monitoring and a register of protected areas. It should also report annually to the Central Water Agency. Moreover, RWDs should concentrate permit-issuing competence for all water abstractions and water works. Contrary to the RWMDs of the 1987 Water Law, the new RWDs have competence for both water quality and quantity management issues. In this context, the RWDs are responsible for collecting and assessing data on water quantity and quality to fulfil their role. Nevertheless, despite the concentration of competence for the management of all water uses to the RWDs (in cooperation with the Central Agency), fears are still expressed that lack of clear text in the 2003 Water Law might encourage the continuation of overlap in competence and activities of different public bodies (UMEWS, 2003; CSEH, 2003b). The RWDs were established via a CMD in 12/2005, by restructuring the pre-existing RWMDs. The RWDs were placed under the administrative and financial mandate of the Secretaries General of the Regions (as previously the RWMDs) and were given no additional income sources of their own (e.g. through water charges). In terms of human resources, it is planned to have a staff of 23 in each RWD, which is slightly higher than the ca.10-15 planned RWMD employees under the 1987 Water Law.

Finally, the 2003 Water Law also required setting up Regional Water Councils to run consultations on the river basin management plans. The broader public should also be informed about the content of the management plan and participate in a process of public consultation according to the WFD. Compared to the Regional Committees of the 1987 Water Law, a broader group of stakeholders can participate, including representatives of the Regions, local self-administration, municipal water supply and sewerage utilities, unions of farmer cooperatives and land reclamation organisations, various chambers (Technical, Geotechnical, Commercial and Industrial), environmental NGOs and management bodies of nature protection areas.[42]

The following Figure 5.4 summarises the main authorities and advisory bodies of the 2003 Water Law.

Figure 5.4   Main authorities and advisory bodies of the 2003 Water Law

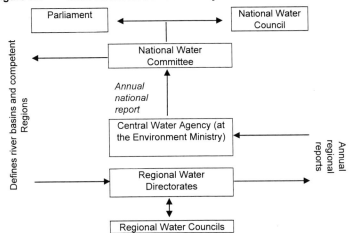

The European Commission (EC) monitors the implementation progress of the EU WFD, which was transposed in Greece by the 2003 Water Law. In 2005, the EC initiated legal action against Greece for inadequacy in transposing and implementing the WFD.

In terms of transposition, the EC issued (after investigation upon its own initiative) a reasoned opinion against Greece for not adopting all complementary legal acts for the full transposition of the WFD (regarding the definition of river basin districts' characteristics, cost recovery in water services and the detailed definition of the content of programmes of measures and management plans), for not transposing important definitions and the environmental objectives of the WFD and for not defining a complete procedure for public information and consultation on the river basin management plans (EC, 2005b). Indeed, certain regulations of the new 2003 Water Law were not concrete and several important issues, such as the establishment of a water-monitoring procedure and the transposition of

---

[42]   The number of participants in the Regional Water Councils could be rather high, especially in case of discussing issues concerning more than one Region. In such cases, the participants of the Regional Water Council may exceed 100 (Giotakis, 2003).

the technical annexes of the WFD, were to be dealt with in additional future legal acts (Interview of Ch. Maniati-Siatou, Director of the Central Water Directorate at the Ministry of Development, In Ydrooikonomia, 17/12/2003).[43]

The NGO community also criticised the content of the 2003 Water Law for several omissions, especially for not explicitly stating several WFD requirements, such as fulfilling the environmental objective of "good water status" by 2015 and "preventing deterioration of the status of water bodies" until the end of 2012 when the programme of measures must be operational (EEB, 2004). Greek environmental groups also criticised the lack of definitions on terms such as "water quantity" and "sustainable water use", the lack of specific commitments for incorporating water policy in other policy sectors and the incomplete reference to the WFD requirements for public participation. With regard to the latter, although the 2003 Water Law mentions participation on a regional level, it does not provide details or time-schedules (Varvaressou, 2004) for a well-planned participatory process.

In terms of practical implementation of the WFD, Greece failed to meet the deadline for providing complete information on the definition of its river basin districts and their competent management authorities by 22/6/2004, resulting in an EC reasoned opinion. Greece also failed to submit reports on the current state of its river basin districts by 22/3/2005 to the EC, thus resulting in an EC letter of formal notice (EC, 2005a). According to Koutalakis (2006), the delay of Greek authorities in designating river basin districts and their competent authorities is related to the lack of a comprehensive national action plan for water management based on sufficient biological, hydrological, economic and administrative data.

5.6 National water regime development

Following the above presentation of water property rights and public water policies in Greece, the next sections characterise the development of the Greek national water regime as a whole. To this aim, use is made of the analytical framework for change in national institutional water regimes as presented in chapter 4/4.1.1. In specific, section 5.6.1 identifies different phases in the Greek national water regime development, while section 5.6.2 reflects on the development of the regime extent and coherence (referring to the relevant indicators presented in chapter 4/4.1.1). Finally, section 5.6.3 discusses the causal factors of external change impetus and institutional context conditions, which have influenced the outcome of attempts to increase integration in the Greek national water regime.

5.6.1 Phases of the national water regime

The development of the Greek national water regime over time can be separated into distinct phases (see Table 5.4). For comprehending the following description of the regime phases, it is useful to refer also to Table 5.2 on the chronology of the main Greek water policies as well

---

[43] At the time of writing, the Presidential Decree needed for the full transposition of the WFD (especially of the WFD technical annexes) had not been adopted yet. It had been sent for written comments to stakeholders (ministries, experts from public and non-public institutions (public scientific institutions, water-related corporations such as the Public Power Corporation, research centres and technical professional chambers) as well as NGOs such as the WWF) and was on hold for signature.

as to Table 5.5 which describes a common pattern of national water regime development based on six European countries other than Greece.

Table 5.4   Phases of the Greek national water regime

Regime phases	Regime change variables	Description
I. Simple to low complex regime (1852-1965)	- Low to medium extent - Medium coherence	- Few water uses regulated and few actors involved - General sanitation rules for certain water uses (health) - Introduction of private property expropriation (facilitating public hydraulic works) - Definition of public domain for surface water (Civil Code)
II. Complex regime (1965-1987)	- Higher extent - Low coherence	- Increase of water uses regulated and of actors involved - 1965 Sanitary Regulation for effluents and wastewater - Competence for water supply and sewerage to local self-administration (Law 1069/1980) - Environmental/nature dimension of water and water quality (for health and environment) in the public domain
III. Complex regime after 1st failed attempt at partial integration (1987-2003)	- High extent - Low coherence	- Water scarcity recognized by policy (1987 Water Law) but quantity still considered separately from quality - 1987 Water Law introduces permits for water abstractions and water works, promotes water policy planning, combines management of surface and groundwater, river basin management, stakeholder participation, defines aquatic ecosystem protection as a water use, but is only partially implemented - Transposition of EU directives brings groundwater protection and diffuse pollution control into public domain
IV. 2nd attempt at integration (phase in development) (2003-...)	Attempt to achieve: - High extent - High coherence	- 2003 Water Law integrates water quality and quantity, emphasizes water environmental protection, cost recovery, makes renewed attempt at holistic policy planning, river basin management and public participation

Phase I/1852-1965: Simple to low complex regime

This phase started as a simple water regime. A few uses were regulated: mainly drainage, flood protection, irrigation, water supply and energy production. This phase also marked the beginning of several hydraulic works (engineering era) of the Greek State to serve drainage, river regulation and irrigation. After the revision of the Constitution in 1911 and the adoption of a law on hydraulic works in 1920, the State was also in a position to expropriate private property for the purpose of public hydraulic works. In 1940, the adoption of the first Greek Civil Code introduced the definition of a public domain for all important surface waters. In the late 1930s, the issue of public health linked to water and sanitation was added to the public domain via elementary sanitary rules. In 1950, the production of energy using water passed on to the hands of a public company, the Public Power Corporation (PPC). Additionally, several acts were passed to regulate the use of irrigation and to set up land reclamation organisations. In general, the first phase of the Greek national water regime was dominated by numerous legal acts related to the use of water in agriculture (irrigation, land reclamation works). The Ministry of Agriculture, being the main consumer of water through irrigation, was the main actor involved in water (quantitative) issues.

In total, this phase developed from a simple to a low complex regime phase, as the public domain was gradually extended to include more water uses along with the appearance of first public concerns for issues of health and water sanitation.

Phase II/1965-1987: Complex regime

In this phase, an increase of water uses regulated by the State and an increase of actors involved in the water policy sector is noted.

In 1965, water quality protection from a health point of view became the focus of attention through the adoption of a national Sanitary Regulation. This introduced regulations on the discharge of urban and industrial effluents, thus restricting the right of industries and human settlements to pollute water. Thus, this phase started with an increased concern about public health and sanitation, which was unfortunately accompanied by the weak enforcement of the relevant adopted policies.

This phase was also characterized by heavy investment in the extension of the domestic water supply system. An important law was introduced in 1980 defining the competence of the local administration and municipal water enterprises for the delivery of water supply and sewerage services.

By the 1980s, the policies adopted by Greece on water resource protection mainly dealt with aspects of public health protection. In other terms, measures until then aimed at reducing the negative impacts of water use on water quality with respect to the human population. Only in the second half of this regime phase, the environmental and nature protection dimension of water management started being introduced into the public domain. Several legal developments, such as the ratification of the international Ramsar treaty on the protection of wetlands (1974), the creation of the Ministry of the Environment (1980) and the transposition of the EU Birds Directive with indirect implications for wetland protection, were first attempts to protect water resources with respect to the environment. In 1986, a national Environment Law was adopted which aimed (among others) at safeguarding good water quality and established the practice of environmental impact assessment studies on works affecting the water environment.

EU pressure even speeded up the development of more specific water quality protection policies. In 1986, a series of common ministerial decisions transposed important EU water quality legislation into Greek law (EU directives on bathing, drinking water and water for fish and shellfish life). At this stage, the national water regime reached a *highly complex* phase with water quality protection gaining a prominent place in the public domain.

Phase III/1987-2003: First failed attempt at partial integration

This regime phase started off with the adoption of the 1987 Water Law, which is considered as the first attempt towards partial integration in the water regime and towards the improvement of water governance structures. On the policy level, water scarcity was recognised as an issue of national significance for the first time. Important aspects of the 1987 Water Law included the enlargement of the public domain for water resources (at the expense of the private domain) with the introduction of water permits for water abstractions and for water works. Water management, thus, came completely to the hands of the State and water use came under public control, except for any pre-1989 established water uses. The 1987 Water Law also promoted water policy planning, a river basin management approach and some stakeholder participation as well as the protection of the aquatic environment from water over-use.

However, the approach to integration in national water policy remained still partial. Although surface water and groundwater should be managed together, water quantity and water quality issues were treated separately by different laws and authorities. In practice, even the required coherent management of surface water and groundwater also remained a principle "on paper" without real management implications. The 1987 Water Law was also criticized for not including adequately the principle of water resource protection from an environmental perspective (Terzis, 1997). Furthermore, as already discussed, most aspects of the attempted reform of the 1987 Water Law were not implemented adequately on the ground. Thus, the national water regime in this phase remained in reality complex and fragmented.

A few other developments in this phase should also be emphasized, namely the transposition of important EU legislation on groundwater and surface water protection from dangerous substances (in 1988) as well as on urban wastewater treatment and on nitrate pollution (in 1997). This EU legislation, in essence, added the protection of groundwater and the control of diffuse water pollution to the public domain in Greece.

Finally, as in other European countries, the debate on an open European market and the liberalisation of water services affected also the mixture of public and private actors involved in Greek water management. In this phase, the water supply companies of the two largest urban areas of Thessaloniki and Athens were partially privatized (in 2001 and 1999 respectively), while the Public Power Corporation was also transformed into a Societé Anonymé in 1999/2000 initiating the deregulation of power production, including hydropower, in Greece.

Phase IV/2003-... (Phase in development): Second attempt at integration

In 2003, a second attempt towards a more integrated national water regime was made under EU pressure for implementing the requirements of the WFD. The new 2003 Water Law was adopted, which for the first time promoted coherence between water quality and quantity issues. The 2003 Water Law made a renewed attempt towards holistic water policy planning, a water basin management approach based on the coordination of central and decentralised authorities as well as towards extended public participation. In the context of efforts to harmonise the national water policy with the WFD, high level discussions took place on the main water problems of Greece and the required action. Discussions and increased interaction among several actors indicate a revived effort to "push" the Greek water regime to more integration. However, until the integrative aspects of the 2003 Water Law are put into practice, the Greek water regime remains highly complex and fragmented.

5.6.2 Development of regime change variables

This section describes the development of the main change variables of extent and coherence for the Greek national water regime.

5.6.2.1 Extent

The regime extent remained low to medium in the first regime phase (1852-1965). In this phase, only few water uses were regulated such as agriculture, drainage, flood protection, energy production and water supply. The regime extent then gradually increased as more water uses were considered by public policies. Especially, in the 1980s and 1990s, the addition of water quality and environmental policies (for both health and environmental protection) onto the agenda led to the consideration of almost all water uses by the national regime. The 1987 Water Law even defined aquatic ecosystem protection as a water use and required the determination of minimum constant vital flows for rivers (and levels for lakes). The WFD transposition into national law in 2003 maintained the high extent of the regime, giving even higher importance to the environment and ecosystem objectives.

5.6.2.2 Coherence of property rights

The internal coherence of property rights can improve, when newly recognised uses become supported by property rights (Kuks, 2004b) in a coordinated manner with respect to existing property rights. Before the 1987 Water Law was introduced, continuous water use rivalries indicated that rights of different users and owners of water were not coordinated. The system of water abstraction permits of the 1987 Water Law was an attempt to overcome inconsistencies within the property rights system. However, the persistence of water use rivalries in several regions to date indicates failure to coordinate rights in practice. So far, the multiplicity of permit-issuing authorities, the lack of intersectoral water planning and lack of real water consumption data made it very difficult to coordinate the allocation of rights to competing water uses. Uncontrolled water use is also commonplace disregarding the terms set down in water permits. Especially rivalries involving the environment cannot be resolved as long as the quantitative rights of the environment to water are not defined. Namely, requirements for minimum vital river flows and lake levels have not been implemented nation-wide. Also the system of rights to pollute water (via effluent disposal permits) remains largely uncoordinated due to the lack of well-functioning enforcement mechanisms. As a conclusion, despite efforts, the internal coherence of the property rights system in the Greek national water regime remained low in practice, up to the latest phase of the water regime.

5.6.2.3 Coherence of public governance

Until the mid-1960s, there were few water uses regulated by policy and a limited number of actors involved (mainly the services of the Ministry of Agriculture). Therefore, coordination of policies and governance was not too complex. This lack of complexity largely explains the medium (and not low) national regime coherence in this first regime phase (see Table 5.4). Later, several additional actors entered the water policy arena (Ministries of Health, of Coordination, of Environment and of Development), decreasing thus the possibility for coordination. Attempts to increase coherence within the public governance system were then made by the 1987 Water Law and later by the 2003 Law. In the following, we describe the main effect of these attempts on the key elements of the public governance system.

Governance levels and scales: In terms of the levels of governance and especially the "fit" between administrative levels and natural boundaries, there was a rise of coherence with the creation of the Regional Water Management Departments (RWMDs) competent for specific water districts (groups of water basins) in the 1990s. However, the RWMDs did not become fully operational. As a result, their role in coordinating water management across administrative levels and water uses was minimal, while water management decisions remained largely centrally dominated. The 2003 Water Law recently gave decentralised water management competence to the new Regional Water Directorates (RWDs). Their competence was, however, delineated on the level of the Regional administrative borders instead of natural river basin boundaries.

Actors in the policy network: Although the actors involved in water management gradually increased, the policy arena has been mainly dominated by closed sectoral policy communities rather than interactive networks. Even the Interministerial Water Committee of the 1987 Water Law did not succeed in coordinating different sectoral ministries. The lack of coordination of sectoral authorities is also aggravated by the fact that no multiperspective national water vision and no river basin management plans have been produced. Regional participation processes involving public and non-public actors are also still almost non-existent, despite attempts to create Regional Water Committees. The 2003 Water Law made a renewed attempt to promote participation nationally and regionally, recognising more explicitly the interests of non-public actors and allowing for the first time the participation of environmental NGOs. Water policy making also seems to be becoming more open to the broader epistemic community[44] with knowledge on water management. Although the making of the 1987 Water Law was criticized for not involving stakeholders, scientific and professional collective groups (Parliament records, 12-19/10/1987), comments of different experts (public scientific institutions, water-related corporations, research centers and technical professional chambers) were invited for the draft 2003 Law (MEPPW, pers. comm., 2005). The 2003 Water Law also extended the participation of epistemic community representatives in consultation bodies on water management. While the 1987 Water Law admitted the participation of limited experts (Technical Chamber of Greece) in the Regional Water Committees, the 2003 Water Law admits a wider spectrum of experts (from various chambers and research centres) both in the Regional Water Councils and the National Water Council.

---

[44] An epistemic community is defined as a network of professionals with recognised expertise and competence in a particular domain and an authoritative claim to policy-relevant knowledge within that domain or issue area. Epistemic communities are considered as important channels through which new ideas circulate from societies to governments (Haas, 1992). In the field of water management, an epistemic community will typically include bureaucrats, experts or scientists with recognised expertise and competence in water management problems.

Perspectives and objectives: There is so far little coordination between the perspectives of different sectoral authorities on both the national and the regional and local level of water management. This is also linked to the lack of water management plans. Rivalries between different uses competing over water resources were recognized and discussed only in a very fragmented manner on the national policy level and mainly in the context of the Joint Interministerial Water Committee of the 1987 Water Law. Furthermore, there is lack of coordination between the perspectives of public and private actors including water users and environmental groups.

Strategies and instruments: Progress in the adoption of integrated water legislation combining quantity, quality and ecological values (which is an expression of a coordinated water strategy) was only made in the latest regime phase (2003 Water Law). As far as management instruments are concerned, so far these are mainly command-and-control tools (permits for water abstractions, water works and effluents disposal). However, given the lack of intersectoral water planning and of effective enforcement mechanisms, even these command-and-control tools have not helped water managers much to avoid water use conflicts. Other types of more indirect and flexible instruments such as full cost-recovery water pricing or participative management instruments have not yet been realised in the national water regime.

Responsibilities and resources for policy implementation: Finally, the implementation of water policy has so far relied on complex institutional arrangements with responsibilities and resources allocated to an array of institutions, making coordination difficult. Some improvement is expected through the coordination of water management by the Regions and the Environment Ministry as required by the 2003 Water Law.

Additionally, organisations responsible for water policy implementation have been inadequately equipped with financial, personnel and authority resources. With regard to authority resources, the Greek water-district RWMDs were given no competence for certain important water management issues, such as water quality (a gap which was bridged with the adoption of the 2003 Water Law), consultation on land use and effluent discharge permits, water infrastructure operation and maintenance or the setting and collection of water charges.[45]

Overall assessment: Bringing together the progress achieved in terms of the five elements described above, no high coherence in the public governance system of the Greek national water regime is achieved so far. The most recent integration attempt of the 2003 Water Law may improve this aspect of coherence, only once its integrative elements are put into practice in the future.

---

[45] According to Blomquist et al. (2005a), the main fields of activity for water basin-scale organisations include planning and/or coordination, infrastructure operation and maintenance, licensing water uses/allocating water supply, water quality monitoring, consultation on land use or new water use/discharge permits and setting/collecting water charges.

In the first phase of the water regime, only few water uses/users were considered by policies and no permits for water use existed. The regime coherence between the systems of property rights and of public governance gradually increased in the first half of the 20th century as public policy started to increase State control over water (e.g. 1920 law on the expropriation of private land for constructing public hydraulic works; 1940 Civil Code defining surface water as "thing in common use"). Later on, systems of permits for effluents (1965 Sanitary Regulation) were also introduced, limiting thus the right of industries and human settlements to pollute water. The 1987 Water Law significantly expanded the public domain on the use (but not ownership) of water by introducing permits for all new water abstractions and water works. The Greek State chose, thereby, the strategy of public control of water use rights by means of public law rather than by means of public ownership. The permit system showed that the State intended to limit uncontrolled water use and guarantee water access to all interested users. However, the private character of groundwater ownership (linked to private land ownership in the Civil Code) has not been eliminated by public policy. Additionally, policy has not intervened yet in the pre-1989 established abstraction rights excluding them so far from the permit system. Thus, the fact that policy has not targeted yet all users with water rights reduces the external coherence between the systems of public governance and property rights. The external regime coherence is also reduced in practice by the lack of well-functioning enforcement mechanisms for authorised abstraction rights as well as authorised effluent disposal rights.

5.6.3 External change impetus and institutional context

5.6.3.1 External change impetus

The first attempt of the 1987 Water Law at a more integrated national water regime was largely stimulated by the recognition on the Greek policy level of the increased problem of water degradation, especially scarcity. The attempt was also stimulated by the need to coordinate the complex national institutional arrangements for water management and "a feeling of obligation" to meet international and European standards. EU policies (mainly new water quality standards) were also strong change impetus for establishing water quality and environmental issues in the regime. The adoption of the 2003 Water Law was also the result of EU pressure to implement the WFD. In view of possible infringement procedures due to inadequate implementation of European law, the EU WFD may continue to act, also in the future, as an external impetus for further changes in the Greek national water regime.

Institutional impetus for the change of national water regimes, in general, may also include national political developments such as federalist trends and the rise of new levels of governance. In the case of the Greek water regime, changes took place in the 1990s in parallel to (but not because of) the national decentralisation reform. Nonetheless, although the decentralisation reform was not a direct impetus inducing change in the national water regime, it did accelerate some of the required changes, especially in terms of the establishment of regional water management authorities.

5.6.3.2 Institutional context

The following mainly discusses the way in which the outcome of the 1987 Water Law, which was a key attempt at a partially more integrated national regime, was undermined by a non-supportive institutional context. Only but a few aspects of the institutional context became somewhat more favourable to integration in the latest phase of the water regime development, which is dominated by the adoption of the 2003 Water Law. Given the very recent adoption of the 2003 Law, a few preliminary remarks only, can be made on the institutional context influencing this latest regime change attempt.

The national institutional context is discussed in terms of three sets of conditions: the set of values in the water sector, the perceptions of water issues and power distribution. Discussions on each of these three sets of conditions elaborate on their relevant indicators which were presented in chapter 4/4.1.1.

Set of values in the water sector

In general, the set of predominant values in the Greek water sector have acted so far as a restraint to attempts to change the Greek national water regime towards more integration.

The integration attempt of the 1987 Water Law was, firstly, hindered by the lack of a traditionally cooperative water policy style. The principles of this Law were undermined by the persistent lack of coordination in water policy and planning among sectoral institutions with water resources competence (MEPPW, MDEV, MIPAD, pers. comms., 2003). This situation is typical of the horizontal fragmentation of policies within the Greek administration. The 2003 Water Law is also bound to suffer from the unfavourable context formed by the so far non-cooperative water policy style. In fact, it is feared that the participatory mechanisms of this Law for national water policy formulation under the leadership of the Environment Ministry, may prove inadequate for eliminating the lack of coordination among different sectoral actors. The coordination problem of the main sectoral ministries with a stake in water (Ministries of Agriculture, of Development and of Environment) is here briefly discussed borrowing "game-theoretical" ideas (see Scharpf, 1997). Considering that these three Ministries have and will continue to have different world views on water management due to the different sectoral interests which they defend, an option for "forcing" cooperation between them might be the creation of inter-dependencies in their resources. Especially, the Ministry of Agriculture seems to play an important "game" role, since agriculture is the main water consumer (by 83%) in Greece. In the case of the 1987 Water Law, no significant resource dependencies existed between the Ministry of Development (national coordination authority for water management) and the Ministry of Agriculture to "force" their cooperation. In the case of the 2003 Water Law, at first sight, the power basis of the Environment Ministry (new national coordination authority for water management) also appears to be too narrow to force cooperation with the Ministry of Agriculture. However, potential resource interdependency may develop from the need to coordinate with one another on the implementation of EU agri-environmental funding instruments and the new Common Agricultural Policy. If this window of opportunity is used, there may be some room in the near future for increased water-related cooperation at least between the Environment Ministry and the Ministry of Agriculture.

Secondly, the low societal awareness of water resource problems and their environmental dimension was not supportive of an increase in integration either. According to Pridham (1994), the prevalence of consumerist values and financial insecurity in the recently modernized Mediterranean EU Member States, including Greece, are generally not favourable to the spread of environmental and resource protection values.

Thirdly, willingness to bring and to keep water resources in the public domain became obvious in Greece only after the introduction of a permit system for water abstractions and water works in 1989. Still, despite the establishment of this public domain for water use, there remains much uncontrolled private use of water (especially of groundwater). This situation has been aggravated by the unwillingness of individuals to restrict their autonomy on water use (especially for groundwater underneath private land) and has prevented the more equitable distribution of water rights.

Altogether, the set of values in the water sector has been quite unfavourable to integration and is also bound to negatively affect the latest integration attempt of the 2003 Water Law. To the unfavourable conditions discussed above, it should be further added that the Greek water sector does not abide yet to the principles of full cost recovery and "polluter pays" in operational water management, as now required by the EU WFD.

Perceptions of water issues

The main paradigms and perceptions of water issues nation-wide have also been quite unfavourable to the integration effort of the 1987 Water Law.

Water problems have been viewed so far largely from the isolated perspectives of different sectoral actors. Sectoral interpretations of water problems have also been linked to a dominant engineering paradigm in water management, both from a research and from a policy point of view. Characteristically, research on Greek water resources has mainly focused on hydrological and engineering studies, usually ignoring policy and economic management instruments. It is also important to point out that, even within the Environment Ministry, competence for the environment is combined with competence for public works. Within this Ministry, the General Secretariat for the Environment has comparatively less resources and expertise compared to that of Public Works. As a result, the Environment Ministry is overstaffed with engineers, while environmental expertise is still lacking and the interpretation of environmental issues remains unavoidably rather technical on the policy level (Spanou, 1995).

The lack of an intersectoral water planning tradition and of an adequate information basis to support water management decisions acted as further constraints to the integration attempt of the 1987 Water Law. As far as the ability to adapt existing institutions to new water functions is concerned, the Greek water sector is plagued by significant administrative inertia which, in fact, affects the entire public sector of the country. There has been especially a marked weakness of authorities to deal with the environmental functions of water systems.

As concerns the ongoing integration attempt of the 2003 Water Law, the "water perceptions" context is not expected to exercise a very positive influence either. In specific, there continues to be lack of intersectoral water planning, single-sector understanding of water problems and lack of a national solid information basis on water resources. With respect to the latter, an exception is the information basis provided by the established regional registers of water permits nation-wide. There also appears to be increased perception of the risks from water degradation and of the need to consider the environmental needs of water ecosystems by the scientific community, NGOs and mass media.

Power distribution

When the 1987 Water Law was issued, power distribution within the Greek administration was unbalanced, especially between administrative levels. Affected by the strong unitary character of the Greek State, its water management sector was highly centralized. Local and regional experience with self-governance was also in general quite limited, since the national decentralization reform (setting up 13 Regions and converting Prefectures to a 2nd level of local self-administration) was initiated later in the mid-1990s. This decentralization reform of the 1990s was indeed, in some ways, favourable to the attempted changes of the 1987 Law, e.g. it supported the creation of all Regional Water Management Departments (RWMDs). However, the top-down character of the Law implies that its reform attempt was rather imposed by the central government instead of being recognized as a priority issue also regionally and locally. Moreover, the human and financial resources made available to the Greek administration remained insufficient to adequately support the changes attempted by the Law. Even the central administration failed to support the Law with continuous political commitment. Commitment to the attempted integration and decentralization reform varied with time and through transitions in the political leadership of the central administration (former MDEV official, pers. comm., 2004).[46] The limited commitment to the implementation of this important national water management law also reflects the priority placed by the

---

[46] In general terms, it is noted that institutions for policy co-ordination and implementation in the Greek public administration are weak and are replaced by continuous political oversight and intervention. Reform initiatives often depend on personalities rather than institutions, thus weakening continuity between changes in governments. In part because of this *ad hoc* and political nature of policy-making, policy implementation can be uncertain and policy effectiveness reduced (Cordova-Novion et al., 2000). As concerns the 1987 Water Law in specific, in the first year after its adoption, several tasks were implemented largely due to the supportive attitude of the Minister of Development in office. His departure from his post in 1989 coincided with a time of internal political turbulence linked to an economic scandal that brought members of the socialist party government (Pasok) before justice. In this unstable context, many 'on-going' governmental issues remained on hold for a long time (former MDEV official, pers.comm., 2004). Pasok, which was in power when the 1987 Water Law was adopted, remained at government until 1989. In 1989-91, opposition parties formed a coalition government without Pasok. In 1991, the more conservative party of New Democracy won the elections and remained in power until 1993. In the next decade (1994-2004), Pasok was again in power until 2004, when New Democracy was re-elected. Against this background, the implementation of the 1987 Water Law was undermined after 1989 and largely depended on the priority given to water issues by each Minister of Development in office.

Greek State on development and growth at the expense of environmental concerns and sustainable natural resource management. In this sense, past policy patterns favouring economic growth and combined with a tight national budgetary policy[47] continued to act as a constraint on change in the institutional water regime towards more integration and resource protection.

Power distribution has also been unbalanced between public and non-public actors. When the 1987 Water Law was adopted, there was no operational forum for citizen and stakeholder participation in water management. Even nowadays, participation is mainly restricted to formal consultations in environmental impact assessment procedures of water-related projects. In general, there was (and continues to be) a low level of institutionalised societal involvement in water and environmental policies. For instance, the impact of environmental NGOs on water policy was until recently limited to nature and wetlands protection issues. Only recently, the role of Greek environmental NGOs in the field of water policy gained the potential of becoming stronger, using the participation chances offered by the WFD.[48]

With regard to the support offered to the national water regime by the environmental policy sector, this has been rather weak. Firstly, the Environment Ministry has had a limited role in water policy itself until recently. Since 1986, it had a role only in water quality issues. Its water "role" was enhanced only when it became national water coordination authority by the 2003 Water Law. Secondly, the Greek environmental policy sector has been in general rather weak. In specific, it is characterised by a legacy of implementation gaps, which are attributed to the prevalence of ideas on the incompatibility of economic development with high levels of environmental regulation (Spanou, 1998).

Regarding the prospects of the reform attempted by the 2003 Water Law, significant institutional obstacles may continue to exist firstly with respect to the lack of water management authorities with substantial power and resources both on the regional and the central level. Secondly, even though some regional water management structures are in place, especially the RWMDs of the 1987 Water Law, continuous commitment (political and financial) of the central government and of the regional authorities is needed for successful water management decentralization. Thirdly, the 2003 Water Law cannot rely on any participatory tradition in water issues, which would ensure a more balanced distribution of power between non-public and public actors. At least, there is now somehow stronger presence of environmental NGOs in the water policy arena than 20 years ago.

---

[47] Economic conditions often shape the willingness to undertake new ventures. As Mitchell (1990) argues, in difficult economic times, governments are frequently less inclined to undertake new initiatives including those linked to integrated water management.

[48] The 2003 Water Law (transposing the WFD) established consultation mechanisms in the form of one national and 13 regional water councils, which also involve non-public actors including e-NGOs. Although it is still early to assess the role of these novel participatory structures, it is postulated that their effectiveness will largely depend on the willingness of the central government to build up trust between stakeholders and institutionalise effective mechanisms of conflict resolution that depart from centralised command-and-control traditions in this policy area. Otherwise, these new structures will fall into disgrace and exacerbate mistrust between stakeholders as in the case of similar initiatives in nature conservation policy (Koutalakis, 2006).

5.7 Concluding discussion and outlook

This chapter characterized the development of the Greek national institutional water regime, concluding thereby on the outcome of regime change attempts towards more integration. In this closing section, key observations on the progress of the national regime towards a more integrated phase are summarised. Subsequently, the observations made on the Greek national water regime are discussed in a comparative manner together with observations on the development of national water regimes in six other European countries. This comparative discussion aims, on the one hand, at further strengthening the validity of the conclusions drawn on the development of the Greek national water regime. On the other hand, it aims at providing a more validated basis for recommendations on how to shift the Greek national water regime towards more integration in the future.

5.7.1 Progress of the Greek national water regime towards an integrated phase

The above analysis showed that the first attempt at a partially more integrated national water regime by the 1987 Water Law was unsuccessful. A change towards more integration was hindered by the unfavourable institutional context prevailing at the time of the attempt. The centralisation of the Greek politico-administrative system combined with the lack of resources of the administration and the top-down style of the integration attempt hindered the set-up of a coordinated balanced network of central and decentralized water management authorities. Integration was also hindered by integration-unfavourable dominant values and perceptions in the water sector, such as the absence of cooperative water policy style and intersectoral planning, the fragmented understanding of water problems by different sectors and the low societal awareness of the environmental dimension of water resources.

A further integration attempt took place through the transposition of the EU WFD (via the 2003 Water Law), whose real effect on the national water regime remains to be seen. In general, formal policies to achieve more integration now seem to be largely in place but a sufficiently integration-favourable context is still missing. From the perspective of institutional path dependency, an integrated national institutional water regime could be adopted with difficulty in Greece, because the regime changes required for a more integrated phase fit very little into the existing Greek institutional structure. In other words, it is considered that the Greek national water regime has a low adaptive potential to institutional requirements for more integration, such as those promoted by the WFD. Although a few institutional context conditions can be considered somewhat more integration-favourable at present than during the time of the integration attempt by the 1987 Water Law, in total the institutional context remains still largely unfavourable.

Ultimately, the path towards more integration in the Greek national water regime has to be supported by efforts to overcome the unfavourable institutional context, hindering the success of integration attempts so far. Relevant recommendations to change the institutional elements of the Greek national water regime and its institutional context in the direction of more integration are made after the comparative discussion of the Greek and other national regimes in the following section.

5.7.2 The Greek & other European national water regimes in comparison

The progress of six other European countries towards more integrated national water regimes is compared in this section with the findings on the Greek regime. The six other European countries have been previously researched using largely the same analytical framework which was used here for Greece (see comparative review of national water regimes in six European countries by Kuks (2004a; 2004c)). The aim is to ensure that the analytical framework has been interpreted consistently for the Greek national regime, reinforcing thereby also the validity of the key regime explanatory variables used. Subsequently, this section can also make recommendations on how to achieve more integration progress in the Greek regime, based also on the findings in other countries.

Regarding the background previous research in other European countries, Kuks (2004a) compared attempts towards integration in the national water regimes of the Netherlands, Belgium, France, Spain, Italy and Switzerland. Therein, national water regimes consisted of water rights and water policies, while regime developments towards more integration were interpreted as shifts towards more coherent and more complete in their extent regimes. Additionally, Kuks (2004c) compared the institutional context conditions which influenced the national water regime changes in these six European countries.

5.7.2.1 Comparative evolution of national water regimes

Identification of a common pattern for national water regime evolution

Kuks' comparative review of six European countries revealed a common pattern of national water regime evolution described in Table 5.5. This common pattern distinguishes between phases of simple regimes, complex regimes and attempts to establish integrated regimes.

Table 5.5   Common pattern for the evolution of national water regimes

Time period	Regime phase	Description of regime phases
1800-1900	Simple regime	Starts with the definition of a public domain in early 19th century (new constitutions, civil codes, expropriation regulations): ownership of continental waters, navigable waters and responsibility for flood protection placed in the public domain
		Private domain or common property domain: right to drainage for agriculture, right to irrigation for agriculture, the right to use water for domestic, agricultural or industrial purposes
		Scarcity first becomes apparent
1900-1950	Low complex regime	Concern for public health and sanitation added to the public domain (construction of sewage and supply systems): mostly local authorities started developing sanitation infrastructure in urban areas
1950-1970	Medium complex regime	Increased concern about public health and sanitation-urban expansion-more systems for public water supply and wastewater treatment
		No restrictions on water demand but some form of redistribution of use rights begins
		First attempts at systematic nature conservation and attention to natural aspects of water management
		Start of communalisation of water rights, State assumes responsibility for guaranteeing access to all users

Time period	Regime phase	Description of regime phases
		Public service to water demands, attempts to protect surface water and attempts to conserve nature added to the public domain
1970-1985	Highly complex regime	Adoption of surface water protection acts to reduce surface water pollution / EU protective policies for bathing and drinking water
		Restrictive regulations on effluent discharges (domestic and industrial uses)
		Growing environmental movement – incorporation of environmental aspects in water management
		Water depletion on the agenda (water overuse as a policy problem)
1985-2000	Attempts at integration	Awareness of crisis of both water quantity and quality
		Integrated policy approaches: quantity and quality, surface and groundwater into one coherent policy perspective
		Expansion of the public domain:
		• Ecosystem aspects
		• Groundwater quality protection and diffuse pollution control
		• Water use and irrigation control
		• Water drainage control
		• Control flood plains and anticipate water risks
		Debate on redefinition of public and private domain
		Debate on open European market – liberalisation of water services
		Clearer attempts at integration included:
		• Adoption of water policy planning
		• Water basin approach
		• Adoption of integrated management and legislation
		• Extended user participation

Source: Based on the comparative review of Kuks (2004a).

From Table 5.5, it can be concluded that the six European national water regimes compared by Kuks evolved from simple and low complex regimes during the 19th and first half of the 20th century towards more complex regimes after the 2nd World War. After the 1950s, water demands increased strongly and various new water use types and use functions were added to the regime extent. This was due to a rapidly growing population related to economic growth, industrialisation and urban expansion. In the 1960s, there was a growing attention for natural aspects of water resources in some countries, followed by the incorporation of environmental and ecological aspects into water management in the 1970s and the 1980s. In the 1980s, besides surface water issues, also groundwater issues started getting into the spotlight. Around 1985, the first attempts towards integrated water management started taking place in most countries. Specifically, in the period 1985-2000, there was an ongoing debate on the redefinition of the public and private domain regarding water resources in several European countries. In the same period, several integrated water policy approaches were adopted integrating water quantity and quality issues, surface and groundwater management and expanding the public water domain for functions such as ecosystem aspects, groundwater quality protection and diffuse pollution control (Kuks, 2004a). Therefore, the development of national water regimes towards more integration was accompanied by an incremental expansion of the public domain with regard to water use. Similarly, Dosi & Tonin (2001) concluded that in EU Member States, there was an increase over time in the regulatory power of public authorities over water use.

In the following, the above cross-country observations are further grounded by giving more detailed information on important milestones in the evolution of the national water regimes of the Netherlands, Belgium, France, Spain, Italy and Switzerland. Except where otherwise cited, most of the information below was summarised by Kuks (2004a). Detailed information on the development of the Greek national water regime, which was given in the previous sections of this chapter, is not repeated.

Regime changes in the Netherlands

In 1985 an integrated approach to water quantity and quality management was adopted, allowing also ecological considerations to enter Dutch water management decision making. This introduced integrated approach formed the basis for the (Third) National Water Policy Plan of 1989, which was the first integrated water policy plan in the Netherlands although water policy planning already started in 1968. Namely, until 1985, separate plans were made for water quality management and water quantity management.

Due to river floods in the early 1990s, an integral vision on the connection between water management and land use planning was also adopted in 1995. This vision proclaims that more space around rivers is needed, especially in response to flood problems in flood plains, and to better anticipate climate change. Water and its natural movement should become key determining factors in land use planning, when water is competing with other spatial claims. After flooding problems in 1998, a state commission developed a long-term vision on water management in the 21st century. Water will thus have a strong claim in spatial planning, because spatial plans will have to be assessed for water risks before their adoption.

There have also been interesting developments in the tasks of the Dutch regional water boards (main regional water management authorities) and representation of interest groups in the boards' councils. Since 1985, the water boards[49] have been allowed to exercise water management according to a so-called 'broad water system approach' (faced also with the combination of tasks in quality and quantity). Concerning representation, until the 1970s, farmers and real estate owners were the most important contributors to the water boards. In the end of 1960s/early 1970s, representation in the water boards was extended due to the water quality management tasks which were officially delegated to the water boards from 1969 and on. As a result, industrial and domestic surface water polluters acquired some seats in the council. Urban interest groups gained increased influence, opposing the traditionally dominant farmers' interests. This created a basis for the adoption of a more

---

[49] The water boards existed before 1985. Some boards even have roots in the Middle Ages, being among the oldest forms of government at regional scale in the Netherlands.

environmental approach to water management. In the 1980s, the central government decided that the water boards should have a constitutional position in administration and that they should also act on behalf of the general interest. The Water Board Act in 1992 introduced general water board taxation for all citizens, while allowing general citizens to acquire water board seats (on the basis of general elections) equivalent to the share of their contribution (Kuks, 2004b). In 2003 it became mandatory to give water boards a say in land use planning to prevent building activities in floodplains (Kuks, 2004a).

Regime changes in Belgium

In 1971 a national Law on the Protection of Surface Waters against Pollution was adopted. This was an attempt to organise water management on a river basin scale for the three main river basins of the country, which cross the Belgian Regions (Flanders and Wallonia). The main idea was to limit emissions and to accelerate the flow of water in order to transport wastewater more quickly to the sea.

Additionally, in the early 1970s, Belgium started to evolve progressively towards federalisation. It became a full federal state in 1993. The management of freshwater became a Regional responsibility while the Federal Government looks after coastal waters. In the 1990s it appears that the complicated process of federalisation delayed an effective approach to water problems. In the context of federalisation the 1971 law failed implementation, despite its ambition.

In the mid-1990s, Belgium took steps to promote more integrated water management. Flanders and Wallonia recognised the need to co-ordinate different water uses at a tributary basin scale and to co-ordinate quantity and quality management. In Flanders, informal basin committees were set up for each of the eleven basins in the region. The committees are composed of representatives of Regional, provincial and municipal administrations and local environmental organisations. In Wallonia, river contracts have been developed in order to integrate stakeholder interests. These are voluntary agreements between stakeholders of a river basin for the purpose of improving the physical and biological qualities of the river. In general, contracts are made between municipalities, but other stakeholders like industry, landowners and local organisations may participate.

Regime changes in France

The French Water Law of 1964 incorporated the concepts of integrated management by river basin, partnerships, and the combination of regulatory instruments and financial incentives based on the polluter pays principle. The 1964 Law focused mainly on surface water. The main innovation of this law was to set up a dialogue between users (industrialists, farmers and local authorities) and government representatives in the Basin Committees of the French Water Agencies. The Agencies were defined in relation to the main river basins aiming at creating financial solidarity between the different users in order to control water pollution.

The Water Law of 1992 created a new legislative framework for water resource management by introducing the concept of ecological planning and management. By proclaiming water as an object of national heritage, the need to protect this resource for itself, both in terms of its quantity and its quality, was recognised for the first time. Therefore, this legislative step is considered as the start of a more integrated water regime in France. The law also provides for a water resource development and management plan for each river basin, together with objectives for the exploitation and protection of water resources, aquatic ecosystems and wetlands. The idea of integrated management was further supported by the creation of new planning and negotiation tools (SDAGE and SAGE),[50] as well as on the creation of new institutions at the local level (the Local Water Commissions composed of local authorities, state public administrative bodies and user representatives).

Regime changes in Spain

As early as 1926, the Spanish Confederaciónes Hidrográficas (River Basin Authorities, RBAs) were created to group all major water users of each river basin, and to allocate the water resources made available by major hydraulic engineering works. Today, there are 9 RBAs for the main interregional basins, 3 intra-regional water authorities and 2 island water authorities for the Canary and Balearic Islands (Maestu et al., 2003).

The 1985 Spanish Water Act brought the use of surface and groundwater into the public domain giving the State the right to control and limit water uses. The Act also aimed at rationalising water uses in harmony with the environment and other natural resources. The Act included the respect of a minimum flow, to assure the availability of common uses and ecological and environmental needs. The Ministry of Environment already defined minimum ecological flows in the river basins of the country. In practice, these are translated as conditions of maintenance of a minimum discharge downstream of new hydraulic works. However, the minimum flows tend not to be maintained when it comes to water supply or irrigation needs (Kallis, 2003).

The 1985 Act also required the formulation of a National Hydrological Plan, to deal with all water resource management issues. The first such plan was completed (but not approved) in 1993. A revised National Hydrological Plan was presented in 2004. In the context of continued Spanish water policy reform and under pressure of citizen-led protests, the newly elected Spanish government in 2004 revised the Hydrological Plan again, announcing that it would cancel a major inter-basin water transfer project (Blomquist et al., 2005b).

Furthermore, the 1985 Act transferred the RBAs from the competence of the Ministry of Public Works to the Ministry of the Environment. Nonetheless, this transfer of competence has not led yet to the transformation of the traditionally technical and development-oriented role of the RBAs (del Moral et al., 2002). The responsibilities of the RBAs today include water resources planning and development, the management of water use rights and emission rights system, including the monitoring and control of water quality and water resources. The

---

[50]   SDAGE: Master plans for water development and management; SAGE: Local plan for water development and management.

central government keeps strong control of the RBAs through the investments that are mainly funded by the Ministry of the Environment (Maestu et al., 2003).

As regards public consultation, stakeholders are represented directly in water policy formulation in the National Council for Water Resources and on a water basin level in the context of the administrative councils of the RBAs and of user's assemblies. Nevertheless, on both the national and the water basin level, the majority consists of traditional users who represent mainly agriculture, hydropower production, water supply and the Autonomous Communities. Environmental groups, users related to recreation and experts from universities are a minority (del Moral et al., 2002).

Finally, a 1999 amendment of the 1985 Water Act allowed water trading, in the sense that concession holders may sell their surplus to other concession holders, in order to achieve more efficiency (demand management). It also imposed a new restriction to the exploitation system: the ecological flow or environmental demand gets a priority over all other uses, except for drinking water supply.

Regime changes in Italy

The Italian Framework Law of 1989 on the creation of Water Basin Authorities is considered as the first legislative effort to develop integrated management of water resources at the level of river basins, taking into consideration both the quantitative and qualitative dimensions and most of the potentially rival uses. This law resulted in the identification of six major national watersheds (covering the most important Italian rivers), each with a special management authority, and 18 inter-regional basin authorities, which are controlled by the Italian Regions involved. The remaining bodies of water were entrusted to smaller authorities under the direct control of the concerned Region. The principal occupation of the Basin Authorities with environmental protection is mostly confined to safeguarding a constant minimum vital flow in watercourses. This provided the Basin Authorities a crucial role in water quantity regulation.

However, the reform of the 1989 Water Law faced difficulties in its implementation phase. Additionally, some important aspects of integrated water management were still missing such as public participation. Water policy continues to be described in legal as well as planning documents as a top-down administrative exercise, with little or no explicit requirement for the involvement of stakeholders in the policy process (Massaruto et al., 2003).

Another important legal development was the Galli Law of 1994 establishing optimum areas for water services. This law deals with water services and their management, allowing Regions and municipalities to raise finance and set user charges. The Galli law also asserts the public ownership of all water resources and sets a hierarchy between various uses of water, giving priority to human consumption.

All in all, although Italy shows attempts that have an integral outlook, these are based on an incomplete integral approach. Attempts at integration occurred at two different and conflicting levels: at the water basin level, through the creation of the water basin authorities, and at the local level, through the optimum territorial areas entitled to administer locally integrated water services. Contradictions between these two levels of integration reflect the major obstacles intrinsic to the process of the Italian State reform, which sees a contrast between the empowerment of the Regions and of the local authorities.

Regime changes in Switzerland

Swiss water policies mainly developed along three different issues, resulting in three rather separate policy communities on flood protection (since the second half of the 19th century), hydropower production (since the establishment of a water use concession system for hydropower in the start of the 20th century) and water quality protection (since the 1950s).

In 1912, the Swiss Civil Code was introduced, which set a unified regulatory system at national level and introduced State property rights to water. Before then, property rights were only regulated at cantonal level.

In 1953, a new article in the Swiss Federal Constitution on the protection of water bodies against pollution was adopted, followed in 1955 by a Federal Law on the Protection of Waters against Pollution. Water quality protection was, thus, intensified by the limitation of wastewater 'discharge rights' as use rights on water.

In 1975, a new article in the Federal Constitution was adopted, which added a water quantity dimension. New restrictions on water uses were added, especially with respect to hydropower production. In specific, residual flows were prescribed in watercourses and streams. The Constitution modification was also an important turning point, because of the introduction of a principle on the 'unity of water management', which implies that water management should deal with the three sectors of water policy simultaneously. From that time on, restrictions could be placed on uses in the interest of other uses. For instance, the abstraction of large water quantities for hydropower could be restricted for reasons of nature conservation or protection of the hydrological cycle.

In 1991, a new Federal Law on the Protection of Waters was adopted, which finally substantiated the principles defined in the Constitution of 1975. The law offered a framework for the integration of sectoral policies, which opens up the way for a more integrated regime. Water uses (like irrigation and hydropower production) and farming activities as diffuse pollution sources started to be considered as target groups of Swiss water policy. This new federal law again imposed an obligation to maintain suitable residual flows for water bodies and established a series of water protection targets, covering also the ecological functions of water bodies. Federal regulations set minimum flows and cantons may then establish more detailed regulations, allowing for economic and ecological factors on a case-by-case basis. These requirements apply when new concessions are granted or existing ones are renewed. For existing concessions (often granted for 99 years) there is no minimum flow requirement, thus efforts can only be made to persuade those concerned to collaborate voluntarily. The 1991 Federal Law, as revised in 1997, also formally introduced the 'polluter pays principle' into Swiss water policy.

Comparative observations based on seven European countries

Table 5.6 illustrates time-wise the evolution of the national water regime in the Netherlands, Belgium, France, Spain, Italy, Switzerland as well as Greece. In fact, all seven countries seem to match the description of the phases in the common pattern of national water regime evolution, described in Table 5.5. However, the transition moments from regime phase to regime phase appear to vary in time in the different European countries.

Table 5.6   Phases of national water regimes in seven European countries

Common pattern of regime evolution	1800-1900 simple	1900-1950 low complex	1950-1970 medium complex	1970-2000 high complex and integration attempts
Netherlands	1814-1891	1891-1954	1954-1969	1969-1985 1985-1995 1995 >
Belgium	1804-1893	1893-1950	1950-1971	1971-1980 1980-1990 1990 >
France	1789-1898	1898-1945	1945-1964	1964-1992 1992>
Spain	1879-1953		1953-1978 (democratisation)	1978-1985 1985-1999 1999>
Italy	1865-1933 land protection	1933-1976		1976-1989 1989-1994 1994-1999 1999>
Switzerland		1912-1953	1953-1975	1975-1991 1991-1997 1997 >
Greece	1852-1965		1965-1986	1986-2003 2003>

Source: Information on countries other than Greece is based on Kuks (2004a) (covering a research period until 2002). Information on Greece is based on the findings of this dissertation (covering a research period until 2005).

The following Table 5.7 summarises key milestone events in the evolution of property rights and public policies in the seven countries comparatively discussed.

Table 5.7   Milestones in national water regimes of seven European countries

Regime milestones	Netherlands	Switzerland	Belgium	France	Spain	Italy	Greece
Redefinition of the public domain	1992	-	-	1992	1985	1994	1987
Adoption of water policy planning	1968	-	1995	1992	2001	1990	1987
Adoption of integral water legislation[51]	1989	1991 (but still incomplete)	-	1992	-	-	2003

---

[51]   Integral water legislation is here interpreted as legislation considering water quantity and quality issues, surface and groundwater and ecological aspects of the water system.

Regime milestones	Netherlands	Switzerland	Belgium	France	Spain	Italy	Greece
Water basin approach in policy	1992: Water boards gain primacy in regional water management	-	1995: Informal basin committees and river contracts	1964: Basin agencies	1926: River basin authorities	1989: River basin authorities	1987: Regional Water Management Departments

Source: Information on countries other than Greece is based on Kuks (2004a) (covering a research period until 2002). Information on Greece is based on the findings of this dissertation (covering a research period until 2005).

According to Table 5.7, a redefinition of the public domain to bring water to State or public ownership took place in most of the seven European countries in the 1980s and 1990s. The Greek 1987 Water Law also defined water as natural good and means for meeting the needs of society, bringing thus water use (but not explicitly ownership) to the public domain. Water policy planning was also adopted in the 1990s in most countries (except for the Netherlands, which was a forerunner on this since 1968). In Greece, water policy planning was required by both the 1987 and 2003 Water Laws but it was not really effectuated in practice (at least by the time of writing this dissertation). Integral water legislation could be adopted in the Netherlands and France already in the 1990s. In Greece, this was achieved on paper by the adoption of the 2003 Water Law. Finally, a water basin approach in policy was adopted in some countries earlier (e.g. in Spain since 1926) and in others later in the 1990s. In general, attempts towards more integrated national water regimes were often linked to administrative arrangements for the devolution of water management responsibilities to regional levels of governance (often also within water basin boundaries). In Greece, a water management approach following water basin boundaries (as parts of larger water districts) was introduced by the 1987 Water Law but was hardly practiced on the ground due to the deficient implementation of most management aspects of the respective Law.

All in all, Kuks (2004a) concluded that only France and the Netherlands managed to develop towards more integrated water regimes around 1990. By 2002, Belgium, Spain, Italy and Switzerland still had a tendency to complex and fragmented water regimes since integration attempts were not sufficiently coherent. As far as Greece is concerned, this dissertation concluded that its national water regime also still remains complex and fragmented.

In fact, the literature is rich in examples of obstacles and delays in the practical realisation of the theoretical ideal of integrated water management. Mitchell (1990) concluded that despite the general acceptance of the concept of integration, progress in actual implementation has been hesitant and unsystematic. At a conceptual (policy) level, integration receives considerable support, but in practice, there are numerous problems in implementing the concept. In part, this is the result of the presence of real obstacles to integration.

Besides several European countries, also in other parts of the world (e.g. Australia and the USA), it has been tried to employ an integrated approach in the management of water, especially within domestic water basins. For instance, the USA and Australia developed over long periods of time highly advanced and resilient institutional resource regimes for integrated water basin management (Shah et al., 2001). However, the results of such efforts in practice have been mixed and experience often showed that it is very difficult to implement comprehensive, multi-purpose management schemes which take into consideration a broad range of hydrological, ecological, social and economic issues and linkages in a single framework (Marty, 2001). In the USA, the trend in the 1980s even moved again steadily away from an integrated approach as various coordinating mechanisms were eliminated (Mitchell, 1990).

According to Mitchell (1990), a major obstacle to integrated water management is the existence of boundary or edge problems that are often a barrier to, and a rationale for integration. Boundary problems are based on the fact that fragmented and shared responsibilities whether between states, between levels of government, among agencies, or among divisions within departments, are always likely to exist. Mitchell (1990) also notes that although legitimisation for the integrative approach is often sought through legislation, a high level of political and bureaucratic commitment is needed in addition to legislation.

The hesitant progress in the practical implementation of integration also reflects a situation where participants in such a process are learning as they proceed, with no obviously correct model to follow. As a result, individuals are usually cautious and follow an incremental strategy in which they move forward slowly (Mitchell, 1990). In practice, integration is often approached incrementally starting out with certain aspects of integration such as combining quantity and quality issues or combining groundwater with surface water, instead of attempting to achieve holistic integration from the beginning. This was also made obvious in the development of national water regimes in the European countries described above.

In his comparative analysis of international experience with integrated water management, Mitchell (1990) noted that early initiatives usually concentrated upon only a few concerns, such as soil erosion, flood control and drainage or hydropower, irrigation and flood control. Subsequent attempts were made to broaden the range of concerns to be addressed. In that respect, many management functions were identified and then allocated to become the responsibility of an implementing agency. Finally, a stage would be often reached when it was realised that it was extremely difficult to implement a wide range of management functions in a comprehensive manner. Almost inevitably a much narrower range of functions tended to be pursued, usually as a result of financial arrangements which favoured some matters over others, or because of professional predispositions to concentrate upon selected activities. In the same time, however, the response was often to consider explicitly the linkages among different issues. To conclude, the incremental nature of dealing with integration is noticeable in practice world-wide, as well as the fact that on the operational level, the ideal of integration seems to be often broken down again to a narrower approach of management.

5.7.2.2 External change impetus and institutional context in comparison

In his comparative review of six European national water regimes, Kuks (2004a; 2004c) also discussed the external regime change impetus (referred to as "triggers" in his analysis) and institutional context conditions influencing regime evolution. A similar analysis with some variations in the indicators used for the institutional context conditions was carried out for the Greek national water regime in section 5.6.3. An overview for Greece and for the six European countries compared by Kuks is given in Table 5.8.

Table 5.8   External change impetus and institutional context in seven European countries

	Nether -lands	Belgium	France	Spain	Italy	Switzerland	Greece
External change impetus							
Problem pressure	+	+	+	+	+	+	+
European water policy	+	+	+	+	+	+	+
Institutional context							
Set of values							
Cooperative policy style	+	-	+	-	-	+	-
Strong environmental awareness in society	+	-	+	-	-	+	-
Community spirit and willingness to restrict individual autonomy (relevance to water access rights)	+	-	+	-	-	+ (at federal level) - (at cantonal level)	-
Adherence to full cost recovery and polluter pays principles	+	-	+	-	-	+	-
Perceptions of water issues							
Common understanding of water problems (and proactive resource protection)	+	-	+	-	-	- (+)	-
Developed water planning and a supportive learning system	+	-	+	-	-	-	-
Ability to adapt existing institutions (and their expertise) to new water functions	+	-	+	-	-	-	-
Power distribution							
Co-governance between central and decentralised authorities	+	-	+	-	-	-	-
Tradition of public participation and debate	+	+	+	-	-	+	-
Strong position of green NGOs	+	+	+	(+)	(+)	+	(recent improvement)

121

	Nether-lands	Belgium	France	Spain	Italy	Switzerland	Greece
Support by a strong environmental policy sector	+	(+)	+	(+)	(+)	+	(recent improvement)

Source: Information on countries other than Greece is based on Kuks (2004c) (covering a research period until 2002) and on Greece on the findings of this dissertation (covering a research period until 2005).

External change impetus

Table 5.8 shows that the main external impetus for national water regime changes was similar in all seven countries including problem pressure from degraded water resources as well as political and institutional impetus, especially from European water policy. However, it must be noted that the degree of adequate practical implementation of EU water policy principles differed on the national level. For instance, the Netherlands and France adopted quite early the "polluter pays" principle and were able to implement and fund urban wastewater treatment as required by EU directives (Kuks, 2004a). Other countries, such as Italy, Belgium as well as Greece, were much slower in the implementation of these and other relevant EU requirements.

Institutional context

As proposed in the analytical framework for national water regimes (see chapter 4/4.1.1), institutional context conditions are separated into three categories: the set of dominant values in the water sector, the perceptions of water issues and the power distribution. Differences in meeting these institutional context conditions can help explain the different progress of the seven European countries of Table 5.8 towards more integrated water regimes, despite the presence of similar external change impetus and their common EU context.

Set of dominant values in the water sector

The conditions of the "set of values" context were, in general, favourable to integration in the Netherlands, France and to some extent Switzerland. The progress achieved in the Netherlands and France in terms of an integrated national water regime can be explained by a strong value placed on community spirit (including willingness to restrict individual autonomy in water access rights), by a cooperative policy style, strong environmental awareness in society and a common adherence to the polluter pays and full cost recovery principles. The partial progress in Switzerland can also be explained by a strong value placed on community spirit (at federal level but not at cantonal level, where the willingness to restrict individual autonomy is almost absent), a cooperative policy style and strong environmental awareness in society, as well as strong willingness to adapt to EU standards and directives due to strong economic and trade relations with the EU (Kuks, 2004c).

Concerning Belgium, Italy, Spain and Greece, the delay of their attempts at more integrated national water regimes can be partly explained by the lack of a traditionally cooperative policy style, weaker environmental awareness, lack of adherence to the polluter pays and full cost recovery principles and less commitment to limiting individual autonomy in water rights. Especially in Spain and Switzerland, the maintenance of private water property rights and long-term water use concessions respectively imposed significant restraints on regime changes to more integration. In Spain's case, there were attempts (especially by the 1985 Water Act) to eliminate private property rights on water and to advance their expiration date. However, these attempts had a limited effect. In Switzerland, water use concessions for hydropower have been granted for periods of up to 99 years and many of these can only be amended voluntarily (Kuks, 2004c). Similarly in Greece, although the public water domain formally increased in 1987 (bringing water use under State control), this was only weakly enforced in practice to reduce individual autonomy in water use. The weakly enforced public domain resulted in inadequate implementation of water use permit regulations and persistent uncontrolled private use of water (especially of groundwater). In countries with a stronger developed public water domain (Netherlands, France, Belgium, and since 1994 also Italy), private property rights and concessions were not as strong a restraint to integration attempts in their national water regimes (Kuks, 2004c).

Perceptions of water issues

On the one hand, the conditions of the "perceptions on water issues" context formed a quite unfavourable background for regime changes towards more integration in Greece as well as in Belgium, Spain, Italy and partly in Switzerland. This was mainly related to weakly integrated (common) understanding of water problems, poorly developed water planning and also the relatively poor ability of these nations to adapt existing water institutions (and their expertise) to an expanding range of water use functions.

A common restraint, particularly in Mediterranean countries, was the continuation of a traditional engineering approach in water management which promoted isolated instead of common understanding of water resource problems. The emphasis on water quantity management (to meet water supply and irrigation needs) led to civil engineering holding sway in water management (European Declaration for a New Water Culture, 18/2/2005).[52] However, also in non-Mediterranean countries, there has been a strong bias in favour of engineered measures due to the approach adopted for flood protection using embankments and channelisation (Kuks, 2004c).

On the other hand, the progress of the Netherlands and France towards more integrated national water regimes can be explained by a largely integrated (and not isolated) understanding of water resource problems, a strongly developed water planning embedded in a wider policy planning tradition and the ability to adapt their existing water institutions (and their expertise) to new water functions added to the water regime. Switzerland is also characterised by a pro-active attitude in resource protection but suffers from the lack of water planning tradition and the isolated handling of different water issues (Kuks, 2004c).

Power distribution

Finally, the "power distribution" context involves different indicators including the degree of co-governance between central and decentralised authorities, the tradition of public

---

[52] European Declaration for a New Water Culture, Madrid, February 18th, 2005, Initiated by the Fundación Nueva Cultura del Agua, Zaragoza: Navarro & Navarro.

participation, the rise of green NGOs as well as the support that can be provided to the water regime by a strong environmental policy sector.

The lack of sufficient co-governance between central and decentralised authorities was a significant unfavourable condition for integration attempts in the water regimes of Belgium, Spain, Italy as well as Switzerland and Greece. This was especially linked to unfavourable State institutional structures (e.g. traditional federalism or lack of coordination between the central government and the regions) as well as ongoing reforms which delayed the formation of a stable administrative set-up.

In Belgium, there was a gridlock in national water regime developments in the period 1970-1993 due to an ongoing process of federalisation. Since this process was completed in 1993, the Belgian Regions started to develop their own capacity to coordinate water policy issues. In Spain, there was also incoherence in the national water regime due to the intervention of national authorities in the water regimes of the autonomous Regions. While the Spanish regionalisation process began in 1978 and allowed Regions to develop their own coordinating role in water management, the National Hydrological Plan of 2001 allowed interbasin transfers between regions, thereby resisting sustainability arguments of some affected autonomous Regions. In Italy, due to a void at the national level until 1986, water management as well as the implementation of European water policy was left to regional and local authorities. Only after 1986, with the birth of the Ministry of the Environment and the subsequent creation of national and regional environmental agencies, has there been an attempt to create a national dimension in water quality policy. However, the national attempts at integration which followed (water laws in 1989, 1994, and 1999) were incoherent, since they still reflected the tensions between the empowerment of the Regions and of the local municipalities which are intrinsic to the Italian State reform (a process of decentralisation which started with the creation of Regions in 1972). Also in Switzerland, the main obstacle to integration efforts seemed to lie in the traditional Swiss federalist system which assigns responsibilities for policy implementation in many fields to the cantons (Kuks, 2004c).

In Greece, there was also lack of effective co-governance between central and decentralised authorities on water. In fact, when the 1987 Water Law was issued, there were no decentralised water management authorities in place at all. The general Greek local and regional experience with self-governance was also quite limited, since a national decentralisation reform was initiated later in the mid-1990s.

Only in the Netherlands and France, the level of co-governance between central and decentralised authorities was evaluated as a quite favourable condition for water regime integration. In both these countries, there was centralised water management to a large extent with the national level having an influential role. In specific, the national level maintained a strong coordination role and also set the framework for regional water management. In this context, Kuks (2004c) also concluded that it is important to have operational decentralised organisations at water basin level (like the Water Boards in the Netherlands and the River Basin Committees and Water Agencies in France) dealing with all aspects of water management and not just water quantity or quality, as was the case with the River Basin Authorities in Spain which only deal with water distribution, or the separate regional institutions which Italy created for quantity management in 1989 and for quality management in 1999.

As far as the tradition in public participation and the empowerment of e-NGOs are concerned, in the Netherlands, water management started to become politicised in the 1960s, resulting in a paradigm change of the traditional engineering approach adopted by the State water authority and the regional water boards. In Belgium, environmental groups became influential in the 1990s. In the Netherlands, France and Belgium, the participation of new users was allowed through planning. In Spain, the entrance of e-NGOs in the policy arena started mainly in the 1990s in combination with the creation of the Environment Ministry in 1996. In Italy, an increasing importance of environmentalists was reported, although their role is still quite minor. In Switzerland, the political system traditionally affords considerable weight to non-public actors through the right of referendum and popular initiative (direct democracy). During the 1960s, 1970s and 1980s, there was also increasing pressure from groups in society to progress with water quality protection (Kuks, 2004c). In Greece, when the 1987 Water Law was adopted, there was no operational forum for public participation in water management and, for considerable time, the impact of e-NGOs on water policy was limited to nature and wetlands protection issues. Even nowadays, public participation is mainly restricted to formal consultation in environmental impact assessment procedures but, at least, there is a stronger presence of e-NGOs in the water policy arena, especially after the adoption of the EU WFD.

The condition of support of a more integrated national water regime by a strong environmental policy sector seemed to be met or partially met in several countries. The Netherlands, Belgium, France and Switzerland all established an environment ministry around 1970s, Italy in 1986 and Spain in 1996. In the case of Belgium, however, environmental policy was delegated to the Regions in 1980 and the ongoing institutional reform delayed the effective operation of environmental departments in the Regions. For the Netherlands, France and Switzerland, it seems that the active role of the national environment ministry was important for the dedication of the national government to environmental issues. In the cases of Belgium, Italy and Spain, the entrance of an environment ministry into the water policy arena induced regime change in terms of the recognition of environmental and ecological aspects of water systems (Kuks, 2004c). In Greece, an environment ministry was set up in 1980 but played a rather insignificant role in water management (being involved only in water quality issues) until 2003. In 2003, the environment ministry was finally assigned also competence in water quantity management issues when it was defined as the new national water authority with regulatory, executive and coordinating competencies in water management.

Summary comparative observations

Although all seven compared countries shared similar external change impetus stimulating their national water regimes to change towards more integration, only the Netherlands and France achieved more progress towards integrated national water regimes around 1990. This difference in progress could be explained by differences in the institutional context which prevailed during attempts at more integration. That is to say, the Netherlands and France were assisted in their attempts by a set of institutional context conditions, which favoured the establishment of more integrative elements in their national water regimes. On the other hand, the countries which made less progress (Greece, Spain, Italy, Belgium and in some aspects also Switzerland) shared a less integration-favourable institutional context (some to a greater and some to a lesser extent). All in all, the fact that the institutional context variables used in this dissertation could explain the integration progress of the Greek national water regime, also in relation to the integration progress concluded for previously researched European countries, reinforces the validity of these variables as meaningful determinants of national water regimes and their changes.

5.7.3 Outlook recommendations on the Greek national water regime

Based on the empirical analysis of the Greek national water regime and its comparison with relevant developments in other European countries, this section attempts to give directions on how the Greek national water regime could shift towards a more integrated phase.

Improvements needed in the institutional elements of the regime

The direction in which the institutional elements of the Greek national water regime should develop to reach a more integrated phase (in the sense of this dissertation) is spelled out in the integration dimensions of the analytical theory-based framework used. Specifically, more progress is required to complete the extent of uses considered by the regime and to achieve higher coherence within and between the regime institutional elements.

The extent of the national water regime still needs to be completed in practice, mainly in terms of the environmental use of water resources (e.g. by respecting policy requirements for the definition of minimum river flows and minimum lake levels).

The need for more coherence in the system of Greek public water governance refers to the need for coordination between administrative levels for water management and natural river basin scales, for the establishment of public participation processes in water planning, for the coordination of perspectives of different sectoral authorities and multiple actors within water management plans, for more resources to implement integrated water legislation and for the simplification of implementation institutional arrangements by concentrating water competence to fewer authorities and by establishing better coordination arrangements.

Additionally, the system of water property rights needs to become better coordinated and coherent. To this aim, the issuing of water abstraction permits should be linked to an intersectoral water planning system, there should be better control of water abstractions according to the conditions set down in abstraction permits and there should be control on the extensive private use of water outside the property rights system. However, it should be noted that strengthening the relevant control and enforcement capacity could prove difficult in Greece, which is characterised by deficient implementation in several policy fields. Even in countries like France with significantly more progress of their regimes in terms of integration, there have been problems in the enforcement of abstraction permits, partly due to the lack of personnel in the territorial water police services of various administrations (Barraqué, 1998).

Finally, there is need to better coordinate the systems of property rights and public governance of the national water regime (external regime coherence). Firstly, there is a need to better coordinate groundwater ownership rights (defined in the Civil Code) and groundwater use rights (defined in national water laws). Namely, the ownership of groundwater remains linked to land ownership, although groundwater use rights can only be allocated by the State. In this case, the more consistent treatment of groundwater in public policy (water laws) and in basic property rights regulations (Civil Code) could be simplified, if private ownership of groundwater is eliminated. Secondly, it is necessary to issue public policies that will regulate the system of those water abstraction rights which were established by law or custom prior to the 1989 permit regulations. Thirdly, the coordination between public policy and property rights could be enhanced in practice by investing more in enforcement mechanisms to control but also to safeguard policy-based rights to abstract or pollute water.

Improvements needed in the institutional context

Section 5.7.1 concluded that to achieve a smoother shift of the Greek national water regime towards a more integrated phase, more effort is needed to improve the institutional context in which regime change attempts take place. Indeed, the institutional context in Greece has so far not been very supportive to integration attempts. Based also on the comparative cross-country discussions of the previous section, it is here concluded that more effort is needed to improve all three core aspects of the Greek institutional context, i.e. the dominant values in the water sector, the perception of water issues and the power distribution.

As concerns the dominant values in the Greek water sector, above all, there should be political effort in achieving a more cooperative water policy style to gradually harmonise rival water use fronts. In this respect, it is important to make use of new opportunities given by the 2003 Water Law, especially the National Water Committee. Political commitment is also needed to safeguard water resources in the public domain, especially through the better enforcement of regulations to control private illegal water use. Additionally, political willingness is needed for adopting and actively pursuing the polluter pays and the full cost-recovery principles. The environmental awareness of Greek society could also be strengthened, despite the still lower economic development of the country compared to other EU Member States, if more awareness efforts are made by the competent environmental administration and non-governmental organisations.

As concerns the prevalent perceptions of water issues, effort and willingness is needed by both public and non-public actors to develop more integrated understanding of water problems and to be open to new paradigms of water protection. The traditional engineering approach, which has also been supported so far by the public administration, should not be the only paradigm guiding Greek water management. Instead, also alternative less engineering-intensive management solutions should be gradually investigated and promoted. Secondly, resources and political commitment should be devoted to the establishment of intersectoral water planning tradition and a nation-wide water information basis suitable to modern water management needs. Finally, the ability of existing water institutions to expand their expertise to a growing range of water use functions and new paradigms in water protection should be strengthened. Again, for this, the commitment of appropriate resources and political effort to overcome administrative inertia on the national, regional and local levels is needed.

As concerns the power distribution in the water sector, especially the level of coordination and co-governance of central and decentralised authorities in Greek water management needs to be strengthened. The 2003 Water Law (similarly to the 1987 Water Law) gives the opportunity to establish a network of Regional Water Directorates to take over water management competence on the regional level. Political willingness and resources need to be committed as soon as possible on the national and regional level, in order to avoid a network of non-functional and inadequately equipped authorities with little ability to communicate regional water management issues to the central level. As far as balanced power distribution between public and non-public actors such as e-NGOs is concerned, new chances to strengthen the relevant institutional context are seen mainly in the 2003 Water Law and the EU WFD (see section 5.5.4 for the new National and Regional Water Councils, which should also involve non-public actors). In general, it is expected to prove difficult to overcome the mistrust of both non-public and public actors in the ability of the government to actually deliver positive results in the area of water policy. The steps made so far have been cumbersome and disappointing. Therefore, significant efforts are needed to boost up expectations (and motivation) for the implementation of new important water policy and the therein required participative processes.

All in all, this chapter provided useful insight into the extent of realising any integration principles in the Greek national water regime so far, as well as explanations for that. The next chapters further explore the realisation of any progress towards a more integrated water regime on the regional level of Greece, in the context of real-life water management processes in two specific basins. The empirical research on the water basin level will also explore in more regional detail several aspects of water property rights and public governance, which have been introduced on the national level of analysis.

6 Selection of two Greek basins as case studies

6.1 Introduction

This chapter describes the selection of two Greek basins as case studies to pursue the research questions of this dissertation on a water basin level. Case study selection took place after screening several candidates, using the criteria defining the case study unit of analysis in chapter 4/4.3.2.2. In brief, the criteria defining the unit of analysis required the selection of cases 1) following water basin boundaries, 2) having such water resource problemat structure, so that it is more likely to find attempts at a more integrated regime, 3) having actor interaction in the context of water use rivalries and 4) involving both economic and ecological uses. Before presenting the outcome of the screening process (section 6.3), the following gives a general overview of the types of the most common water resource problems in Greece (section 5.3), in order to place the selection of two basins into a broader context.

6.2 Types of common water resource problems in Greece

The following observations on the most common water problems in Greece are based on the description of the status and uses of Greek water resources in chapter 5/5.3. Distinction is made between key water categories, such as rivers and lakes, and between problems of water quantity and water quality. As previously mentioned, the most significant water problem of Greece is quantitative, while water quality is in general good with pollution problems of mainly local character.

In the category of rivers, there are only 45 rivers classified as perennial in Greece, despite the numerous recorded streams. Greek rivers mainly suffer from problems of water overabstraction, usually leading to rivalries of upstream-downstream territorial character. Problems of intense water pollution are only encountered in the few large transboundary rivers entering northern Greece. In these cases, rivalries are again of upstream-downstream character between downstream Greece and its upstream neighbours.

As mentioned in chapter 5/5.3, there are also several lakes (almost 60) in the Greek water landscape, mostly located in the wetter northern and western Greece. Greek lakes are often affected by both water quantity and water quality problems. Water quantity problems are due to overabstraction directly from the lakes or from their interconnected aquifers for agricultural and industrial purposes. Lakes are also often the subject of local rivalries over water pollution, due to the accumulation of pollutants from nearby urban settlements, industrial centres and agricultural areas. Often, water abstraction and pollution problems are shared by the lakeshore users with similar use benefits (problems with reciprocal distribution of externalities). However, there are also cases where water use rivalries gain a more upstream-downstream character, if important users are dispersed in the whole basin and not only on the shores of the lake.

As concerns Greek aquifers, pollution is rising due to overabstraction which causes saline intrusion in coastal areas and on islands. However, Greek aquifers are mainly under significant pressure in terms of scarcity. Indeed, groundwater abstraction in Greece is significant, satisfying 40% of total water consumption. Agriculture, which is the main water user in Greece, is served to 35% from groundwater sources (Kouvelis et al., 1995). Similar to lakes, aquifer overabstraction can affect quite symmetrically users with similar use benefits. There are, however, also cases where rivalries of a more upstream-downstream nature arise, due to the specific aquifer shape (e.g. inclination, flow direction).

Finally, Greece has a very long coastline and numerous islands facing their own specific water resource problems. Coastal waters are mainly affected by local pollution problems, when they are close to important urban and industrial centres. The islands face intense water scarcity problems (in terms of their aquifers), which are however of small scale and mainly seasonally related to the rise of tourism in the summer.

All in all, the most common water degradation problems in Greece have led to a variety of water use rivalries in many parts of the country. In cases of water scarcity, rivalries appear both between homogeneous users (mostly between irrigation communities in the summer) as well as between heterogeneous users. In urban areas, territorial conflicts usually arise between water users in the urban area and remote water-richer areas which serve as water source for the cities. Conflicts also develop between drinking water supply in the urban centres and local farmers. In agricultural areas, conflicts arise between water use for irrigation and water use for water supply, tourism and the conservation of the environment. In areas dependent on tourism (particularly the Aegean islands), conflicts arise mainly in the summer between drinking water supply and irrigation. In cases of decreased water quality, industries or urban agglomerations come into conflict with supporters of health, the environment, fisheries and recreational interests of the local population.

6.3 Selection of case studies

For the selection of two case studies which would be suitable for the research purposes of this dissertation, seven different candidate water basins were screened. All candidates were inland basins of important rivers as well as of important lakes, with water problems often present also in their interconnected aquifers. It was decided not to consider islands or degraded coastal waters as candidate case studies. Firstly, islands and coastal waters cannot always be clearly related to the spatial unit of a single basin (e.g. many islands consist of several small basins, if any, while coastal systems may be affected by the outflow of several basins). Secondly, islands and coastal waters often demand other management strategies than inland waters, tailored to fit their specific characteristics.

For the purpose of screening, preliminary information was collected for each candidate case, based on available literature and a few personal communications with local water experts. Table 6.1 summarises key information on all candidate cases, according to the different case study selection criteria defined in chapter 4/4.3.2.2.

At a first stage, two candidate cases were screened out (case of Prespa Lakes and of Strymon basin), because their water use rivalries were not sufficiently intense and clear. The absence of intense rivalries was considered as suboptimal for observing any potential attempts at more regime integration urged by the pressure of conflict.

At a second stage, the screening was completed in favour of two basins, which include important lakes and small stream systems: the basin of Vegoritida and the basin of Mygdonian. Indeed, most other screened-out candidate cases concerned mainly rivers or rivers connected to lakes. Preliminary information gathered on the candidate cases showed that there were relatively more actor interaction and more cooperation efforts in the two selected basins of lakes than in the non-selected basins of rivers. This *a priori* information on the existence of some cooperation efforts in the selected "lake" basins led to the expectation that there might be relatively more chances to find attempts at more regime integration there than in the other candidate basins.

In a way, the presence of relatively more interaction and cooperation efforts in stressed "lake" basins than in stressed "river" basins provides preliminary support for the assumption put forward in chapter 4/4.3.2.2, that upstream-downstream water problems with unidirectional externalities (as usually encountered in river systems) have lower cooperation potential than water problems with reciprocal externalities (as more frequently encountered in shared lakes or shared aquifers). Table 6.1 includes preliminary conclusions drawn on the distribution of externalities from the main water problems in all candidate cases, on the basis of the limited information selected during screening. Indeed, for most examined candidate cases of rivers (including the case of the artificial Lake Plastira which mainly supplies water to a downstream river system), a unidirectional distribution of externalities was assumed, while candidate cases of mainly important lakes seemed to be characterised by predominantly reciprocal externalities.

The two selected basins of Vegoritida and of Mygdonian are situated in north-western Greece and in northern Greece respectively (see Figure 5.3).

Both basins are at the lower end of large-sized basins[53] (ca. 2000 km²). Both are characterised by clear problem pressure and intense water use rivalries. In practice, the most intense water use rivalries were observed in specific parts of the two basins in question, which thus became the main focus of research. The main rivalries identified are central to the descriptive part of the case studies in chapters 7 and 8, since the storyline of these rivalries is interlinked to the water management process.

The two case studies selected are quite similar in terms of their water resource problems and use rivalries. This similarity could increase the possibility of observing similar water management developments and of coming to potentially converging research conclusions. Given the analytical limitations of research in only two case study areas, similar conclusions from two similar case studies could be analytically stronger and easier synthesised than distant conclusions from two very different case studies. Both selected case studies of the

---

[53]   For the size characterisation, the definition of the WFD Annex II (1.2.1) System A has been used. According to this definition, a large catchment equals 1000 – 10000 km².

Vegoritida and the Mygdonian basin involve important water quantity pressures, covering thus the most significant water problem of Greece (which is that of quantity). Both, however, also involve water quality problems of local character, which are intensified by scarcity. Although water pollution problems are not considered typical for the majority of Greek water systems, it is believed that the examination of two cases combining quantity and quality problems will allow interesting insights into the degree of integration of quantity and quality aspects in Greek water management.

Except for similarities, there are also certain interesting differences between the two selected case studies. Water use rivalries in the Mygdonian basin are more diffuse (involving several disperse polluters and abstractors) than in the Vegoritida basin, where rivalries greatly focused for considerable time on the activities of a single industrial actor (the Public Power Corporation). Another important difference concerns the administrative splitting of the basins. The Vegoritida basin is split between three Prefectures (also pertaining to two different Regions), while the largest part of the Mygdonian basin lies in the jurisdiction of only one Prefecture.

Figure 6.1   Location of the Vegoritida and Mygdonian basins

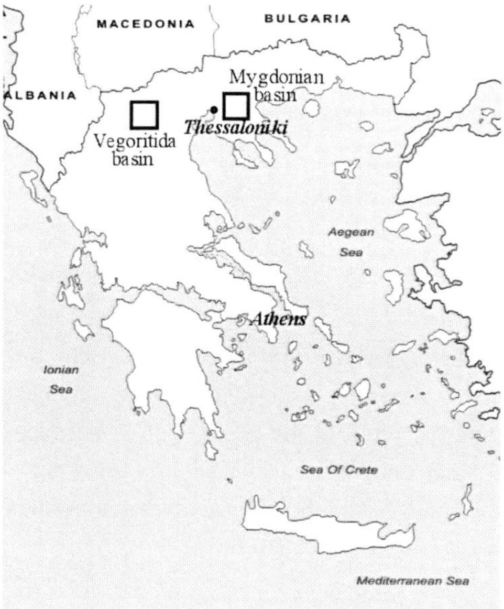

Table 6.1   Summary information on seven candidate case studies

Screening criteria	Water resource problem structure			Rivalries as object of strong actor interactions	Rivalries of economic vs. ecological uses
	Clear problem pressure & rivalries	Type of water use rivalry	Distribution of externalities*		
Vegoritida basin (incl. Lake Vegoritida)	Yes	Energy production abstractions vs. irrigation, tourism, fisheries, environment Discharge of pollutants vs. tourism, fisheries, environment	Reciprocal for abstractions Reciprocal for wastewater discharge among two Prefectures sharing Lake Vegoritida Unidirectional for wastewater discharge among three Prefectures sharing the basin (upstream Prefecture of Kozani pollutes stream Soulou ending up into the lake, at the expense of the two downstream lakeshore Prefectures)	Yes	Yes (partly)
Mygdonian basin (incl. Lake Koronia)	Yes	Agricultural and industrial abstractions vs. environment, fisheries Discharge of pollutants vs. environment, fisheries	Reciprocal for abstraction Reciprocal for wastewater discharge	Yes	Yes
Basin of River Strymon (incl. artificial Lake Kerkini)	Yes, but rivalries not intense	Irrigation and flood protection storage in lake vs. lake and downstream river environment Irrigation upstream vs. irrigation downstream *Note*: Transboundary nature of basin upstream	Unidirectional negative externalities from upstream abstractions Unidirectional positive from upstream flood control	Partly	Yes
Basin of artificial Lake Plastira (with water transfer to neighbour basin)	Yes	Initially, energy production downstream of artificial lake vs. irrigation further downstream Later, irrigation abstractions from lake for use downstream vs. lake landscape, ecotourism	Unidirectional for abstractions; no evidence for important cooperation efforts	Weak	Weak
Basin of River Aliakmon	Yes	Initially, energy production upstream vs. irrigation downstream	Unidirectional for abstractions; no evidence for important cooperation efforts	Partly	Yes

133

| Screening criteria | Water resource problem structure | | | Rivalries as object of strong actor interactions | Rivalries of economic vs. ecological uses |
	Clear problem pressure & rivalries	Type of water use rivalry	Distribution of externalities*		
		Later, rivalry among irrigators downstream Energy production and irrigation vs. water supply and environment downstream			
Basin of Prespa Lakes	Partly (problem pressure not strong anymore; rivalries not intense)	Nature vs. irrigation Discharge of pollutants in neighbouring country vs. environment *Note*: Basin is transboundary	Reciprocal for abstractions among Greek lake settlements Reciprocal for wastewater discharge	Partly	Yes
Basin of River Nestos	Yes	Energy production upstream vs. environment downstream Irrigation upstream vs. environment downstream Minor conflict between energy production upstream vs. irrigation downstream *Note*: Basin is transboundary	Unidirectional for abstractions; no evidence for important cooperation efforts	Yes	Yes

* Only externalities among homogeneous parties, i.e. users with similar benefits from the same resource use, are considered.

7 Institutional water regime of the Vegoritida basin

7.1 Introduction to the case study on the Vegoritida basin

Water basin description

The water basin of Vegoritida is situated in north-western Greece. It is shared by two administrative Regions, the Region of West Macedonia (including the Prefectures of Florina and of Kozani) and the Region of Central Macedonia (including the Prefecture of Pella).[54] Lake Vegoritida is the largest water body of the basin and its surface is divided between the Prefecture of Pella and the Prefecture of Florina (see Figure 7.1).

Figure 7.1   Prefectures sharing the water basin of Vegoritida

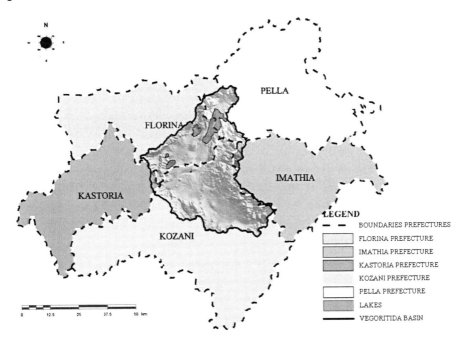

Source: Map produced by Dr.Martha Stefouli, Institute of Geology and Mineral Exploitation, Greece.

The hydrological basin of Lake Vegoritida is surrounded to its largest extent by mountains (Sioutis, 1989). The lake has only natural underground outflow (of $49*10^6 m^3/y$) towards a neighbouring basin (Paraschoudis et al., 2001). In the past, Lake Vegoritida also had artificial surface outflow. This was achieved via a tunnel transferring water to a hydropower plant in a neighbouring basin of the Pella Prefecture (1955-1990) and via a water pumping station abstracting water for steam-electricity plants in the Florina and Kozani Prefectures (1976-1997) (Sioutis, 1989).

There is seasonal surface water inflow into Lake Vegoritida from streams and torrents. The longest stream of the basin is the stream Soulou (Sioutis, 1989), which flows from south to

---

[54]   The basin also includes minor parts of the Prefectures of Kastoria and Imathia.

north close to the city of Ptolemaida and mouths into the southern part of Lake Vegoritida. In 1954, part of stream Soulou was converted into an artificial channel during the drainage works of the neighbouring Sarigiol basin, which was previously marshland (PPC General Mines Directorate, 2005). After these drainage works, stream Soulou connected the basin of Vegoritida with the Sarigiol basin and extended the basin of Vegoritida by 516 km² (Paraschoudis et al., 2001).

As a result of further land reclamation works in the 1960s, Lake Vegoritida also became the final recipient of water from 3 smaller nearby lakes (Lake Chimaditida, Lake Petron and Lake Zazari, which previously formed a distinct basin of 355 km²). The four lakes were artificially connected via ditches and formed one common hydrological basin. Since then, there is artificial water inflow from Lake Zazari into Lake Chimaditida, from Lake Chimaditida into Lake Petron and finally, from Lake Petron into Lake Vegoritida (Paraschoudis et al., 2001).

The current total surface of the Vegoritida basin (see Figure 7.2) is 1922 km² (Paraschoudis et al., 2001), including all artificial connections to neighbouring basins which took place since the 1950s and 1960s.

The surface of Lake Vegoritida has been subject to frequent fluctuations, at first due to natural mechanisms, especially climatic variations.[55] In 1885-1955, the lake surface used to vary between 50-70 km². Researchers characterise a level of 541-542 masl (meters above sea level) as the normal level of Lake Vegoritida under natural conditions. Since 1955, Lake Vegoritida showed a level drop of over 30 m (from 541 masl in 1955/56 to 509 masl in 2001) due to human-induced processes. As a consequence of the level drop, the lake surface was reduced from 60 km² (in 1955) to 28-30 km² (in 2001) (Paraschoudis et al., 2001).

Demographics and employment

The total population in the Vegoritida basin is 95,000. The largest urban centre is the city of Ptolemaida with a population of 27,000. Other settlements are smaller with a population between 500-1,000, except for Amyndeo with 3,400, Arnissa with 1,800 and Perdika with 1,900 (Koklis, 2003). Due to the location of the basin close to the Greek north-western borderline, it is politically important to maintain the local population at a stable level. Because of this, it has often been argued that there is little room for reducing the local income for the sake of water protection measures (Planet Northern Greece S.A & AN.FLO. S.A., 2000).

---

[55] The climate of the region is characterized by high precipitation and the lowest temperatures in Greece (Reference to the climate of the city of Ptolemaida, which is located in the basin (Sotiropoulou et al., 1992)).

Figure 7.2   The water basin of Vegoritida

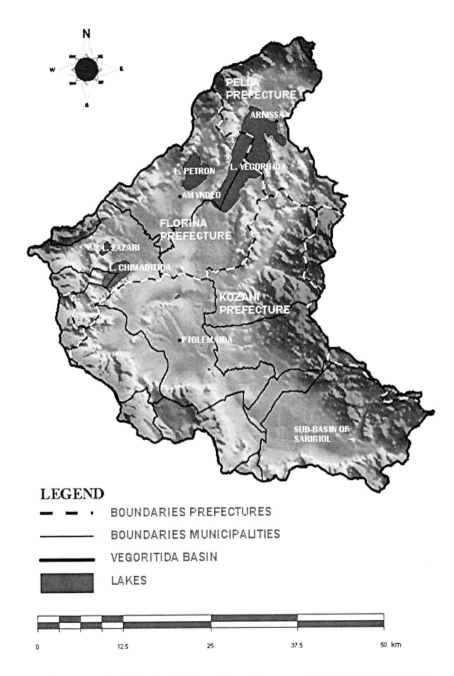

Source: Map produced by Dr.Martha Stefouli, Institute of Geology and Mineral Exploitation, Greece.

The agricultural sector is the most important economic sector both in terms of land use and of employment for the local population (AN.FLO. S.A., 1998). Additionally, the Public Power Corporation (PPC) offers employment to many inhabitants in the energy production sector. The employment situation of lakeshore communities was different in the past, when most lakeshore inhabitants were fishermen. For instance, in 1965, there were 80 fishermen families only in the lakeshore community of Agios Panteleimon, close to Amyndeo. Today, only 6 fishermen remain there due to the lack of fish (Interview Community Agios Panteleimon, 22/1/2005).

Methodological issues of the case study

The reasons for selecting the basin of Vegoritida as a case study were elaborated in chapter 6. The methodology used for case study data collection in this case study was described in chapter 4/4.3.2. Here, additional information is given on the focus of the specific case study and the sources of data used.

In the Vegoritida basin, many water users compete over water for industry, agriculture and urban use. The case study focuses on the water resource problems of Lake Vegoritida, which has been the main source of conflict in the basin. Lake Vegoritida is one of the most important lakes of Greece supporting multiple water uses for humans. Although the chronologies of two different water use rivalries are described in separate (in sections 7.6.2 and 7.7.2), the case study is analysed as one process, because the rivalries in question and the related changes in water regime elements are highly interdependent. The time period covered by the case study starts in the 1970s, when first protests on the decline of water resources in the basin began, until approximately mid-2005.

A wide range of written documents and primary data were collected in the field and personal interviews and communications with actors in the process were carried out (see Annex III and Annex IV).

Structure of case study

Section 7.2 gives information on the use of water in the basin, emphasising the main pressures on water resources. Section 7.3 describes the key water resource problems and the main resulting water use rivalries. Section 7.4 describes milestone events in the case study process, which are common for both key water use rivalries dealt with in the case study. After the presentation of the main actors involved in the water management process of the Vegoritida basin in section 7.5, sections 7.6 and 7.7 deal with the two key water use rivalries in the basin on dropping water levels and on decreased water quality. In specific, both sections 7.6 and 7.7 start with the description of water property rights of users involved in these rivalries. The description of rights serves as the basis for the later assessment of changes in the property rights component of the water basin regime. Subsequently, the chronology of events of each rivalry, the actors involved and their interrelationships are described. The actor analysis serves as a useful source of interaction observations for the subsequent evaluation of the regime change and implementation processes.

Sections 7.8 and 7.9 are dedicated to the analysis of the case study findings according to the analytical framework for change in institutional water basin regimes (see chapter 4/4.1.2). In specific, section 7.8 evaluates the process of making changes to the water basin regime and section 7.9 evaluates the process of implementing any new incentives (measures and policies) resulting from the regime change process.

7.2 Key water uses and their impacts

The current uses of water resources in the Vegoritida basin include:

- Irrigation using lake water, the aquifer and streams, especially stream Soulou.

- Industrial use: Lake Vegoritida and stream Soulou receive industrial effluents. Especially stream Soulou receives wastewater from the Prefecture of Kozani (steam-electricity plants of Agios Dimitrios and Kardia, the Southern Mine and industrial units in the area of Ptolemaida including the Ptolemaida Public Slaugtherhouse) (PPC General Mines Directorate, 2005). Industries, especially the PPC, also abstract water from the basin aquifer and from stream Soulou. In the past, the PPC abstracted water directly from Lake Vegoritida for use in a hydropower plant and its steam-electricity plants;

- Urban use: Water for drinking is abstracted from the basin aquifer and springs. In the past, Lake Vegoritida was also used directly for water supply (for 4,500 people) of certain lakeshore communities (Goulios, 1990). In 1979, the water of Lake Vegoritida was characterised as unsuitable for drinking by a common decision of the Prefectures of Florina, Kozani and Pella (decision 1900/79).

  Lake Vegoritida also receives untreated urban wastewater from Amyndeo (first discharged into Lake Petron) and other lakeshore settlements. Stream Soulou receives wastewater from the city of Ptolemaida (treated) and nearby settlements (partially untreated) (PPC General Mines Directorate, 2005).

- Fisheries (professional and leisure).

- Recreation and tourism, although bathing in Lake Vegoritida does not take place to the extent it used to until the 1970s. In 1960-1970, there was a trend for tourism development in the area around Lake Vegoritida. A club for water sports was also very active in 1976-1985 when the lake level was still high enough (pers. comm. Association MESNA of Arnissa, 27/1/2005).

- Finally, Lake Vegoritida forms an important wetland system together with the lakes of Zazari, Chimaditida and Petron, which are a vital habitat for flora and fauna and a determinant of the local climatic conditions.

Table 7.1 gives information on the annual water consumption rates of the main water uses in the Vegoritida basin.

Table 7.1   Water consumption in the Vegoritida basin in the 1990s

Use	Irrigation	Industrial use by the PPC	Drinking water
Annual water consumption	101*10^6 m³	Water from the Vegoritida basin: 16*10^6 m³ in the Amyndeo mine 3-12*10^6 m³ in the South Field mine of the Kozani Prefecture 10.6 *10^6 m³ for the Ptolemaida steam-electricity plant (groundwater and stream Soulou) 2.9*10^6 m³ for the Amyndeo steam-electricity plant (groundwater) Water from the southern neighbouring Aliakmon basin: 50-60*10^6 m³	15*10^6 m³
Source of water	Groundwater: 70% Springs and surface water: 30%	Groundwater and surface water	Groundwater and springs

Source: Paraschoudis et al. (2001); PPC General Mines Directorate (1994a); PPC General Mines Directorate (2005).

Agriculture

Impacts on quantity of water resources

Abstractions for irrigation take place to 30% from surface waters (Lake Vegoritida, streams and ditches) as well as springs. 70% of irrigation water comes from groundwater abstracted in private wells (Paraschoudis et al., 2001). There are only a few small collective irrigation networks in the basin but these are not in the immediate vicinity of Lake Vegoritida. These collective networks are managed by Local Organisations for Land Reclamation and use either surface water (torrents, Lake Zazari) or groundwater or both (Interview Florina Pref., 26/1/2005).

In the part of the basin in the Pella Prefecture, there are several "collective" wells managed by groups of land owners. "Collective" wells are connected to an underground pipe system and every farmer of the group has an outflow in his field. "Collective" wells belong to groups of 10-20 farmers, who share operation and maintenance costs and engage a person for distributing water to the farmers. Although there are similarities to Local Organisations for Land Reclamation, "collective" wells are a private arrangement and operate in a more flexible way (Interview LVPS, 22/1/2005; Interview Agricultural Cooperative of Arnissa, 25/1/2005).

In 2001, farmland covered approximately 1/3 of the Vegoritida basin (58,350 ha). Irrigated surface of farmland amounted to ca. 2,000 ha in the period 1955-1975, which increased to ca. 8,000 ha in 1975-1995. In the year 2000, 15,500 ha were irrigated with a clear tendency to increase even more (Paraschoudis et al., 2001). Especially in the Florina Prefecture, there was an increase of the irrigated surface from 26.96% of its total agricultural surface (in 1990) to 40.1% (in 1998). This was due to the increase of irrigated crops, especially maize (by 64.7% from 1990 to 1998) and sugarbeet (by 596.7% from 1990 to 1998). In fact, many crops in the Vegoritida basin are water-demanding crops such as maize and alfalfa. In the

Florina Prefecture, wheat and maize are the dominant crops, while in the Pella Prefecture, trees and wheat dominate (Koklis, 2003).

Two main types of irrigation techniques are used in the basin. First, spraying irrigation in the form of artificial rain is used to irrigate crops such as maize and alfalfa. Secondly, systems of drip irrigation and sprayers are used to irrigate tree crops, mainly apples and peaches (Planet Northern Greece S.A – Planet S.A., 1998). Considering the dominance of maize and alfalfa in the basin, it can be concluded that the largest part of irrigated surface is irrigated with the technique of artificial rain leading to great water losses. Only a small part of cultivated crops (tree crops) is irrigated with drip irrigation or sprayers.[56]

Impacts on quality of water resources

The use of fertilisers on lakeshore farmland endangers the quality both of Lake Vegoritida and of the aquifer. Inorganic fertilizers are the most damaging, especially nitrogen and phosphorus fertilizers (IGEKE & D.A.P/AUTH, 1999).

Moreover, animal farms in the basin cause significant pollution of the aquifer and surface waters. The total population equivalent of the numerous animal farms of the basin is approximately 170,000 p.e., in terms of their water pollution potential. The animal farms in the basin do not treat their effluents (despite their obligation to do so). Their effluents are either used as field fertilizers or are disposed in the proximity of the farms directly into surface waters or underground (Iosifidis & Avgitidis, 2000).

Industry

Because of the absence of industrial activity until the 1950s, the basin of Vegoritida was not affected by degradation and pollution of its natural environment. This picture was reversed when large industrial units started settling in the basin since the 1950s.

In 1965, the industry AEVAL for the production of nitrogen-fertilisers was set up and remained operational until the 1990s (PPC General Mines Directorate, 1994a).

Additionally, following the 2nd World War, intensive research started for the exploitation of large lignite reserves found in the Vegoritida basin and extensive lignite mining activities started in the 1950s. Since 1960, four lignite mines have been set up by the PPC in the Vegoritida basin, in the Prefectures of Kozani and Florina.[57] Lignite from the mines is mainly used as fuel for 4 steam-electricity plants, which were set up within a radius of 12 km from the mines with a total installed capacity of 4,438MW (in 2004) (PPC S.A., 2005). Until 1975, only one steam-electricity plant operated in the Vegoritida basin near Ptolemaida. The other

---

[56] Water loss with the water-demanding technique of spraying irrigation or artificial rain reaches 35%. On the contrary, drip irrigation and sprayers limit water losses to ca. 10%. In practice, water quantity needed with spraying irrigation is 500 m³/sm, while with drip irrigation or sprayers it is 250-300 m³/sm (Planet Northern Greece S.A – Planet S.A., 1998).

[57] Currently, lignite reserves of the Vegoritida basin which can be exploited reach 2700 M tones equivalent to 400 M tones of oil, worth more than 55 billion $. The rate of lignite mining rose rapidly from 1960 until 1975. Lignite extraction increases with an annual rate of 20%. Nevertheless, research on the impacts of lignite mining on flora, fauna and climate of the area is inexistent (Koklis, 2003).

three steam-electricity plants were constructed later on: the steam-electricity plant at Kardia (Kozani Prefecture) in 1974/75, at Amyndeo (Florina Prefecture) in 1986/87 and at Agios Dimitrios (Kozani Prefecture) in 1990 (Paraschoudis et al., 2001).

The Public Power Corporation (PPC) extracts lignite in this area of Greece as a priority, because of the large availability of water resources which are necessary for the operation of its steam-electricity plants. According to data of the PPC, the lignite reserves in the basin will be exhausted around 2065. Currently, energy produced in the Region of West Macedonia covers 75% of national needs, characterising thus the Prefectures of Kozani and Florina as National Energy Centre.[58]

Nowadays, the range of industries in the basin includes the following: lignite mining, energy production, milk industry, slaughters, wine, mineral water and refreshment production.

Impacts on quantity of water resources

The AEVAL industry abstracted water from four wells, which were conceded to the Municipality of Ptolemaida after the industry's closure in the 1990s (Paraschoudis et al., 2001).

The main industrial pressure on the water quantities of the basin, however, came from the PPC activities. From 1955-1990, the PPC abstracted large quantities of water from Lake Vegoritida for a hydropower plant operating in a neighbouring basin of the Pella Prefecture. Additionally, the operation of the four steam-electricity plants in the Florina and Kozani Prefectures required the consumption of large water quantities for cooling. Until 1975, the only steam-electricity plant at Ptolemaida was provided with cooling water from surface streams and wells. The three additional plants set up later were provided with cooling water directly abstracted from wells around the lignite mines, from stream Soulou and from Lake Vegoritida via a pumping station at Agios Panteleimon close to Amyndeo. Since 1997, the needs of the steam-electricity plants are met by water transferred from a PPC dam-reservoir on the River Aliakmon (located southern of the Vegoritida basin in the Kozani Prefecture) as well as water abstracted from the aquifer (e.g. both the plant of Ptolemaida and of Amyndeo abstract water from wells with an annual abstraction of $6.5*10^6$ m³) (Paraschoudis et al., 2001). 30-50% of water transferred from the Aliakmon basin ends up into the basin of Vegoritida, after its use by the steam-electricity plants (Paraschoudis et al., 2001).

Additionally, for the operation of the PPC lignite mines, the level of the aquifer must drop and be kept lower than the excavation level in order to allow the extraction of the lignite. This is accomplished by operating numerous wells around the mines. Mainly two of the PPC mines in operation significantly affect water resources in the basin: the Amyndeo Mine and the Southern Mine at the Kozani Prefecture. The abstracted water is used to reduce dust production in the mines, to cool the nearby steam-electricity plants (Paraschoudis et al., 2001) and for irrigation and water supply of local communities. Remaining abstracted water

---

[58] Environment Ministry Decision 26295, Official Government Journal B 1472/09.10.2003, on "Approval of a Regional Framework for Spatial Planning and Sustainable Development of the Region West Macedonia".

in the mines is ultimately discharged into the stream Soulou or Lake Vegoritida via the ditch connecting lakes Chimaditida and Petron. In winter, when irrigation stops, most water abstracted in the mines ends up into Lake Vegoritida. In total, $20*10^6$ m³ of water is pumped annually around the lignite mines. These abstractions disturb the hydrological balance of the wider basin, even if 30-50% of the water is finally returned to the basin system after use (Paraschoudis et al., 2001). In 1988-1993, water abstracted at the Southern Mine caused a 7m drop of the aquifer. The environmental impacts of the mines on the aquifer also include the decrease of discharge or even drying out of nearby natural springs and wells. Also, the humidity of the ground decreases in the area influenced by the aquifer level drop (PPC General Mines Directorate, 1994b). Wells around the lignite mines are destroyed once the extraction of the lignite is completed.

Impacts on quality of water resources

Pollution from industrial sources originates mainly in the Prefectures of Kozani and Florina. In the part of the basin of the Pella Prefecture, there are no significant industrial discharges.

The steam-electricity plants produce effluents and suspended ash which are driven into Lake Vegoritida via the stream Soulou. The effluents of the steam-electricity plants consist of wastewater, untreated cooling water, suspended solids and substances included in lignite (Amanatidou, 1997). The water abstracted in the lignite mines and then discharged into surface waters is also loaded with suspended solids (earth, lignite dust) (PPC General Mines Directorate, 1994a). The nitrogen-fertiliser industry (AEVAL) also used to significantly pollute Lake Vegoritida with large quantities of different nitrogen compounds (Amanatidou, 1997). Finally, also other industrial activities in the basin (slaughterhouses, milk industry and others) contribute to the pollution of surface waters.

Urban use

Impacts on quantity of water resources

Drinking water in the basin stems to 50% from mountainous springs and to 50% from wells. Only abstractions for the water supply of the city of Ptolemaida from 6 wells amount to 3-$3.5*10^6$ m³/y. Another $7*10^6$ m³/y are transferred to a neighbouring basin to supply the city of Kozani (Paraschoudis et al., 2001).

Impacts on quality of water resources

By the end of this case study, the state of urban wastewater treatment in the Vegoritida basin was far from satisfactory.

Lake Vegoritida receives the untreated wastewater of approximately 6,000 permanent residents of lakeshore settlements. The situation worsens during winter when a large number of ski tourists (ca. 5,000 during peak weekends) visit the area, thus resulting in a total of 11,000 population equivalent dumping wastewater into the lake (LVPS, 2002).

60% of the population in the basin is served by sewage networks, to their greatest part old and incomplete. Another 15% is served by combined networks for stormwater and sewage. 25% of the population is not served by any sewage network at all and disposes wastewater into septic tanks or cesspits, despite the prohibition of underground wastewater disposal at least in the Florina Prefecture by decision 555/90 (Iosifidis & Avgitidis, 2000). Wastewater collected in domestic cesspits is ultimately dumped untreated into Lake Vegoritida by freight trucks (LVPS, 2002).

The city of Ptolemaida is the only settlement with a wastewater treatment plant which operates since 1994 and was funded by the Cohesion Fund. Tertiary treatment was put in operation in 11/1999. According to data of 2000, the treatment plant meets the requirements of the EU Urban Wastewater Treatment Directive for organic load and suspended solids, but not for phosphorus and ammonium (Iosifidis & Avgitidis, 2000).

Two further settlements on the shores of Lake Vegoritida were equipped with natural wastewater treatment facilities in 2001. In the early 1990s, also a treatment plant for the towns of Amyndeo, Xino Nero and Petres was constructed funded by the 1st Community Support Framework but was never put into operation. In 2005, wastewater from these towns was still being disposed untreated into Lake Petron and ultimately into Lake Vegoritida.

Fisheries

The annual average fisheries production of Lake Vegoritida decreased in 1971-1980 by 55% compared to the decade 1961-1971. In 1981-1991, fisheries production further decreased by 46%. The production decrease was linked to the lake pollution and the lake level drop, which caused the destruction of lakeshore habitats important for fish reproduction (IGEKE & D.A.P/AUTH, 1999). In specific, the presence of ammonia caused a toxic state for the fish population of the lake which led in the past to incidents of mass deaths in its southern shallow part. Mass deaths could be observed usually in the beginning of spring when fish naturally reproduced in the shallow parts of the lake (Gravanis et al., 1997).

Environmental value

As early as 1977, lakes Vegoritida and Petron were characterised (by decision of the Ministry of Culture) as places of special natural beauty and, for each lake, a protection zone of a radius of 50 m was defined (Planet Northern Greece S.A & AN.FLO. S.A., 2000).

The wider area of lakes Vegoritida and Petron is also a proposed Natura 2000 site[59] (Planet Northern Greece S.A & AN.FLO. S.A., 2000). Vegoritida was also defined as an Important Bird Area on a European level. In 2003, 11,500 aquatic birds were recorded on Lake Vegoritida.[60] 162 species of birds were recorded in the system of both lakes Vegoritida and Petron (Municipality of Amyndeo, 2004).

Despite its nature conservation value, Lake Vegoritida was not included in the 27 natural protected areas designated nation-wide by the Environment Ministry in 2002 according to the Greek Environmental Protection Law (L. 1650/86). In order for the system of lakes Vegoritida

---

[59] Natura 2000 includes natural areas in the EU member states that will become Special Areas of Conservation of nature, according to EU directives 92/43 and 79/409.

[60] From letter of HOS to the Environment Ministry and Minister of Macedonia-Thrace, Subject: Problems in Lake Vegoritida, 30/6/2003.

and Petron to be designated as protected area, a Specific Environmental Study (SES) must first be carried out to propose specific zones of protection (see also Box 7.1). Once designated as a protected area, the system of the two lakes can be managed under a nature protection regime (involving a management body and a management plan).

Box 7.1   Designation of protected areas based on the 1986 Environment Law

> The framework Environment Law 1650/86 required the issuing of a Presidential Decree for the declaration of an area as a protected area. Law 1650/86 introduced the term of the "Specific Environmental Study" (SES), which is a kind of management plan. An SES is a document "which contains the specifications needed for the documentation of the protected area's importance, the accession of the area in one of the five protection status defined by the law and the usefulness of the proposed management measures". The SES accompanies the Presidential Decree and is a prerequisite for its issuing. Therefore, the SES is a prerequisite for the declaration of a site as a protected area (Fotiou, 2001).
> Once the SES is accredited by the Environment Ministry, the competent ministries can issue the Presidential Decree, after having taken into account the opinion of the local Prefectural council(s) of the area considered for designation. Sometimes, after the SES is prepared and before the Presidential Decree is issued, a Common Ministerial Decision is prepared which protects the site during the period needed for the preparation of the Presidential Decree (a bureaucratic procedure that is more time consuming than a Common Ministerial Decision) (Fotiou, 2001).
> Law 3044/2002 defined that protected areas can be designated directly via Common Ministerial Decisions instead of Presidential Decrees.
> The administration and management of protected areas of the Natura 2000 network are assigned to Management Bodies according to L. 2742/1999, as amended by L. 3044/2002. Management Bodies are legal entities of private law and of public benefit character, based within the protected areas and supervised by the Ministry of the Environment.

7.3   Key water resource problems and water use rivalries

This section identifies the main rivalries between water uses and users in the Vegoritida basin, which will be the center of attention for describing the chronology of the water management process in this case study. Water use rivalries emerged in relation to the two main water resource problems in the basin, namely the dropping water levels and the rising water pollution. Although these two types of water resource problems are interlinked, the series of events of the resulting rivalries could be largely separated in sections 7.6.2 and 7.7.2.

Rivalry of dropping water levels

The most important hydrological problem of the Vegoritida basin is the drop of the level of Lake Vegoritida (Sioutis, 1989) due to human activities. The following paragraphs and Table 7.2 give an overview of the intensity of water abstractions and their impact on the lake level since the 1950s.

Table 7.2   Water abstractions in the Vegoritida basin in 1955-2001

Period	Total abstractions	Main water uses	Impact on Lake Vegoritida level
1955-1975	$2630*10^6$ m³	- 90% direct abstraction from the lake for the PPC (mainly for hydropower) - 10% aquifer and surface water abstraction for irrigation	Drop by 16 m (from 541 to 524 masl)
1975-1995	$1800*10^6$ m³	- 38% direct abstraction from the lake for the PPC (mainly for the steam-electricity plants) - 62% aquifer abstraction for irrigation - very dry weather in 1985-1990	Drop by 14 m (from 524 to 510 masl)

Period	Total abstractions	Main water uses	Impact on Lake Vegoritida level
1997-2001	264*10^6 m³	- 138% abstraction for irrigation - 38% positive contribution to the water basin balance from the River Aliakmon (return water of the PPC steam-electricity plants)	Drop by 1.7 m (from 510 to 508.3 masl)

Source: Paraschoudis et al. (2001).

1955-1975

In 1955, a hydropower plant was constructed by the PPC at the location Agras in the Pella Prefecture, in a basin neighbouring the Vegoritida basin. For its operation, until 1975, large water quantities (at a rate of 10-3,000*10^6 m³/y) were transferred from Lake Vegoritida through a tunnel constructed at the lakeshore location of Arnissa. In the period 1955-1975, 2,242*10^6 m³ of water, which amount to ca. 70% of the initial lake volume, were abstracted from Vegoritida and transferred to this hydropower plant (figures based on a data series provided by staff of the PPC Agras hydropower plant).

Abstraction intensity for the hydropower plant varied according to the energy needs of Greece, political and economic circumstances. The Agras plant was one of the first projects of the PPC and it initially greatly contributed to national energy production. From 1955-1960, large amounts of lake water were abstracted in the context of great efforts to increase electricity supply nation-wide after the 2nd World War and the Civil War. In 1965-1967, there was again transfer of large water amounts to the Agras plant. The largest abstraction took place following a military putsch in 1967, probably for political and economic reasons. Again in 1973-1975, there was intensive abstraction for the hydropower plant linked to the oil crisis of that period (Paraschoudis et al., 2001).

The impact of these abstractions on the Vegoritida level quickly became obvious. From 1955-1975, there was a 16 m drop of the lake from 541 to 524 masl (see Figure 7.3).

In the meantime, the extensive lignite mining activities which started in the 1950s and the gradual set up of steam-electricity plants in the Kozani and Florina Prefectures started exercising some additional pressure on the water resources of the Vegoritida basin.

Figure 7.3   Annual level of Lake Vegoritida in 1895-2005 and abstractions for hydropower

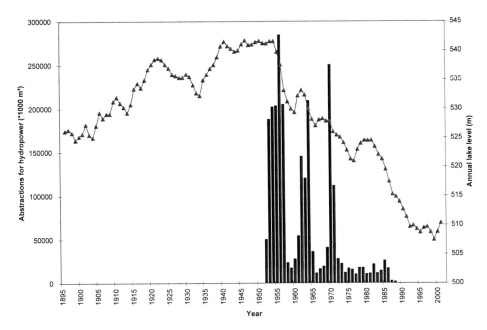

Source: From a data series provided by staff of the PPC Agras hydropower plant in 2005.

Furthermore, as the lake dropped, it gradually exposed much fertile land which was previously covered with water. Already in the mid-1960s, the first exposed land started being cultivated in the eastern and south-western side of the lake where the waters were shallower. Since the early 1970s, there was also a turn to irrigated water-demanding crops such as maize and alfalfa. Nevertheless, in this period, abstractions from the aquifer for irrigation purposes were still kept relatively low compared to consumption rates in the coming decades (Paraschoudis et al., 2001).

1975-1995

After 1975, direct abstractions of the PPC for use at the hydropower plant were significantly reduced due to the reduced ability to operate the transfer tunnel at a low lake level. From then on, the Agras plant did not greatly contribute to the national energy potential (pers. comm. PPC, 2/12/2005). In 1977, the lake level dropped approximately to the level of the entrance to the Arnissa tunnel (515.5 masl) and water had to be mechanically pumped into the tunnel, mainly over the summer with abstractions at a rate of $10\text{-}30*10^6$ m³/y. After its use at the hydropower plant, water was transferred further downstream to irrigate the plains of Skydra in the Pella Prefecture[61] (Paraschoudis et al., 2001).

---

[61]   Being in the public sector, the PPC had to generally fulfill multipurpose works when constructing hydropower plants to serve also other uses such irrigation, flood protection etc. The irrigation network of the Skydra plains in Pella served ca 55,000 ha of fields.

However, in this period, large water quantities were needed for the operation of the - in the meantime - four PPC steam-electricity plants in the Kozani and Florina Prefectures. Abstractions for this purpose took place from the basin aquifer, stream Soulou and Lake Vegoritida. Water abstraction from the lake was possible via the pumping station constructed in 1974 at the lakeshore location of Agios Panteleimon, close to Amyndeo, abstracting $14*10^6$ m³/y (Paraschoudis et al., 2001).

Concerning agriculture, irrigated surface reached in this period 8,000 ha, of which 3,000 ha were located on the immediate lakeshore zone.

All in all, Paraschoudis et al. (2001) concluded that the lake drop in this period was mainly due to dry weather and the direct abstractions of the PPC from the lake. The large increase of irrigated surface and the related overexploitation of the aquifer were considered to have a smaller effect. Actually, during the irrigation period, the aquifer dropped by 3-4 m lower than the lake level. Due to the parallel lake drop, the lake was only in direct communication with the impermeable layers of its bottom. Therefore, there was no immediate response of the lake level to the changes in the aquifer levels. This implies that although the lake and the aquifer tend to become balanced in the long-term, the lake level was much more directly and rapidly affected by direct abstractions rather than by aquifer abstractions.

1995-2001

In this period, instead of abstracting water from the basin, the PPC started contributing its return water from the stream-electricity plants ($25*10^6$ m³/y), which were fed with water from the neighbouring Aliakmon river basin since 1997. However, in this period, large water quantities were consumed in the Vegoritida basin for irrigation, since the cultivation of irrigated crops increased from 8,000 ha to 15,500 ha. The lake level in 1997-2001 was, nonetheless, quite stabilized and it was predicted that, in the long-term, the lake level should rise by 0.30 m/y if no direct abstraction from the lake took place (Paraschoudis et al., 2001). There is however fear that a new dry climate period, which may naturally occur every 4-5 years, combined with the current overconsumption in irrigation would be catastrophic for water levels in the basin (Interview Agricultural Cooperative of Arnissa, 25/1/2005).

All in all, the drop of water levels had a number of impacts, which also indicate the resulting water use(r) conflicts (Planet Northern Greece S.A – Planet S.A., 1998):

- Financial impact on farmers of the basin, which had to abstract water from deeper levels due to the drop of the aquifer, especially in the Pella Prefecture where the aquifer is deeper by nature.
- A "positive" impact for farmers was the exposure of fertile land whose cultivation started already in the mid-1960s.
- Financial impact on urban users, which gain drinking water from the dropping aquifer (rivalry mainly between urban users and the PPC lignite mining activities).
- Financial impact on the PPC which had to invest in water transfer works from the basin of Aliakmon to secure cooling water for its steam-electricity plants.

148

- Inability to supply water to the PPC hydropower plant at Agras and the downstream irrigated Skydra plains in a neighbouring basin.
- Aesthetical degradation of the surrounding landscape due to the lake drop resulting in reduced potential to develop tourism.
- Disturbance of the ecological balance of Lake Vegoritida due to its reduced volume and surface and due to increased levels of water pollution. Specific ecological impacts included the loss of fish habitats and the disappearance of certain bird species, such as the mallard (*Anas platyrhynchos*) and the pelican.

Especially the activities of the PPC were initially an important driver for the quantitative degradation of water resources in the basin. As a result, the actor interaction process over the dropping water levels started with a clear anti-PPC campaign of the lakeshore communities in the 1980s. The lakeshore communities were motivated by their interest to preserve the lake and maintain its landscape value, fisheries and tourism development, but also by their interest in safeguarding water for irrigation. The "irrigation front" was supported both by representatives of the local communities and by local agricultural associations. This conflict between the PPC and local society lingered on until the end of the case study due to disagreement over responsibility for the large drop of the lake. An additional conflict between farmers, the local population and the PPC is related to the PPC mining activity in the basin, which impacts the irrigation and water supply sources from the aquifer in close proximity to the mines. This type of rivalry is also bound to increase in the future since the lignite reserves will not be exhausted before 2065.

Rival fronts over dropping water levels became less anti-PPC focused towards the end of the case study. In the 1990s, the PPC reduced and then completely stopped abstractions from Lake Vegoritida. Thus, the main problem for the lake and aquifer levels became overuse in irrigation. However, there was no direct opposition of the lakeshore communities against farmers' overuse of water for irrigation. In the very end, the lakeshore population wanted to maintain agriculture. Nevertheless, farmers' demands to cultivate exposed land came into conflict with plans to define an ecologically sustainable lake level. There was a clear voice of opposition expressed by local associations and environmental NGOs against the cultivation of exposed land which contributes to further lake level drop and degradation of the lake ecosystem.

A peripheral conflict also grew since the early 1990s between neighbouring lakeshore villages over rights to cultivate exposed land. Lake Vegoritida actually became famous when a violent conflict broke out between two neighbouring communities over the ownership and cultivation of exposed land in 1998 (Bousbouras, 2004). The conflictual situation over concessions to exposed land is very much linked to the lack of land use planning in the basin.[62] Due to the lack of official lakeshore delimitation, there was no overall prohibition to use the shore of Lake Vegoritida.

All in all, the first rivalry of the case study focuses on issues of water levels, and specifically on the level of Lake Vegoritida and the aquifer of the basin. Despite the focus on Lake Vegoritida, it should be kept in mind that the long-term management aim in the basin is the co-management of the level of all four interconnected lakes. By the end of this case study, only the (minimum and maximum) level for Lake Chimaditida was defined (AUTH & EKBY, 2001). However, at least discussions had also started over the definition of a lake level for Lake Vegoritida. Gradually, all sluices connecting the four lakes will be controlled by the Division for Land Reclamation of the Florina Prefecture (in cooperation with local municipalities and local land reclamation organizations), so that a more coordinated managed system can be achieved (Interview Florina Pref., 26/1/2005).

Rivalry of the discharge of pollutants vs. fisheries/recreation/environment

The rise of different human-induced pressures in the Vegoritida basin increased surface water pollution levels due to increased incoming pollution loads and the reduced volume of water. Pollution became noticeable already in the 1960s, after the set up of the AEVAL nitrogen-fertiliser industry, the first PPC steam-electricity plant, the lignite mines and the growing population of the basin. As a result, Lake Vegoritida showed in the past tendency to become eutrophic but its quality started to improve again since 1989. From eutrophic, it was characterized in 2001 as oligotrophic-mesotrophic, its overall quality being satisfactory and stable. However, the water quality improvement should not be reason for reassurance since several human-induced pressures which degraded the lake quality in the past continue to exist (Planet Northern Greece S.A – Planet S.A., 2001). Moreover, other surface waters in the basin did not show the same degree of improvement as Lake Vegoritida. Lake Petron, which mouths into Lake Vegoritida, showed continuous quality degradation since 1988 with increased levels of conductivity, SO_4, Na, Mg and hardness. Lake Petron was characterised eutrophic due to urban wastewater and agricultural effluents. Lake Chimaditida showed degradation since 1995/96 due to fertilizers and pesticides but its quality still remained within acceptable levels. The stream Soulou, which mouths into Lake Vegoritida, has high concentrations of nitric nitrogen (above the limit of 10 mg/l) indicating pollution by urban and animal wastewater, as well as fertilizers and pesticides (Paraschoudis et al., 2001).

In total, water pollution sources in the basin can be summarized as follows (LVPS, 2002; Paraschoudis et al., 2001; Planet Northern Greece S.A – Planet S.A., 2001):

---

[62] In the entire basin, there is only a general policy document on spatial planning approved in 2003, namely the Environment Ministry Decision 26295 (Official Government Journal B 1472/09.10.2003) on "Approval of a Regional Framework for Spatial Planning and Sustainable Development of the Region West Macedonia". The only specific spatial planning study with proposals for land uses and restrictions on the lakeshore was completed for the Pella Prefecture, which covers the biggest part of Lake Vegoritida but only a small part of the wider Vegoritida basin (Planet Northern Greece S.A & AN.FLO. S.A., 2000).

- Industrial effluents, especially from the steam-electricity plants, the lignite mines and the public slaughter premises of Ptolemaida discharged into stream Soulou.
- Urban wastewater from settlements in the basin, although pressure in this respect slightly reduced after the operation of a wastewater treatment plant for the city of Ptolemaida since 1994.
- Pesticides and fertilizers, which are rising in importance as a pollution source.
- Effluents of animal farms.

Water quality degradation in the basin had a number of impacts, which also indicate the resulting water use(r) conflicts:

- Reduced recreational value of Lake Vegoritida (e.g. for bathing) and reduced potential to develop tourism.
- Prohibition of using Lake Vegoritida as drinking water source.
- Disturbance of ecological balance and the environment of the lake.

The main resulting rivalry of use(r)s since the early 1970s developed between lakeshore communities (with claims over good water quality to safeguard fisheries, human health and recreation) and large polluters (mainly the PPC, AEVAL, urban centres as well as the authorities for not enforcing pollution control measures). Later on, also aspects of environmental protection were added to the rivalry as the EU entered the picture with its own policy requirements for good water quality and wetland protection. The rivalry over water pollution in the basin was also the basis for most legal complaints in the chronology of the case study events.

7.4 Milestones in the water management process

7.4.1 The beginning of local protests

Against the background of increasing pressure on both the quantity and quality of water resources, the lakeshore communities of Lake Vegoritida started to protest since the late 1970s regarding pollution and since the late 1980s regarding the extensive lake drop.

Social pressure was initially exercised on local politicians, Prefectures and some times on Ministries. In 1988, an Action Committee for the Salvation of Lake Vegoritida was set up by several lakeshore actors (communities of Arnissa, Agios Panteleimon and Perea, agricultural associations and fishing associations) to take action against the dropping water levels and pollution. The Action Committee aimed at safeguarding fisheries, bathing and the aquifer to supply local irrigation wells. Further triggered by prolonged scarcity in the late 1980s and allegations of PCB leakage from the PPC facilities, local protests reached a peak in 1990.

In the meantime, the only action taken by the authorities was the issuing of Prefectural decisions since the late 1970s to control effluent treatment and disposal. Prefectural decisions to control water abstractions and the drilling of wells were mainly issued in the 1990s, after the adoption of the national 1987 Water Law.

The lakeshore communities were unhappy with the inactivity of authorities to deal with the problems of Lake Vegoritida and its wider basin. Realising that their protests on a local level were not having a real effect, they sent a complaint to the EU in 1987 and again in 1990 about the increasing degree of pollution and the drop of Lake Vegoritida. As it will be illustrated in section 7.7.2, the resulting EU infringement procedure focused mainly on the pollution issue. Greece was taken to the European Court of Justice (ECJ) in 1995 and it was convicted in 1998 for not complying with the EU directive on pollution from hazardous substances for Lake Vegoritida and stream Soulou.

7.4.2 Lake Committee of the Region West Macedonia

The first attempt to deal with the problems of the Vegoritida basin in a more holistic approach was made in 1994 by the Region of West Macedonia. The Region was concerned about the situation in the basin because it was under two-fold pressure: on the one hand from the EU and the ECJ and, on the other hand, from the local population. Especially the local lakeshore association MESNA (educational and cultural association of the community of Arnissa) was actively trying to impose restrictions on the PPC (Interview Planet, 8/2/2005; Interview RWM, 17/2/2005).

After a first informal meeting with local actors, the Region decided that it had to take action on the water problems of Lake Vegoritida both from a quality and quantity perspective. In February 1994, by decision of the Secretary General of the Region, a Committee was set up to examine in a scientific way the current status of the lake, describe the extent of the problem and propose solutions (Interview RWM, 17/2/2005). The Committee consisted of 5 sub-committees on 1) policy planning, 2) industrial use of water, 3) agricultural use of water, 4) qualitative characteristics of the environment and 5) the evaluation of existing studies. Participants were representatives from a wide administrative and professional spectrum, including also water users: Region of West Macedonia, Region of Central Macedonia, the three Prefectures sharing the basin, presidents of the local communities, the local non-governmental association of Arnissa MESNA favouring the protection of the lake, the Ministry of Macedonia and Thrace, local agricultural cooperatives, representatives of the PPC, university experts, research institutes and the Technical and Geotechnical chambers of Greece. The Region West Macedonia favoured broad participation in this Committee, believing that this way the Committee's outcome and proposed solutions would be more easily and broadly accepted. One of the aims was thus to make as many actors as possible realize that management of the lake was necessary (Interview RWM, 17/2/2005).

Discussions in the Committee were largely based on a review of previous studies carried out by the Region on Lake Vegoritida and its basin. After 1.5 year of work, the Committee came to a common set of conclusions on the causes and impacts of the quantitative and qualitative lake degradation and made proposals for certain measures. One of the proposals emphasised the need for an integrated study on the qualitative and quantitative management of water resources in the basin. The Committee also recognised the need for a technical advisor to coordinate all necessary action (Interview RWM, 17/2/2005).

The Committee participants considered that the Committee was a positive experience in raising their awareness and that it fulfilled its objective to a satisfactory degree. A disadvantage was that it was not legally binding to implement the Committee conclusions through follow-up activities. Unfortunately, the Secretary General of the Region who was involved in the activation of the Committee left his post shortly afterwards, lowering thus personal engagement to coordinate the follow-up and ensure the practical application of measures proposed. Thus, even though the Committee succeeded in sensibilising its participants, its positive effect was short-term since its conclusions were not put into practice and its participatory mechanism was dissolved. At least, the Committee contributed in securing funding from the Region West Macedonia for a research programme on the stability of water levels and the purification of Lake Vegoritida (Interview RWM, 17/2/2005).

7.4.3 Investing in research: Programme on the stability of water levels and purification of Lake Vegoritida

In 5/1998, the Region West Macedonia assigned a technical consultant to the coordination and execution of a programme on the "Stability of water levels and purification of Lake Vegoritida". The role of the technical consultant was two-fold: a) to propose a basic set of projects and actions to be carried out as part of the programme and b) to monitor and coordinate the practical execution of the different projects. The funding of 1 billion Greek drachmas (ca. 3 million Euro) came from the Region's Operational Programme.

There were two main factors pushing for the initiation of this programme by the Region. First, pressure from the local society was large. It was necessary to deal with the drop of the lake and the associated social unrest over concessions for the cultivation of exposed land. Secondly, the programme was related to the open case before the ECJ on the degradation of Lake Vegoritida (Interview RWM, 15/2/2005). In fact, the initial intention of the authorities was to carry out a study on the entire system of the four interlinked lakes of the basin, but the study focused only on Lake Vegoritida as a priority because of the ECJ case (Interview Florina Pref., 26/1/2005).

The initial plan was also for both Regions of West and Central Macedonia, which territorially share Lake Vegoritida, to co-finance this programme (Interview Planet, 8/2/2005). The Region West Macedonia would fund activities in the Prefecture of Florina and the Region Central Macedonia would fund similar activities in the Prefecture of Pella. This way also spatial integration on a basin level would be achieved (Interview RWM, 15/2/2005). However, the Region Central Macedonia dealt with the proposed programme in a less coordinated manner than the Region West Macedonia. Under pressure by local interests, the Region Central Macedonia proposed to use its funds for infrastructure works, such the construction of a road around Lake Vegoritida, instead of environmental interventions. Technical consultants to the Region Central Macedonia advised against road construction at that stage due to the lack of official lakeshore delimitation and lack of knowledge on the behaviour of the lake and the broader aquatic ecosystem. Amidst uncertainty and disagreement over the activities to-be-funded, funding on behalf of the Region Central Macedonia was cancelled

(Interview Planet, 8/2/2005). As a result, the Region West Macedonia was the sole funding authority and the Florina Prefecture was defined as the programme executive authority. The result of the inability to achieve funding coordination and cooperation of the two Regions was that action within the programme was taken only on one side of the lake, even though similar quantity and quality pressures occurred on both sides (Interview RWM, 15/2/2005). In practice, many studies carried out in the context of the programme included data analysis and evaluations only in the Region West Macedonia and not in the entire basin.

A second attempt to coordinate the two Regions of the basin (and their respective Prefectures) took place towards the end of the lake stability and purification programme in 2000. It was realized that the measures proposed to be taken in the territory of the Region West Macedonia should also be taken in the Region Central Macedonia. Actions such as restrictive measures on water use, public awareness campaigns, a proposed agrienvironmental programme, a pilot programme to control nitrate pollution as well as the creation of a management body should be embraced by all administrative entities around the lake to have meaningful results in practice. To this aim, a joint coordination committee, advisory to the lake stability and purification programme, was set up upon the initiative of the Minister of Macedonia and Thrace and the support of the programme technical consultant. However, although this committee convened a few times, it was not easy to operate in practice. Progress was hindered by the fact that there were no practical ways to implement coordinated action. There was no administrative authority having competence over the entire basin to undertake coordination (Interview Planet, 8/2/2005).

The lake stability and purification programme itself was supervised by an administrative Committee, which was set up by decision of the Prefect of Florina. The Committee participants included representatives of the Region of West Macedonia and of selected divisions of the Florina Prefecture (Interview Florina Pref., 26/1/2005). This Supervisory Committee had the technical mandate to supervise the work of the programme technical consultant and to monitor progress of the various projects carried out. The Committee had no policy-making competence. In the context of this Committee, the Florina Prefecture representatives showed motivation to understand the water problems in the basin and to contribute to the improvement of the water resource situation. Actually, the Committee often acted beyond its mandate as technical supervisor and discussed issues of practical management interventions, e.g. measures on the legalization of wells, on a register for wells, on the prohibition of new wells and restrictive measures on direct abstraction from the lake (Interview Planet, 8/2/2005).

In this context, the Committee helped the Florina Prefecture staff to clearly perceive the water problems of the basin, understand the programme scientific results and recognize the need for water management. The Region West Macedonia also perceived the depth of the water resource problem and this influenced several of its future actions. Since the lake stability and purification programme, the Region proactively included projects on Vegoritida in its funding programmes (Interview Planet, 8/2/2005). The Supervisory Committee also promoted a kind of a group learning process. Within the Committee, representatives from different Florina Prefecture divisions learned to cooperate with one another. Additionally, representatives of the Florina Prefecture and the Region West Macedonia were trained in technical issues, such as modern techniques of water management and issues of national and European legislation on the environment, especially water.

Finally, the Committee served as a platform for informal consultations and discussions of the different positions of local actors on different technical projects, such as a proposed water transfer scheme to enrich the Vegoritida basin (Interview Planet, 8/2/2005).

Programme of lake stability and purification in practice

At the outset of the Lake stability and purification programme, it was realized that there were inadequate up-to-date data on the water balance of the basin and on water pollution sources (Planet Northern Greece S.A – Planet S.A., 2001). Therefore, many assigned projects focused on the evaluation of previously available data and new data collection. 85% of the available programme funds were spent in studies (Interview Florina Pref., 26/1/2005). All in all, the programme involved a series of short-term actions and several activities setting the basis for holistic basin management in the mid-term and long-term. The actions consisted in 3 main groups of measures (Planet Northern Greece S.A – Planet S.A., 2001):

- Measure 1: Stability of the lake level and the wider basin (projects on the management of irrigation water and on the measurement of basic parameters of the lake and aquifer)

- Measure 2: Monitoring and improvement of surface water and groundwater quality (projects on the treatment of effluents and the measurement of quality parameters of the lake). This included also the installation of two pilot systems of natural wastewater treatment for lakeshore communities.

- Measure 3: Development works and complementary projects to improve the environment of the basin. This included proposals for selected technical works around the lake and public awareness actions.

Table 7.3   Proposed actions of the stability and purification programme of Lake Vegoritida and their implementation

Project or activity	Implementation progress
Measure 1 : Stability of the lake level and the wider basin	
Study on the restructuring of water-demanding crops	Completed
Investigation of possibilities of water transfer from other basin	Completed
	It concluded that no water transfer from other basin should take place
Gradual replacement of irrigation with drip irrigation	Not yet completed (2005)
	Pilot activity, not linked to any subsidy programme for the replacement of irrigation techniques
Hydrogeological study of the basin -Definition of water balance	Completed

Project or activity	Implementation progress
GIS system of the Vegoritida basin	Completed
	Its operation is pending (2005)
Morphological mapping and geophysical study of the lake bottom	Completed
Measure 2: Monitoring and improvement of the quality of the surface and groundwater	
Pilot natural treatment plant of the community of Vegora	Completed
Pilot natural treatment plant of the community of Faraggi	Completed
Study on the operation of the treatment plant of Amyndeo	Completed
	In 2005, the study was being updated again and the assignment of complementary construction works was pending
Study and works for the protection and restoration of the wetlands of Chimaditida and Zazari	Study completed
	Works to raise the dykes and replace the sluice awaiting funding
Study on the treatment and disposal of urban wastewater and industrial effluents in the basin of Vegoritida	Study completed
	Funding application submitted to the Cohesion Fund but was not approved
Pilot programme for wise use of fertilizers	2-year pilot programme completed
Monitoring quantity and quality parameters via telemetric transmission	Partly completed
Programme of periodical in situ measurements of the lakes	Completed
Measure 3: Development works and accompanying action	
Pier of Agios Panteleimon	Completed
Channelisation of part of stream Soulou	Not executed
Maintenance of tunnel connecting lakes Petron-Vegoritida	Not executed
Awareness and information campaign on the lake stability/purification programme results	Not executed (due to disagreement between involved parties on an authority level)
Programme of employment training	Not executed
Study on a Management Body for the water resources of the Vegoritida basin	Completed

Status: March 2005. Sources: Interview Florina Pref., 26/1/2005; Interview RWM, 15/2/2005.

All in all, most assigned studies concluded to several proposals for further action, but only few practical measures were taken on the ground, confirming in a way local communities, which were initially sceptical about the programme effectiveness (Interview of community president of Agios Panteleimon, Newspaper "Macedonia", 9/5/98). Details on progress made on selected proposed measures are given in the chronologies of the two case study rivalries (in sections 7.6.2 and 7.7.2). A few proposed measures which proceeded to practical implementation were mainly relevant to the water quality problems of the basin.

Here, it is worth discussing in some detail the outcome of the study on a management body for water resources of the Vegoritida basin (part of Measure 3). Considering that, at that stage, no management body had been set up according to the Environment Law 1650/86 and not even the necessary SES had been assigned for the designation of a nature protected area, the specific study proposed the creation of an alternative management body by activating mainly local actors. Even though such a body would not be equivalent to a management body of the 1986 Environment Law, it was seen as beneficial to set up a non-profit organisation of a scientific profile to promote activities in the basin with the support of local actors and the Region West Macedonia. This proposal, however, was not put into practice because there was no local actor willing to take on the leading role (Interview Planet, 8/2/2005). At least, in the context of the Lake stability and purification programme, the Region West Macedonia managed to secure funding for the SES, which is necessary for initiating the process of designation as nature protected area under the Environment Law 1650/86 (Interview RWM, 15/2/2005). Unfortunately, although the SES was in procurement in 2005, there were objections to the tender procedure resulting in its cancellation. By the end of the case study, it was uncertain if and when the SES procedure would be open again (Interview RWM, 17/2/2005).

7.5 Actors in the process of the Vegoritida basin

The actors involved in the water management process of the basin included actors of the public administration, water users, non-governmental groups and external experts. Table 7.4 presents the main relevant actors from the different levels of the administration giving information on their water management competence. Table 7.5 lists the main water users, non-governmental groups and experts involved in the process. More specific information on their roles in the two rivalries of the case study is given in sections 7.6.3 and 7.7.3.

Table 7.4   Actors of the administration (Vegoritida basin)

Administrative levels	Actors	Main competence in water management	Water policy fields
EU	European Commission European Court of Justice	Source of funds for the Regional Operational Programmes, for agricultural subsidies and for wastewater treatment plants Judicial competence on the implementation of EU directives	All
National		Definition of national framework legislation	
	Ministry of Environment	National competence for the implementation of EU and national environmental legislation: - EU water quality and environmental directives (including UWWTD, Nitrates Directive, Bathing Directive) - 1986 Environment Law (national network of water quality monitoring stations, designation of nature protected areas, approval of EIAs of heavy industry such as PPC) - 2003 Water Law	Spatial planning Environmental protection Water management (> 2003)
	Ministry of Development	National competence for the implementation of the 1987 Water Law National competence for the regulation of industries (Direct) supervision of PPC (until its 2001 semi-privatisation)	Industry Energy production Mining

Administrative levels	Actors	Main competence in water management	Water policy fields
		Competence in mining regulations (including the Mining Code)	Water management (1987 – 2003)
	Ministry of Agriculture	Competence in national water legislation relevant to agriculture, prior to the 1987 Water Law, including earlier regulations on authorisations for wells	Agriculture
		Management of agricultural subsidies	
		Policy-making and funding of agrienvironmental measures	
	Ministry of Economy	Role in concessions of lake exposed land for cultivation (exposed land is officially State property)	
Regional		Application of national legislation and responsibility for planning on the regional level	All
	Region West Macedonia		
	Regional Managing Authority	Management of the Regional Operational Programme: Financing, programming and execution of works in the Region	
		It secured funding for the lake stability and purification programme, for wastewater treatment plants, for the SES to promote the designation of the basin as nature protected area	
	Regional Water Management Department (RWMD)	Main authority responsible for the implementation of the 1987 Water Law and 2003 Water Law	
		- 1987 Water Law: Water management competence on the entire Vegoritida basin, although part of it lies in the administrative Region Central Macedonia; issued permits for water use and water works for energy production and for multiple use of water; Approved all permits for water use and water works, before they could be issued by other authorities;	
		- 2003 Water Law: Responsible for issuing all water permits; water management competence on the Vegoritida basin shared with Region Central Macedonia (still uncertain whether basin will be co-managed by both Regions)	
		- Proposes restrictive regulations for quantitative water protection to be issued by all Prefectures in the basin	
	Regional Directorate for the Environment and Spatial Planning	Approval of certain EIAs (depending on type of project)	
		Competence in environmental and nature protection (1986 Environment Law; Natura 2000 network)	
	Region Central Macedonia		
	Regional Managing Authority	Management of the Regional Operational Programme: Financing, programming and execution of works in the Region	
	Regional Water Management	It exercised all regional duties of the 1987 Water Law for the Vegoritida basin until 1997, when the Region West	

158

Administrative levels	Actors	Main competence in water management	Water policy fields
	Department (RWMD)	Macedonia took over responsibility	
		Under the 2003 Water Law, it may share competence in managing the Vegoritida basin with the Region West Macedonia	
Prefectural			
	Prefecture of Pella	Regulatory and administrative competence, responsibility for planning on the Prefectural level	All
	Prefecture of Florina	Elected Prefects give final approval for sanctions in case of non-compliance with Prefectural regulations	
	Prefecture of Kozani	Prefectural Council is consulted on the approval of EIAs for projects within the Prefecture	
		Prefecture approves certain EIAs (depending on project type)	
	Prefectural divisions of land reclamation or agriculture	Prepare restrictive regulations on water quantity protection to be issued by Prefect	
		Under the 1987 Water Law, they issued permits for water use and works in irrigation, animal farms, aquaculture (this competence was transferred to the Regions by the 2003 Water Law)	
	Prefectural divisions of health, of environmental protection, of industry	Share competence in issuing regulatory decisions on water quality protection, in issuing disposal permits of wastewater and effluents and in checking compliance	
		Under the 1987 Water Law, the divisions of industry issued permits for abstractions and water works in industry (this competence is transferred to the Regions by the 2003 Water Law)	
Municipalities and communities		Competence for the delivery of water supply, sewerage and wastewater treatment services	Water supply Wastewater treatment
		Under the 1987 Water Law, the Union of municipalities and communities issued water permits for domestic use, mainly drinking water supply (competence transferred to the Regions by the 2003 Water Law)	
		Local self-administration and rural police forces have responsibility for compliance-check of terms of water use permits and wise use of water in their territory (assisted by the permit-issuing authorities) (Presidential Decree 256/89)	
Local organisations for land reclamation		Responsibility for management and distribution of water in collective irrigation networks (which serve a small part of the Vegoritida basin)	Agriculture

Table 7.5   Water users and non-governmental groups (Vegoritida basin)

Actor	Description
Users	
Public Power Corporation (PPC)	PPC is the largest electricity producer in Greece. In 2004, it generated 96% of the total electricity in the country. It is also the second largest lignite producer in the EU
	The PPC was established in 1950 as a legal entity of private law but it pertained to the public sector and was owned by the Hellenic Republic. Being part of the public sector and directly linked to the central State, it has had a rather autonomous presence in the Vegoritida basin. In January 2001, it was incorporated as a Societé Anonymé (PPC S.A.), under the Liberalisation Law of the Electricity Market (2773/1999). The Hellenic Republic holds 51.12% of the company's share capital (PPC S.A,. 2005), therefore the PPC is still partly under public control

Industries	Industries participated on rare occasions in the basin process and always on an individual basis (not through representative unions)
Farmers and their representatives	Farmers in their entirety are not represented in the process by any strong representative organizations
	The agricultural cooperatives could play a role in organizing farmers, but so far they only had a role in negotiating prices for products
	Only few agricultural cooperatives had an active role during the case study process, especially the agricultural cooperative of Arnissa (in the Pella Prefecture)
NGOs and associations	
Local associations	The lakeshore population was to a certain extent expressed in the process by the local municipalities and communities
	Additionally, it was expressed by locally active associations, especially the MESNA association of Arnissa and the Lake Vegoritida Preservation Society (in short, Vegoritida Society).
	MESNA was the first actor mobilised for the protection of the lake both on its own as well as through the Action Committee for the Salvation of the Lake. MESNA enjoyed broad support in Arnissa and mobilized the public through publications in its local newspaper.
	The Vegoritida Society was set up by citizens of lakeshore municipalities. In 2005, it numbered almost 300 members. It was set up out of the need for a non-governmental organization focusing on the lake and expressing the lakeshore population. MESNA was a cultural organisation and action for the lake was only part of its wide range of activities. Although the Vegoritida Society does not have access to decision-making, it managed to become a reliable discussion partner for the public authorities (Interview LVPS, 1/2/2005). Its principle goals are the lake protection from pollution, the protection, preservation, and development of the lake and surrounding area, the rational use of the lake water for continuation of its principle role for irrigation of surrounding fields and tourist development (LVPS, 2002). The activities of the Society include meetings with key decision-makers in public administration as well as information seminars for the local population on the lake
Hellenic Ornithological Society	This is an important national NGO, which became involved towards the end of the case study process concerning ecological aspects and the definition of the level of Lake Vegoritida
External experts	Consultants, universities, National Institute of Geology and Mineral Exploitation, Technical and Geotechnical Chambers of Greece

7.6 Rivalry on dropping water levels

7.6.1 Property rights

As already mentioned in chapter 3/3.3.2.2, in the empirical analysis of this dissertation, property rights are discussed mainly with relevance to use and ownership rights to water. The present section does not refer extensively to water ownership rights in the Vegoritida basin, mainly because private water ownership (e.g. of groundwater underneath private land) gained less importance nation-wide after water use became subject to use permits (in 1989 and onwards). Therefore, mainly the water use rights, and less the water ownership rights, of the key users (industry and agriculture) are of interest in the following. The following Table 7.6 gives an overview of the main use rights concerning the quantitative use of water in the basin.

Table 7.6   Rights to quantitative water use in the Vegoritida basin

Use	Users	Use rights	Relevant policies and regulations
Living environment	Flora and fauna	No right defined (no minimum lake level defined according to the 1987 Water Law)	1987 Water Law

Use	Users	Use rights	Relevant policies and regulations
			2003 Water Law
			1986 Environment Law
Irrigation	Farmers	*Surface water:*	1987 Water Law
		> 1989: rights on the basis of permits for water use and water works are required	2003 Water Law
		Local organisations for land reclamation hold use permits in as far as they use surface water	Law 1988/1952 on wells
		Not all private users have rights according to law (unauthorised abstractions from Lake Vegoritida and stream Soulou)	Prefectural regulations and restrictive measures on water protection
		Groundwater:	
		< 1989: authorisation to drill private wells and gain electricity was required (wells connected to the power network prior to 1987 are considered legal)	
		> 1989: rights on the basis of permits for water use and water works are required; pre-1989 abstractions were maintained	
		In the Florina Prefecture, there are 890 legal wells (690 with use permits and 200 pre-1989 wells without permit) (Florina Prefecture written communication, 11/11/2005)	
		In the Pella Prefecture, there are 69 legal wells (Paraschoudis et al., 2001) (ca 2/3 of those are pre-1989 wells)	
		In the Kozani Prefecture, there are 54 legal wells (Paraschoudis et al., 2001)	
		There are many private users illegally using groundwater (without use permits) but their number is difficult to estimate. Most illegal wells are in the Florina Prefecture (estimated to be 180-200 wells)	
Industry	PPC and other industries	*PPC (surface and groundwater):*	1987 Water Law
		< 2001: PPC needed no specific authorisation for its surface and groundwater abstractions, except for use permits for any new water uses after 1987. In practice, some groundwater abstractions after 1987 took place without a use permit (PPC does not hold use permits for any of its wells in the basin, although it should have issued use permits for wells drilled after 1987)	2003 Water Law
			Decision of the Ministry of Agriculture of 1968 (26278/3128/701) on drilling wells for several uses where restrictive measures apply
		> 2001 (PPC semi-privatisation): permits for all water uses and water works of the PPC are required	
		Other industries (groundwater):	Prefectural regulations and restrictive measures on water protection
		< 1989: authorisation to drill wells was required	
		> 1989: rights on the basis of permits for water use and water works are required; pre-1989 abstractions were maintained	
Hydropower generation	PPC	*Surface water:*	Law 1468/50 (PPC set-up)
		> 1950: exclusive water use right for hydropower	

Use	Users	Use rights	Relevant policies and regulations
		generation held by the PPC nation-wide	1987 Water Law
		> 1989: permits required for any new water use but PPC maintained competence to define itself the quantities of its abstractions and submit this to the central government	2003 Water Law
		> 2001 (PPC semi-privatisation): permits for all water uses and water works of the PPC are required	
Mining	PPC	*Groundwater:*	210/1973 Mining Code
		> 1973: right to abstract groundwater for mining purposes given without further authorisation except for the mining license (210/1973 Mining Code)	1987 Water Law
			2003 Water Law
		> 1987: any new use of groundwater after abstraction requires water use permit; in practice, there may be implementation inconsistencies with policy, since groundwater abstracted in some mines is further used (e.g. for water supply) without specific use permits	
Tourism/ leisure		No right defined	
Fisheries	Fishermen	No minimum lake level defined	
Water supply	Municipalities and communities	*Groundwater and springs:*	1987 Water Law
		> 1989: permits for any new water use required; in practice, most municipalities and communities did not apply for water use permits even for new wells drilled after 1989 (implementation deficit of 1987 Water Law); uses prior to 1989 were maintained	2003 Water Law
			Prefectural regulations and restrictive measures on water protection
		> 2003 (2003 Water Law): permits for all water uses and water works are required	

General regulatory framework

The implementation of the national 1987 Water Law took off on the level of the Prefectures of the Vegoritida basin mainly after 1989, i.e. after the issuing of national regulations for a system of permits for water abstractions and water works. As a result, since 1989/1990, the 3 Prefectures of the Vegoritida basin issue on a regular basis decisions to restrict water uses and to regulate the water permit issuing procedure on their territories according to the 1987 Water Law.

As a rule, any abstractions which began after 1989 without a permit are illegal. Any water abstraction, which began prior to 1989 according to custom or law, is maintained (pre-1989 established rights). Prior to 1989, authorisations from public authorities were only needed for constructing wells and connecting them to the power supply system. Therefore, wells drilled prior to 1989 and authorised for connection to the power supply network are considered legal and can be replaced if damaged. By 2005, no accurate estimations of the amount of water used through pre-1989 established rights in the Vegoritida basin existed. Wells drilled prior to 1989 without authorisation for connection to power supply (and thus, operating on generators) are considered illegal by the Prefectural authorities. In order to connect such wells to the power supply network at present or replace them with new ones, the same restrictions apply as for new wells (Interview Florina Pref., 26/1/2005).

In the Pella Prefecture, regulations to protect groundwater were issued as early as 1982 following regulations of the Ministry of Agriculture, preceding the 1987 Water Law of the Ministry of Development. These early regulations of the Pella Prefecture defined a procedure of authorisations for wells. Later, the regulations became stricter in terms of the minimum distance allowed between wells and the surface of private land needed to drill a new well. These regulations proved very useful for preparing the Pella Prefecture for the later implementation of the 1987 Water Law (Interview Pella Pref., 8/3/2005). In the Kozani Prefecture, similar restrictive measures for the protection of groundwater were also issued since 1972 but they were mainly relevant to the subbasin of Sarigiol (pers. comm. Kozani Pref., 29/11/2005).

Restrictive measures prior to 1989 were related to the rural electricity supply programme (nation-wide), according to which power supplied for agricultural use was cheaper than for domestic or industrial use. Therefore, a financial incentive was given to farmers to apply for power supply connection and, in the same time, for authorization of their wells. Otherwise, farmers were left with the expensive option of using gasoline to operate their wells. Furthermore, at that stage, wells were to a large extent still drilled by the administration itself (by the regional land reclamation units of the Ministry of Agriculture situated in the Prefectures) rather than private companies. This system, in a way, allowed the administration to plan the drilling of new wells and to control the water quantities consumed. When drilling a new well, the competent authority of the Pella Prefecture calculated the maximum optimal flow rate of the aquifer at that location. Then, it approved as much power supply capacity as necessary to abstract enough water for the irrigation of the respective field. If, later, there was need for irrigation water on a nearby field, the authority simply extended the power supply capacity at the existing well instead of drilling an additional well. Because of this collective planning, there are several "collective" wells in the Pella Prefecture (Interview Pella Pref., 8/3/2005).

The introduction of the 1987 Water Law did not affect much the regime of surface water abstractions, which mainly concerned communal water use and were maintained as preexisting uses. It significantly affected though the regime of groundwater abstractions. Specifically, the introduction of the 1987 Water Law discontinued the previous authorisation system for wells. The new system of permits of the 1987 Water Law established a more elaborate procedure based on more information about the fields to be irrigated, but it also established a system of personalised use permits which could not be extended and modified by the competent authority to cover the future needs of nearby land owners. Moreover, after 1990, wells were mainly drilled on a private basis and not by the authorities (Interview Pella Pref., 8/3/2005).

Initially, restrictive measures issued by the Prefectures of the Vegoritida basin were uniform for the entire territory of each Prefecture. In 1993, after a prolongued scarcity period in the Vegoritida basin, a first proposal was made by the Ministry of Development to introduce specific stricter measures for the Vegoritida basin in all 3 basin Prefectures.[63] As a response to the proposal of the Ministry of Development, the Pella Prefecture issued the first specific restrictive measures for the Vegoritida basin in 1994/95, followed by the Kozani Prefecture in 1995 and last by the Florina Prefecture in 1998 (see overview of restrictive measures in Annex V).

The specific restrictive measures for the Vegoritida basin were at start different in content in the three Prefectures. Already in 1995/1996, under the coordination of the Region West Macedonia, the three Prefectures agreed informally to follow a common approach on specific restrictive measures around Lake Vegoritida. However, although the Pella Prefecture and Kozani Prefecture issued strict specific measures for the Vegoritida basin, the Florina Prefecture continued issuing less strict measures until 1998 (Interview Pella Pref., 8/3/2005). Moreover, practical implementation of the restrictive measures issued by the Florina Prefecture was rather weak. On the contrary, implementation of restrictive measures in the Pella Prefecture was very strict (Interview Pella Pref., 8/3/2005). These differences in water protection regulations were often a source of conflict between the Prefectures of Pella and Florina.

Towards the end of this case study, the specific restrictive measures for the Vegoritida basin became more harmonized among the basin Prefectures, especially after the Florina Prefecture also started issuing specific measures for the Vegoritida basin in 1998. In 2005, the Florina and Pella Prefectures issued identical restrictive measures on the basin level. Agreement on a common approach for restrictive measures was reached through joint meetings also attended by the RWMD of the Region. In this context, the RMWD served as a coordination platform by proposing since 2004 common policy measures for all three Prefectures of the Vegoritida basin based on the results of the lake stability and purification programme (Interview RWM, 15/2/2005). Only the Kozani Prefecture maintained a different position aiming for less strict measures on its territory of the basin. The Kozani Prefecture disagreed with the proposed restrictive measures of the RWMD, because it would have to extent the application of stricter measures to a greater part of its territory.[64] Thus, the Prefect of Kozani disregarded the RWMD proposal (against the legal provisions of the 1987 Water Law)[65] and issued less strict restrictive measures. The RWMD regards these measures as invalid and non-compliant with the law (Interview RWM, 15/2/2005).

---

[63] Already in 1992, a committee set up by Ministry of Development drafted a report on the "Drop of the Lake Vegoritida level: consequences and proposals", which was the basis of the 1993 proposal of the Ministry of Development to the 3 Prefectures.

[64] The RWMD re-defined the boundaries of the Vegoritida basin for policy purposes on the basis of the findings of the lake stability and purification programme.

[65] According to the 1987 Water Law (art. 11, para 4), the Kozani Prefecture is obliged to follow the RWMD proposals. The Prefect either follows the RWMD proposal or does not issue any restrictive measures. Then

Rights of industry

Energy production is the industrial activity with the greatest impact on the water resources of the basin. The Public Power Corporation (PPC) was founded as a corporation in the property of the Greek State. Thus, the PPC was entirely part of the public sector until its conversion to a Societé Anonymé in 2001.

Prior to the adoption of the 1987 Water Law, all PPC abstractions took place without any specific permits. Activities planned by the PPC were submitted to the central government in the context of approving the overall PPC development plans. Therefore, State approval, also for planned abstractions, was given in the context of the strategic national planning for energy production.

Concerning the use of Lake Vegoritida for hydropower generation (from 1955 to 1991), the Law setting up the PPC gave the PPC an exclusive right to construct, operate and exploit hydropower plants nation-wide using the water of rivers and lakes to the extent necessary for accomplishing its purpose (Law 1468/50, article 2(1)).[66] Even though a similar clause did not exist for the production of steam-electricity, the PPC abstractions from Lake Vegoritida for its steam-electricity plants (from 1976 to 1997) took place without specific permits. When the 1987 Water Law introduced the requirement of water permits in 1989, the PPC (as a public sector entity) was exempt from the obligation to acquire permits for water works but not from the obligation to acquire permits for new water uses. Although no surface water abstraction was initiated by the PPC within the Vegoritida basin after 1989, there was continuous activity concerning groundwater abstractions. A few wells were put into operation for the needs of the steam-electricity plants, for which no water use permits were issued. Although this was in some cases made known, the competent authorities did not take action against the PPC. All in all, although water use permits were legally required for public water users such as the PPC and municipalities, this requirement was usually overlooked in practice.

After the PPC conversion to a Societé Anonymé in 2001, the PPC became obliged to apply for permits for both water uses and water works. Temporary water permits are already issued for wells drilled for a new PPC steam-electricity plant in the Florina Prefecture, very close to the Vegoritida basin (Interview RWM, 15/2/2005).

Wells for the PPC mining operations were also drilled without specific water permits both prior and after the 1987 Water Law. According to the Mining Code 210/1973, concessions given for mining operations allow the drilling of all necessary wells without further permits. However, use rights of water abstracted from the mining wells are not completely clarified. According to the 1987 and the 2003 Water Law, if further use of the water abstracted takes place, a water use permit should be issued. The EIA studies of the PPC on its mining

---

the RWMD has the right to ask for the intervention of the Ministry of Development and the restrictive measures are issued directly by the Minister.

[66] Even after the adoption of the 1987 Water Law, the PPC could define its own water needs for hydropower generation to be approved in the context of the national water development programmes.

activities note that no water is used in the production process of the mines. However, abstracted water is in some cases used for the water supply of the buildings of the mines and washing vehicles and engines. Therefore, water use permits may be needed for the water quantities further used, if the permit regulations are strictly applied.

Rights of irrigation

In the Vegoritida basin, only a limited area is irrigated via surface water collective irrigation networks. These networks are managed by Local Land Reclamation Organisations and hold permits for their abstractions. Most water for irrigation in the basin, however, comes from numerous private wells, several of which do not hold a permit. There is no exact estimation of the total number of wells in the basin due to the unknown number of illegal ones (Interview Florina Pref., 26/1/2005). Most sources converge to a total number of ca. 1200 wells (legal and illegal ones) for irrigation (estimation is based on Paraschoudis et al. (2001) and written communications with Florina Prefecture on 11/11/2005).

The number of wells in the Vegoritida basin rose dramatically in the 1980s and especially in the 1990s. On the one hand, this was due to technological innovation in the drilling of wells by introducing new easier techniques (air drilling) (Interview Pella Pref., 8/3/2005). On the other hand, public schemes were introduced aiming to increase the income of farmers, which also increased the cultivation of irrigated crops (improvement plans and crop subsidies) (Interview Florina Pref., 26/1/2005).

The multitude of wells is unevenly distributed across the Vegoritida basin. A 1994 survey showed that there were 90 wells on the Pella Prefecture territory and 830 wells on the Florina Prefecture territory of the basin.[67] Except for the fact that the cultivable flatland of the basin is much more extended in the Florina than in the Pella Prefecture, there are additional factors which contributed to the uneven distribution of wells in the two Prefectures. Firstly, the Pella Prefecture adopted a system of permits for wells to protect groundwater since 1982 and implemented it very consistently with the help of a well-operational competent division. As a result, only few wells could be drilled after the 1982 regulations. On the contrary, the Florina Prefecture only established a Prefectural division competent in these matters much later and adopted the first water use restrictive measures in 1990 (Interview Pella Pref., 8/3/2005). Secondly, natural conditions favoured a much higher number of wells in the Florina Prefecture. The table of the aquifer on the lakeshore of the Pella Prefecture is deeper than on the lakeshore of the Florina Prefecture. Therefore, for farmers in the Pella Prefecture, it was expensive to drill deep wells illegally and operate them on gasoline (as mentioned, cheap electricity was provided only by the authorities). On the contrary, on the lakeshore of the Florina Prefecture where much land was exposed from the dropping lake, it was much easier and cheaper to drill wells (legally or illegally) in the shallow aquifer. As a result, the

---

[67] Conclusions of Lake Committee established by the Secretary General of West Macedonia (Decision 1067/27-07-1994), Meeting of the Lake Committee on 9-10/12/1994 in Arnissa and Florina.

number of drills rose dramatically in the Florina Prefecture especially in the 1990s, facilitated by the introduction of new drilling techniques (Interview Pella Pref., 8/3/2005).

Illegal water use for irrigation and farmers' claims over exposed land

Illegal abstractions for irrigation take place widely in the Vegoritida basin. Water is illegally abstracted from the aquifer (mainly in the Florina Prefecture) and to a smallest extent from surface waters, especially from Lake Vegoritida.

In a strict sense, illegal water use also includes abstractions with expired water permits, i.e. permits whose 10-year validity period expired. In order to renew a permit, it is necessary to control the initial terms set for the abstraction (type of cultivation, water consumption, water quality suitability for irrigation, validity of land rental certificates). By the end of the case study process, lack of personnel capacity of the competent Prefectural authorities kept the procedure of renewing water permits non-operational (Interview Pella Pref., 8/3/2005; Interview Florina Pref., 26/1/2005).

Illegal wells are also largely linked to the regime of cultivating exposed land. The land use of the exposed lake surface is not defined on a planning basis. Such land was nevertheless conceded for use to farmers or is used illegally for cultivation to contribute to the local income (Planet Northern Greece S.A. – Planet S.A., 2001). In fact, the cultivation of exposed land is a very widespread practice all around the lake. Approximately 80% of the lake inhabitants cultivate exposed land illegally to a small or large extent (Interview Agricultural Cooperative of Arnissa, 25/1/2005). To irrigate exposed land, farmers who are normally given annual cultivation concessions (if any) drill wells without any authorization (Interview LVPS, 22/1/2005). Thus, a vicious circle between the cultivation and irrigation of exposed land leads to unsustainable water management and the further drop of water levels in the Vegoritida basin.

The issue of exposed land is highly political. It is often a topic of discussion between elected politicians (ministers, local politicians), who are reluctant to bear the political cost of removing illegal farmers from exposed land. The situation is aggravated by the fact that there is no systematic control and inspection of illegal cultivations. In essence, there is no control authority. The Land Services of the Prefectures are not active in inspecting illegal cultivations due to the lack of personnel (Interview Municipality Amyndeo, 16/2/2005).

In 1998, an intense conflict over concessions for exposed land broke out between two neighbouring communities in the Florina Prefecture (Agios Panteleimon and Vegora). Farmers from both communities cultivated land without having any titles of property. When the Ministry of Economy formally conceded 310 ha to Agios Panteleimon, intense protest came from Vegora (Macedonian Press Agency: News in Greek, 16/2/1998)[68] leading to

---

[68] Some argue that the conflict of the two villages over land had actually started long before the drop of the lake level in the context of an ethnic identity rivalry. Namely the villagers of Agios Panteleimon are part of the local population. The villagers of Vegora are immigrants settled in the area by the central government after the exchange of Greek-Turkish population in 1923. For the local population, the arrival of immigrants shook their entire regime of land property. The villagers of Agios Panteleimon argue that they were treated unfairly

confrontation of the two communities. Finally, the conflict was resolved by signing an agreement between the two communities, whereby Agios Panteleimon conceded 120 ha of the 310 ha for indefinite use to Vegora.

Table 7.7   Use of land and water on the exposed surface of Lake Vegoritida

Cultivation regime of exposed land	Type of crop allowed	Use of water allowed	Use of land and water in practice
Illegally cultivated	No crop cultivation allowed on public land (exposed land belongs to the State, specifically the Ministry of Economy)	No drilling of wells is allowed	Land is illegally cultivated and wells illegally drilled or irrigation water is pumped directly from the lake
Cultivated with temporary concessions (annual)	Only annual crops may be cultivated	No drilling of wells is allowed according to the Prefectural restrictive measures	Temporary concessions (annual) were given both in the Pella and Florina Prefectures In the Florina Prefecture, temporary concessions are given each year to local communities, which then concede land to farmers In the Pella Prefecture, temporary concessions were given in the past but were not renewed To irrigate this land, farmers drill wells illegally (especially in the Florina Prefecture) or use lake water (both in the Florina and Pella Prefectures) In practice, if farmers are given annual concessions once, they stay on the land even after the concession expires. Some farmers (especially in the Pella Prefecture) plant trees as a means to establish "permanent rights" on the exposed land Also, land temporarily conceded for cultivation used to be eligible for crop subsidies. As a result, farmers preferred to cultivate subsidized water-demanding crops such as maize
Cultivated with permanent concessions and/or property titles (ownership of land)	No restriction	Owners have the right to drill wells, if they meet the conditions defined by the Prefectural restrictive measures	Land property titles were only given for land which was exposed quite early by the initial lake drop in the 1960s

because the immigrants were given more land to cultivate from the State (Newspaper "Eleftherotypia", 15/3/1998).

168

7.6.2 Chronology of events

The water levels rivalry in the basin was until 1997 very much focused on the activities of the PPC. Towards the end of the case study process, other issues started gaining importance, especially the coordination of authorities over common restrictive measures to protect water quantity and discussions on the definition of a level for Lake Vegoritida. The following describes the events of the rivalry chronologically since the 1970s.

Lakeshore communities against the PPC

In 1977, Lake Vegoritida dropped to the level of the entrance to the Arnissa tunnel at 515.5 masl. Already then, the lakeshore community of Arnissa started complaining to its Prefecture (the Pella Prefecture) about the drop of the lake due to the PPC abstractions. The community was concerned about the impacts on the aquifer level and the operation capacity of the local irrigation wells, while in the same time, water abstracted from the lake benefited farmers of another basin of the Pella Prefecture (in the plains of Skydra downstream the Agras hydropower plant; these plains received irrigation water from the PPC water transfer scheme after water was used in the hydropower plant).[69] In the same period, however, the PPC was receiving requests from lakeshore communities and farmers to keep decreasing the lake level in order to gain more fertile exposed land (phone comm. PPC, 18/11/2005). These requests were especially intensified after strong precipitations in 1979 caused the lake to suddenly rise, flooding exposed land which was already being used as farmland (pers. comm. PPC, 2/12/2005).

In 1978, a PPC study on the water balance of Lake Vegoritida confirmed concerns about the impacts of the dropping lake. The study proposed a maximum lowest lake level at 521 masl, which would allow water abstraction for the Agras hydropower plant and also avoid adverse impacts on the aquifer level of the basin and the operation of irrigation wells. The study also concluded that PPC abstractions from the lake in the period 1955-1977 caused a lake drop of average 0.90 m yearly. The study proposed to use also alternative water sources to avoid further drop of the lake in the period 1977-1990[70].

The community of Arnissa remained alarmed about the decrease of the lake and the aquifer level throughout the 1980s[71], while in 1987, a coalition of lakeshore communities complained to the EU about the increasing lake pollution and the lake drop. In the meantime, also the Pella Prefecture was concerned about the lack of lake water to irrigate the plains of Skydra (the heart of the Pella Prefecture economy). The threat of deficient irrigation water was brought to the attention of the Pella Prefecture by a PPC letter on 2/1988 emphasising that "in 1991, the irrigation of the Skydra plains may be impossible given the current technical

---

[69] Letter of community of Arnissa to the Environmental Protection Committee of the Prefecture of Pella, Subject: Protection of Lake Vegoritida from pollution, 18/7/1977.

[70] Letter of PPC Directorate of studies and constructions of hydropower works (Mimikou, M) to Department of pre-studies of hydropower energy and environment, Subject: Conclusions of the Study of Water Balance of Lake Vegoritida, 25/11/1978.

[71] Minutes of meeting of the Community Council of Arnissa on 12/9/1988, Subject: Pollution of Lake Vegoritida.

works". To deal with the situation, the Pella Prefecture met in 10/1988 with members of the Ministry of Agriculture, the PPC, the National Institute of Geology and Mineral Exploitation, the Mayor of Edessa and the Local Land Reclamation Organisation of Skydra. One of the solutions proposed to safeguard water for Skydra was the installation of pumps at the entrance of the Arnissa tunnel to enhance water pumping from Lake Vegoritida, in case necessary.

In 8/1989, the agricultural cooperative of Arnissa was activated and requested from the Ministry of Agriculture to take action against the lake drop emphasising that the endangered local irrigation wells had been an expensive agricultural investment. The agricultural cooperative proposed a water transfer scheme to enrich the lake (with financial contribution from the PPC) and stated the determination of the lake inhabitants to hinder the installation of water pumps at the Arnissa tunnel to further irrigate the fields of Skydra in another basin. While the lakeshore communities had paid large amounts to drill and operate their irrigation wells, the farmers of the Skydra plains were receiving lake water "for free". The lakeshore communities further resented the fact that the decision to enhance water transfer from the lake by installing pumps was taken at a meeting without their participation. Therefore, it becomes obvious, that, at that stage, a territorial conflict over irrigation water evolved.

The PPC, from its perspective, attempted to stay out of this confrontation. It did not accept responsibility for the lake drop claiming that a) the drop of the lake was due to the natural water outflow of the lake underground and b) lake abstractions at the Arnissa tunnel did not serve the PPC but the irrigation needs of Skydra.

In 1990, the lake dropped down to the entrance of the Arnissa tunnel. In order to hinder the planned installation of further pumps, a massive mobilisation took place at Arnissa with the participation of several lakeshore communities. As a sign of protest, lake inhabitants sealed off the entrance to the Arnissa tunnel with stones, which was an event with wide press coverage and an important communicational impact. There were two important factors which induced this act of protest of the lakeshore population against the further transfer of water for the PPC and the irrigation of a neighbouring basin. First, weather conditions were especially dry. Secondly, there was social unrest because of news on alleged PCB pollution of the lake from the PPC industrial premises (Interview LVPS, 1/2/2005).

Finally, the lakeshore communities met with members of Parliament and the Prefect of Pella and decided to allow water abstraction for the Skydra plains in the summer of 1990, under the condition that alternative arrangements would be made for the irrigation of Skydra in the following year and that funds for restoring Lake Vegoritida would be allocated. Indeed, the Ministry of Agriculture agreed that, due to dry weather, abstractions from the lake should cease to avert an ecosystem imbalance and avoid protests of the lakeshore communities. To meet the irrigation needs of Skydra, the Ministry of Agriculture ordered the drilling of wells at Skydra to provide groundwater for irrigation. In 1991, these wells were drilled to serve the Skydra plains but no funds for restoring Lake Vegoritida were given although promised (Interview LVPS, 1/2/2005).

In 1991, the PPC abstractions at the Arnissa tunnel ceased altogether. It was not cost-effective for the PPC anymore to pump water from the lake because of the high operational cost of this water transfer given the low lake level (Interview LVPS, 1/2/2005).

Although confrontation over abstractions at the Arnissa tunnel in the Pella Prefecture was settled, the lakeshore communities had already started fighting against the PPC over its abstractions for its steam-electricity plants in the Florina and Kozani Prefectures. The Action Committee for the Salvation of the lake proposed to halt all PPC abstractions from the lake and to search for alternative water sources for the PPC steam-electricity plants.[72] Farmers also became involved when, in 1992, the farmers' association of the lakeshore community of Vegora in the Florina Prefecture complained to the Ministry of Agriculture, the Florina Prefecture and the PPC about the lake drop. The farmers' association requested the return of lake water abstracted for the steam-electricity plants back to the lake (via stream Soulou) after its use. The PPC replied that, except for the fact that it only abstracted a minimum amount of water annually entering the lake, the return of flows would involve costly technical works and would cause the protest of other communities which were using this water for irrigation further upstream.

First actions by the authorities in the 1990s

After the introduction of the 1987 Water Law, the escalation of local protests in the early 1990s and the initiation of EU infringement procedures against Greece, some action started being taken by the local and regional authorities of the basin.

In 1994, first specific restrictive measures to protect water resources in the Vegoritida basin were adopted by the Pella Prefecture, followed by the Kozani Prefecture in 1995 and by the Florina Prefecture in 1998. As mentioned in section 7.6.1, these specific restrictive measures for the Vegoritida basin gradually converged among the three Prefectures towards the end of the case study process.

In 1994/1995, the water levels issue was also discussed by the participatory Lake Committee set up by the Region of West Macedonia. Main conclusions were that the drop of the lake and the aquifer was due to PPC abstractions directly from the lake and the aquifer, abstractions of the AEVAL nitrogen-fertiliser industry, abstractions for irrigation as well as natural outflow and dry weather. Some of the main proposals made were to: cease PPC abstractions from the basin of Vegoritida and investigate alternative water sources, transfer water from other basins, install water meters, replace water-demanding crops with less water-demanding ones, change to drip irrigation and sprayers, stabilise the lake at 520 masl and prohibit concessions for the cultivation of exposed land on the lakeshores. The PPC also participated in this Committee with the aim to clarify its contribution to the water abstractions in the basin (phone comm. PPC, 1/3/2005). Although the PPC accepted its contribution to the quantitative problem, it argued that also farmers were to blame. The PPC also argued

---

[72] Proposals of the Action Committee for the Salvation of Lake Vegoritida to the Inter-Prefectural meeting of 6/6/1990 in Florina.

that, at that stage, it actually had no alternative but to abstract water from the lake for its energy production activities (Interview RWM, 17/2/2005). Since the 1970s, the PPC began searching for alternative water sources outside the Vegoritida basin. There were plans to provide water to two steam-electricity plants from the PPC reservoirs in the River Aliakmon basin of the Kozani Prefecture already in 1981-82, while Lake Vegoritida would only supply one steam-electricity plant (PPC Directorate for Studies and Constructions for Hydropower Generation 1979). Nevertheless, the planned water transfer from River Aliakmon was not realised until 1997.

PPC resorts to water transfer from a neighbouring basin

In 11/1997, the PPC stopped abstractions from Lake Vegoritida altogether. The termination of the abstractions was not the result of an official top-down decision imposed upon the PPC (phone comm. PPC, 3/3/2005). The decision was taken by the PPC itself in the context of its own strategic management planning and imposed by the dry climatic conditions. Because of the continued lake drop, the PPC had to continuously extent its pumps to reach water in the lake. The PPC calculated that it could continue abstracting water from Lake Vegoritida for only another 3-4 years, but then it would need an alternative water source. Therefore, the long-term solution of water transfer from the southern neighbouring basin of the River Aliakmon was given (Interview LVPS, 1/2/2005). Although at that stage the Prefect of Pella had sent a request (in 9/1997) to the Ministries of Development and of Environment to intervene for the termination of the PPC abstractions, pressure by local authorities and local farmers was a less important factor affecting the PPC decision to stop abstraction.

Thus, since 1997, the PPC steam-electricity plants receive water from a PPC artificial reservoir on the River Aliakmon.[73] To some extent, the steam-electricity plants also continue to use groundwater from the Vegoritida basin either from their own wells or from wells operating at the lignite mines. For instance, the steam-electricity plant of Amyndeo uses water from the River Aliakmon to 65-70% and from wells to 30-35% (Interview Director of PPC Amyndeo steam-electricity plant, Newspaper "Ethnos tis Kiriakis", 1/10/2000).

Definition of a level for Lake Vegoritida and emergence of nature demands

In the context of the lake stability and purification programme, a comprehensive hydrogeological study was carried out examining the past and present water balance of the basin. Based on scientific criteria, the study made management proposals, most of which were adopted by the RWMD of West Macedonia (set up in 1997) and thereafter by the Pella and the Florina Prefectures in their 2005 restrictive measures, including the prohibition of direct abstractions from Lake Vegoritida, prohibition of new wells in the basin and allowance of replacement of existing wells only if further increase of the irrigated surface is not involved (at least, until the lake reaches 515.5 masl).

---

[73] It should be noted that some experts were against the water transfer from the basin of Aliakmon into the basin of Vegoritida, because water volumes of different origin and different quality are ultimately mixed in Lake Vegoritida (Interview State Chemical Laboratory, 14/2/2005; Interview Ornithological Society, 25/2/2005).

The hydrogeological study also proposed to prohibit any activity on the lakeshore until the lake reaches 515.5 masl and to define the zone between 509 masl (lake level in 2001) and 515.5 masl (at the entrance to the Arnissa tunnel) as "Zone of Protection of Vegoritida".

In 11/2004, a common meeting of the Pella and Florina Prefectures took place under the coordination of the RWMD of West Macedonia to discuss the proposed maximum lake level at 515.5 masl. In principle, the Prefectures informally agreed over the proposed level, but several complementary studies and decisions were still needed before any final official agreement could be reached.

The issue of the lake level definition was especially complicated by the unclear property regime of exposed land around Vegoritida. Although several efforts took place by local authorities and non-governmental organizations to deal with the issue of exposed land, it remained unresolved until the end of the observed case study process. In the Pella Prefecture, concessions to cultivate exposed land were last given in 1984, while temporary annual concessions given in 1990 and 1993 were not renewed (Newspaper "Proini", 18/2/1997). In 1997, the Prefect of Pella proposed administrative action to evict farmers illegally cultivating exposed land on the territory of Pella. The president of the community of Arnissa, however, rejected the eviction of farmers and requested instead permanent concessions to keep young farmers in the area (Newspaper "Proini", 20/2/1997). In 2003, the Municipality of Vegoritida (including the community of Arnissa) issued a decision to create a lakeshore zone, as a means of prohibiting the cultivation of exposed land illegally (Newspaper "Politis Florinas", 10/4/2003). In practice, a road was provisionally opened up around the lake but without resources to guard the entire area.

The unclear regime of the use of exposed land also appears difficult to reverse, if one considers that the legal system on land property provides a legal leeway for the establishment of ownership over land (up to 1 ha), if this is occupied for more than 10 years and a small financial contribution is paid by the claimer. Being aware of this legal leeway, many farmers around the lake who cultivate exposed land (illegally) for several years, already submitted applications to the authorities to claim property titles (Interview Agricultural Cooperative of Arnissa, 25/1/2005). Therefore, there is fear that the definition of a lake level, which would result in the flooding of much exposed land, could lead to social unrest and compensation claims by local farmers who cultivate exposed land for more than 10 years. Farmers who cultivate such fields for more than 10 years consider themselves owners and not simply users of such land (Interview RWM, 15/2/2005). Thereby, an agreement on exposed land property and the lake level is expected to be difficult. Until the end of the case study process, even new requests to cultivate exposed land continued. The Region West Macedonia and local authorities could only reassure farmers that the issue was being examined but environmental issues first needed to be considered given the proposal of the Vegoritida area as an EU Natura 2000 site. Except for the issue of claims over exposed land which still needs resolving, the official definition and delimitation of the lakeshore zone is also required before the lake level can be defined.

Furthermore, discussions on the proposed lake level at 515.5 masl were criticised for being exclusively use-oriented and mainly considering the water needs of agriculture. In essence, the lake was perceived as an irrigation storage reservoir (Interview Ornithological Society, 25/2/2005). Before the final definition of the lake level, also the environmental needs of the lake ecosystem need to be officially considered. In this respect, the Hellenic Ornithological Society became involved in 2003 in the issue of the lake level definition and the illegal cultivation of exposed land. The Ornithological Society sent a letter to the Environment Ministry reporting on the adverse impacts of the cultivation of exposed land on the bird populations of the lake. Subsequently, the Environment Ministry demanded that prior to any further concessions for exposed land, a study should be carried out on the viability and sustainability of the Vegoritida ecosystem. This study would examine the natural fluctuation of the lake, the habitats created at different lake levels and the impacts on the viability of the ecosystem. The aim of the study would be to define a sustainable lake level to provide safety for the wetland. This study would also involve defining a maximum and minimum level for the management of the lake. If a maximum level could be defined, exposed land above it could theoretically be given to cultivation, as requested by the local population (Newspaper « Macedonia », 7/10/05). The Specific Environmental Study (required by the 1986 Environment Law to designate the basin as protected area) could also be based on this ecosystem viability study on issues of lake level definition and management. However, despite talks between the Florina Prefecture and the Region West Macedonia to fund such an ecosystem viability study, such a study was not yet officially assigned by the end of the observed process (Interview Ornithological Society, 25/2/2005; Phone comm., 18/11/2005). Thus, the proposed lake level at 515.5 masl remained a temporary proposal, which should be supported by various other studies and procedures to delineate the lakeshore zone and lake level.

Fate of proposed measures to reduce water use in irrigation

As part of the lake stability and purification programme, a study on the impacts of agriculture on the environment was carried out. This study investigated possibilities of replacing irrigated crops with non-irrigated ones in the Florina Prefecture part of the basin. By comparing financial gains from the cultivation of different crop types, the study concluded that financial gains from irrigated crops, such as maize, were much higher than from non-irrigated crops. Therefore, the study concluded that a change of irrigated crops to non-irrigated ones on a significant scale would significantly reduce the income of farmers and impact social and economic aspects of local livelihood (Planet Northern Greece S.A – Planet S.A., 2001). An exception to this was the cultivation of grapevine, which is not water-demanding but nevertheless very profitable and was proposed for the partial replacement of maize cultivations (IGEKE & D.A.P/AUTH, 1999).

In parallel to the above study on the replacement of crops, discussions took place with the Ministry of Agriculture on the possible introduction of an agrienvironmental programme for the protection of water resources in the basin. Such a programme could indirectly and rapidly lead to the replacement of crops bearing in mind factors of the market. It was specifically proposed to introduce an agrienvironmental programme within a zone of 5 km around the lake offering incentives for farmers to change to fallow land and subsidies to change irrigated crops such as maize into less water-demanding ones. At first, relevant talks between basin actors and the Ministry of Agriculture came rapidly to an end without the desired outcome (Interview Planet, 8/2/2005). Later on, in 2005, an agrienvironmental measure was proposed (for the basin of lakes Vegoritida, Petron, Chimaditida, Zazari) by the Ministry of Agriculture to reduce water consumption and non-point pollution (as a follow-up to the earlier request by the Prefectural and Regional authorities of the basin and the evidence given by the lake programme). The relevant CMD activating this proposed agrienvironmental measure was finally issued in August 2006. The aim of the measure is to subsidise farmers for the reduction of water consumption for irrigation and of nitrogen fertiliser application on fields cultivated with maize, alfalfa and sugarbeet. Farmers entering the measure should commit themselves (for five-year periods) to an annual reduction of water consumption by 30% and of fertiliser application by 10%. The methods to be followed include fallow land, alternating their crops with non-water-demanding crops and creation of permanent uncultivated farmland.

The lake stability and purification programme also involved a pilot study on the gradual replacement of irrigation techniques from spray artificial rain irrigation to drip irrigation. Unfortunately, by the end of the case study period, the Ministry of Agriculture did not respond with any institutional instrument to go ahead with such a replacement of irrigation systems. Unless subsidies are provided to farmers for the costs of introducing drip irrigation, farmers in the basin will not give up spray systems for which they have already made investments (Interview Florina Pref., 26/1/2005).

Chronology summary

The rivalry on water levels in the Vegoritida basin started off with social unrest and local protests against the water consuming behaviour of the PPC, initially for hydropower generation purposes and subsequently for steam-electricity generation and lignite mining operations. In the end of the 1980s, communities on the shores of Lake Vegoritida turned to the European Court of Justice raising their stakes in the rivalry against the inactivity of the Greek State. Moreover, further mobilisations on the local level in 1990-1991 achieved a halt to water transfer from the lake for the irrigation needs and the operation of a PPC hydropower plant in a neighbouring basin. Subsequently, the rivalry of lakeshore communities against the PPC continued over lake abstractions for the PPC steam-electricity plants within the Vegoritida basin, until these abstractions also ceased when the PPC turned to a water transfer scheme from another basin (in 1997).

Triggered by pressure from the ECJ (which convicted Greece in 1998 for the pollution of Lake Vegoritida and stream Soulou) and local social pressure especially related to claims for the cultivation of lake exposed land, the Region West Macedonia initiated in 1998 an extensive scientific programme on the stability and purification of Lake Vegoritida. As a result of this programme, discussions started among authorities in the basin over the definition of a level for Lake Vegoritida at 515.5 masl towards the end of the case study period. In this context, also nature conservation and ecosystem demands emerged with the intervention of the Environment Ministry and an environmental NGO. By the end of the case study, policy developments to protect the quantity of water resources in the basin remained mainly on the Prefectural level.

7.6.3 Actor analysis

Table 7.8 gives an overview of the roles played by the main actors involved in the rivalry on dropping water levels, their perceptions over the resource problem, their objectives and resources.

Table 7.9 illustrates key relationships between the main rivalry actors, focusing thereby on the development of more or less cooperation over the studied process. In few cases, there has been some improvement of cooperation, for instance between the Pella and the Florina Prefectures or between the PPC and the RWMD of West Macedonia. In many cases, however, no significant improvement or even a decline in cooperation was noted. The main actor interactions, where conflict was maintained or developed anew, were between the PPC and several local actors (lake municipalities/communities and farmers impacted by scarcity), between farmers and the Florina Prefecture and between the Kozani Prefecture and the RWMD of West Macedonia (see also emphasis in framed cells of Table 7.9). Explanations for these conflictual relationships are given in later sections (7.8.2.3 and 7.9.2), which discuss actor interactions and their influence on the process and outcome of regime change and implementation. Sections 7.8.2.3 and 7.9.2 also make some suggestions on how the cooperation potential between key actors could improve. Additionally, the recommendations on changing the regime contextual conditions and the regime institutional elements towards a more integrated phase in chapter 9 have indirect relevance for the improvement of actor cooperation.

Table 7.8 Vegoritida rivalry on dropping water levels: Problem perceptions, roles, objectives and resources of actors

Actors	Problem perception & roles played in the process *(detailed institutional competence in* Table 7.4*)*	Objectives & resources
EU	**Role:** Indirect role by exercising pressure on the Greek authorities through the ECJ case on pollution	**Resources:** Legal, economic
Ministry of Environment, Physical Planning and Public Works (MEPPW)	**Problem:** It does not agree to further concessions for the cultivation of exposed land before the lake level and the lakeshore zone are defined through scientific studies	**Objectives:** Avoid infringement procedures due to the inadequate implementation of EU nature protection legislation (Lake Vegoritida is part of a proposed Natura 2000 site)
	Role: In 2005, it demanded that prior to any further concessions for exposed land, a study for the viability and sustainability of the lake ecosystem should be carried out. It also demanded that the lakeshore zone	**Resources:** Political

Actors	Problem perception & roles played in the process *(detailed institutional competence in* Table 7.4*)*	Objectives & resources
	should be officially delimited	
Ministry of Agriculture (MINAGR)	**Role:** In 1990, it intervened for resolving the water conflict between farmers of Lake Vegoritida and farmers of a neighbouring basin (Skydra plains) In 2006, it issued an agrienvironmental measure for protection of lakes in the Vegoritida basin	**Resources:** Political/policy-making, economic
Ministry of Development (MDEV)	**Role:** It received complaints about the impacts of the lake abstractions of the PPC	**Objectives:** Protect interests of industrial users, including the PPC
Regional West Macedonia - Managing Authority (RWM/MA)	**Problem:** Lack of coordination between the two Regions sharing the basin was a major drawback in bringing the process forward Water quality and quantity are interconnected and must be managed together The Florina Prefecture should become more active and use the lake stability and purification programme results to impose measures on users **Role:** It activated and coordinated the first participatory Lake Committee in 1994 (motivation of its political leadership)	**Objectives:** Relax pressure from the EU and the local population Make progress in the legal designation of the basin as nature protected area Proceed with the formal definition and delimitation of the lakeshore zone and the lake level **Resources:** Economic, planning power
Regional Water Management Department of West Macedonia (RWM/RWMD)	**Problem:** The past implementation of water restrictive measures was inadequate in most of the basin The Kozani Prefecture does not comply with the RWMD proposed restrictive measures for the Vegoritida basin The proposed lake level at 515.5 masl seems illusionary to many parties involved **Role:** It promoted a common water policy approach for all three Prefectures of the basin following scientific proposals of the lake stability and purification programme	**Objectives:** Achieve acceptance and implementation of the common proposed restrictive measures by all three Prefectures of the basin Resolve disagreement with the Kozani Prefecture **Resources:** Political/regulatory but inadequate human resources
Regional Directorate for Environment and Physical Planning of West Macedonia (RWM/RDEPP)	**Problem:** There is lack of adequate control of the environmental terms approved for activities which overabstract water Interventions of elected politicians complicate the resolution of the problem, especially the issue of exposed land, whose resolution is vital for sustainable water management	**Objectives:** Proceed with the designation of the basin as nature protected area **Resources:** Political/regulatory, technical/EIA approval, but inadequate human resources
Pella Prefecture (PoP)	**Problem:** Lax policies of the Florina Prefecture on water resource protection resulted in numerous wells and were a main driver for water scarcity and the lake drop in the basin (Director of Division for Land Reclamation in the Pella Prefecture, Newspaper « Politis Florinas », 9/7/1996) PPC is mainly responsible for the lake drop	**Objectives:** Strict water protection measures should be implemented by the Florina Prefecture Promote the creation of a management body for the basin **Resources:** Political/regulatory, technical

Actors	Problem perception & roles played in the process *(detailed institutional competence in* Table 7.4*)*	Objectives & resources
	and also significantly impacts the basin aquifer through its mining activities	
	The absence of a management body hinders a holistic approach to deal with the lake problems	
	Role: It actively protected groundwater since early 1980s	
	In 1997, it was involved in activities to stop the PPC lake abstractions	
Florina Prefecture (PoF) – especially referring to its Division for Land Reclamation	**Problem:** The complex problems of the basin largely result from the lack of coordination of sectoral policies, such as agricultural and water policy	**Objectives:** An agrienvironmental measure should be adopted for the basin
	Water problems in the basin are highly political, especially the issue of exposed land (pressure of farmers on politicians)	Strengthen the institutions of the Prefecture to be able to implement the results of scientific work through decisions of the Prefectural Council
	The lake stability and purification programme had no policy validity to impose management measures	**Resources:** Political/regulatory, technical (but not enough use made if sanctions are taken to courts), but inadequate human resources
	The PPC should compensate local society. The Special Development Fund paid by the PPC to the affected local authorities is not earmarked for projects to enrich water resources. Instead, the Fund is used for different purposes, such as road construction and cultural activities[74]	
	Role: It served as executive authority to carry out the lake stability and purification programme	
Prefecture of Kozani (PoK) – especially its Division for Land Reclamation	**Problem:** Water administrative management in the basin worsened since the set up of the Region (difficulty in communication between the Kozani Prefecture and RWMD)	**Objectives:** Implement less strict restrictive measures than those proposed by the RWMD
		Resources: Political/regulatory, technical
Municipalities and communities (Mun/comm)	**Problem:** PPC was the main driver of the lake drop. Exposed land is a significant problem	**Objectives:** Limit the PPC abstractions in the basin and receive financial compensation from the PPC (Special Development Fund for all affected lakeshore communities)
	Municipality of Amyndeo is concerned about the impact of new PPC mines on its water supply. It resents the drilling of wells by the PPC before the approval of the relevant mine EIA. The PPC does not even comply with approved environmental terms for existing mines	Agios Panteleimon aims at lake restoration for the return of fisheries and tourism
	Water basin problems cannot be resolved without an integral management body	**Resources:** Legal (legal mobilisation locally, regionally and on the EU level), political pressure (on local MPs at Parliament to bring issues to the upper policy arena)

---

[74] Although the Special Development Fund of the PPC to the local authorities should be invested in activities on "infrastructure, development and environmental protection", local communities take advantage of this to build roads and water supply systems. So far nothing has been done on the environment (Interview LVPS, 22/1/2005).

Actors	Problem perception & roles played in the process *(detailed institutional competence in* Table 7.4*)*	Objectives & resources
	Role: They were strongly mobilised to protect the lake, initially on an individual basis and then as coalition	
	Local communities together with local associations and private individuals filed a complaint to the EC culminating to the ECJ case	
	Municipality of Amyndeo was active in promoting the ecological and touristic value of the lakes Vegoritida and Petron	
Agricultural Cooperative of Arnissa (Agr.Coop.Arnissa)	**Problem:** Problem causes include the cultivation of very water-demanding crops in the Florina Prefecture irrigated with artificial rain techniques, the lack of strict water protection policy in the Florina Prefecture, the cultivation of the exposed land and the PPC continued overexploitation of the aquifer	**Objectives:** Promote sustainable water management and irrigation in the context of its integrated crop management system **Resources:** Technical (influence on its farmer members), legal (involved in denunciations to local courts, EU)
	Role: It supported the campaign of local associations to protect the lake and was involved in protests against the PPC	
	It supports the introduction of a subsidy scheme for non-irrigated crops	
	It urges farmers to invest in less water-demanding techniques (sprayers or drip irrigation) in the context of an integrated crop management certified system (for peach and cherries).	
Farmers	**Problem:** The PPC is responsible for the drop of the aquifer in the irrigation wells. Some farmers realise that if the lake water reserves are further stressed, their production will be adversely impacted	**Objectives:** Variable are objectives (large, heterogeneous group) Some aim for short-term financial profit by cultivating and irrigating illegally exposed land.
	Other farmers see the potential rise of the lake level as a problem, because it would flood the exposed land they already cultivate	Others prefer to protect water resources, so that they have enough water to irrigate **Resources:** Social and political pressure (also for exposed land)
	Role: Some were involved in protests against the PPC to stop abstractions. Some, at times, asked the PPC to continue abstractions, so that they could cultivate the exposed land	
Public Power Corporation (PPC)	**Problem:** The PPC denied sole responsibility for the lake drop. The drop was also due to underground water loss through natural processes. Additionally, a significant role was played by the change	**Objectives:** To secure energy production and further develop the energy production potential of the region close to Lake Vegoritida **Resources:**

---

[75] Letter of PPC Managing Councilor (Nr. Γ.ΔΝΣ/2618) of 23/7/2002 to Ministry of Development, Subject: Reply to questions 4257/20.06.2002, 4270/20.06.2002 and 4201/19.06.2002 of Members of Parliament N.Tsiartsioni, I. Tzamtzi and P.Fountoukidou, to questions 8657/1306/20.06.2002 and 8659/1308/20.06.2002 of Member of Parliament I.Tzamtzi and to question 33/02.07.2002 of Member of Parliament G.Karasmani to the Greek Parliament on Lake Vegoritida.

Actors	Problem perception & roles played in the process *(detailed institutional competence in* Table 7.4*)*	Objectives & resources
	of crop types from wheat to water-demanding crops such as maize. Thus, farmers are also responsible for the lake drop and mainly for the aquifer drop through irrigation. Farmers have also had inconsistent behaviour. In the 1970s, they requested from the PPC to continue abstraction so that they could gain more fertile exposed land, while, in the 1990s, they demanded from the PPC to stop abstraction so that they could cover their irrigation needs (phone comm. PPC, 18/11/2005) Since it ceased abstracting water directly from the lake, the PPC positively contributes to the water basin balance through return flows from the River Aliakmon[75] The PPC also recognized the potential impacts of its mines on the water resources of the basin (impact on irrigation of nearby fields and the water supply of settlements) (PPC General Mines Directorate, 1998) **Role:** It was the target of campaigns of the lakeshore population to protect the lake since the 1970s It had a rather autonomous presence in the constellation of basin actors in the process. Arguing that issues of national energy security are at stake, the PPC refused to be influenced by local protests in its strategic planning The PPC recently makes an effort to improve its environmental profile, since its conversion to a Societé Anonymé in 2001	Political: Political support of the central government, at least until its semi-privatisation in 2001, and political support of the Florina Prefecture in its energy expansion objectives in the area (e.g. a new steam-electricity plant was inaugurated in 2/2004 near the city of Florina, very close to Lake Vegoritida, enjoying the political support of the Florina Prefecture whose top priorities include the sector of energy production (Newspaper « Politis Florinas », 14/11/98) Economic: The PPC uses its financial resources to compensate for the impact of its activities in the basin. A Special Development Fund was introduced by law in 1996, whereby 0.4% of the annual PPC revenue is paid to local municipalities affected by the PPC activities Social support: The PPC is one of the larger employers in the basin, a fact which contributes to significant social support of the PPC by certain parts of the local population
Local associations: Lake Vegoritida Preservation Society (LVPS) and the association MESNA of Arnissa	**Problem:** From the perspective of the LVPS, the main problem hindering holistic basin management is the split up of administrative competence and the absence of a management body Secondly, the PPC is not yet sensibilised for more sustainable water use. On the policy level, restrictive measures on water use proposed by the RWMD on the prohibiting of new wells in the basin should also apply to the PPC (Interview LVPS, 22/1/2005) Thirdly, farmers became aware of water scarcity since the early 1990s when the lake drop affected the aquifer in their wells. However, farmers expect (financial) assistance from the State to change to less water-demanding agricultural practices and they do not have a strong representative organisation yet to voice their demands	**Objectives:** To impose restrictions on the PPC To promote the creation of a management body for the lake To officially define the lake level at 515.5 m. The LVPS calculated that areas which would be flooded by a lake rise to 515/516 m are illegally cultivated fields and no issue of farmers' compensation claims would interfere. The LVPS proposed to proceed to legalization of cultivated exposed land above 515.5 masl, whereby cultivation could continue with the installation of sustainable irrigation systems **Resources:** Legal (local courts, EU), social support, technical/cognitive (public information campaigns) but LVPS lacks knowledge of policies and network of partner organisations

Actors	Problem perception & roles played in the process *(detailed institutional competence in* Table 7.4*)*	Objectives & resources
	(Interview LVPS 22/1/2005)	
	Role: MESNA was a main driving force within the local population of Arnissa to start protests against the lake drop	
	The LVPS continued the work of previous local associations in promoting the protection of the lake on behalf of local society. The Society has a rather active role and ecologically conscientious members. Although it cannot participate in decision making, it influences people holding key posts	
Hellenic Ornithological Society (HOS)	**Problem:** Cultivation of exposed land results to the destruction of natural vegetation, habitats for birds and the increase of pollution into the lake	**Objectives:** Increase public awareness on the wetland values of the basin
	The local population only engaged in protecting the aesthetic value of the lake but not its natural values	Study and conserve lake bird populations by maintaining high water levels in spring for the creation of wet meadows
	The PPC water transfer from the basin of the River Aliakmon is unsustainable	Support initiatives to resolve the issue of cultivation of exposed land (e.g. via a study on the viability and sustainability of the lake ecosystem)
	Role: The HOS was involved in scientific research on bird populations of Lake Vegoritida in cooperation with the Municipality of Amyndeo	**Resources:** Technical/cognitive (public awareness), political pressure/lobbying
	It requested the intervention of the MEPPW in the issue of cultivation of exposed land pointing out relevant environmental issues	
External experts	**Role:** Conducted studies and research projects on the water resources of the basin	**Resources:** Technical/cognitive
	They proposed new water management approaches and solutions (mainly technical) to water resource problems	

Table 7.9 Vegoritida rivalry on dropping water levels: Relationships among key actors in the process

Actors	Administration									Users			NGOs		HOS	Experts
	EU	MEPPW	MINAGR	MDEV	RWM/MA	RWM/RWMD	PoP	PoF	PoK	Mun/comm	Agr.coop. Arnissa	Farmers	PPC	LVPS/MESNA	HOS	experts
EU	▨													⌂/0		
MEPPW	(↓)/(↓)	▨												⌂/0	0/⌂	
MINAGR		0/↓	▨	0/↓			↓/0	↓/0		↓/0	↓/0	↓/0				
MDEV			+/+	▨	+/+											⌂/0
RWM/MA				+/+	▨	+/+	++/+	+/+	+/−							
RWM/RWMD				+/+	+/+	▨	+/+	(+)/+	+/−	++/+	++/++	0/(−−)	0/+			⌂/0
PoP		+/0			+/+	+/+	▨	−/+		++/+	+/+	+/+	−/0	+/+		
PoF		0/↓	+/0		++/+	(+)/+	−/+	▨		+/+	++/++	0/(−−)	−/0	0/+	0/+	⌂/0
PoK					+/−	+/−			▨							
Mun/comm			+/0		+/+	++/+	++/+	+/+		▨	++/++	+/0	−/−	++/++	0/+	
Agr.coop. Arnissa			+/0			++/+	+/+	++/++		++/++	▨	+/+	−/0	++/++	0/+	
Farmers				+/+		0/+	+/+	0/(−−)	0/(−−)	+/0	+/+	▨	−/−	−/−		
PPC					0/+	0/+	−/0	−/0		−/−	−−/0	↓/−	▨	−/0		
LVPS/MESNA					0/+	0/+	+/+	0/+		++/+	++/++	0/+	−/0	▨		
HOS								0/+		0/+					▨	
Experts				⌂/0	⌂/0			⌂/0								▨

Note (a): The table should be read as the relationship of the actor in the vertical column to each of the actors in the horizontal rows. Blank cells indicate no particular relation observed. Actor abbreviations are explained in Table 7.8.

Note (b): Symbols in the table are explained in the legend below. The presence of symbols left and right of the slash (/) aim at indicating noteworthy changes in actor relationships over the observed process. If left and right symbols are identical, no significant change in the actor relationship was observed. If symbols are different, a certain change took place. For instance, symbols (++/+) indicate close collaboration at first, which was reduced over time. Symbols (−−/+) indicate an initial conflict which was followed by certain improvement (formal interaction or sporadic collaboration).

LEGEND OF SYMBOLS:

++	close collaboration (problem-solving interaction)	⌂	information supply or exchange
+	only formal relationship or sporadic collaboration	€	financial contribution
0	no noteworthy interaction observed	→	One-sided pressure
−−	conflict		

7.7 Rivalry on decreased water quality

7.7.1 Property rights

Ownership rights are not really relevant to the pollution of water resources. Therefore, this section focuses on an overview of use rights affecting water quality in the basin (in Table 7.10), referring also to the relevant regulatory framework.

Table 7.10   Use rights affecting water quality in the Vegoritida basin

Use	Users	Use rights to water	Relevant policies and regulations
Living environment & nature	Flora and fauna	Ecosystem aspects only briefly dealt with in the CMD of 2001 on a "Special pollution reduction programme of Lake Vegoritida and stream Soulou"	CMD of 2001 on a "Special pollution reduction programme of Lake Vegoritida and stream Soulou"
Discharge of pollutants	PPC & other industries	Rights defined by effluent disposal permits Most but not all of the 18 industrial units in the basin hold effluent disposal permits (see Annex VII for details): - 12 units hold permits (1 of these must extend permit for additional effluents) - 3 units hold temporary permits (3 steam-electricity plants of the PPC) - 3 units (small units of dairy and wine production) hold no permits (illegal effluents disposal) The CMD of 2001 on a "Special pollution reduction programme of Lake Vegoritida and stream Soulou" required the revision of the industrial effluents permits in the basin, based on the new adopted quality objectives. By the end of the case study period, there was no progress in this respect	Interprefectural decision 1900/79 Florina Prefecture decision 555/90 EU Urban Wastewater Treatment Directive CMD of 2001 on a "Special pollution reduction programme of Lake Vegoritida and stream Soulou"
	Municipalities, communities	Rights defined by effluent disposal permits Main urban polluters with permits are the city of Ptolemaida for its wastewater treatment plant and Amyndeo for its old sewerage network Most communities hold no effluents permits	Interprefectural decision 1900/79 Florina Prefecture decision 555/90 EU Urban Wastewater Treatment Directive
	Farmers	Rights for animal farms defined by effluent disposal permits. Most farms in the basin do not hold permits (illegal effluents disposal). All new animal farms must apply for specific effluent disposal permits according to the new Sanitary Regulation Y1β/2000/95, which requires collection of effluents and disposal in septic tanks or cesspits No rights defined for crop farms regarding their emissions of certain pollution loads	Florina Prefecture decision 555/90 Sanitary Regulation Y1β/2000/95 CMD of 2001 on a "Special pollution reduction programme of Lake Vegoritida and stream Soulou" set standards for pesticides and nutrients in Lake Vegoritida and stream Soulou (in the CMD, it was not yet judged necessary to further regulate diffuse pollution) Codes of Good Agricultural Practice
Tourism/ recreation	Local inhabitants	Rights to water of good quality set by policy standards	EU Bathing Directive CMD of 2001 on a "Special pollution

Use	Users	Use rights to water	Relevant policies and regulations
(bathing)			reduction programme of Lake Vegoritida and stream Soulou" set aim to safeguard aesthetic value and bathing
Fisheries	Fishermen	Rights to water of good quality set by policy standards	EU Water Directive for Fish CMD of 2001 on a "Special pollution reduction programme" set aim to safeguard fisheries as use

The regulatory framework for the protection of water quality in the Vegoritida basin consists of laws and regulations issued by the central government (largely following EU water quality policies) and by the local Prefectures.

The intended water use of Lake Vegoritida and stream Soulou was initially for "drinking water" as defined by an interprefectural decision of 1976 (decision 2667/76). Later, this intended use was revised. According to an interprefectural decision of 1979 (decision 1900/79), the quality objective of stream Soulou and Lake Vegoritida was for "bathing and any other use except for drinking". In 1987, another interprefectural decision (decision 10032/87) set quality standards according to the EU directives on quality for bathing, fish life, shellfish and drinking water which were transposed nation-wide in 1986.

Except for these early general interprefectural decisions, the Florina Prefecture also issued its own Prefectural decision (555/90), which specified in detail the emission limits applying to the disposal of wastewater and other waste into Lake Vegoritida (from polluters on its territory). This decision defined the emission limits for urban settlements, animal farms and industrial effluents within the Florina Prefecture. In detail, decision 555/90:

- Prohibited the direct disposal of any type of wastewater and effluents into lakes, rivers, streams and torrents without prior treatment.
- Prohibited underground disposal of effluents and wastewater.
- Defined maximum allowed quality standards for wastewater before their disposal into a sewage network or surface water.
- Required secondary treatment for settlements with more than 2000 p.e. and primary treatment for settlements with less than 2000 p.e. The decision emphasized the construction and modernization of wastewater treatment plants by private businesses, municipalities and communities.
- Defined maximum emission limits for wastewater and effluents from certain types of industry.

The Kozani Prefecture, which also hosts significant sources of pollution in the basin, issued no specific decision for the disposal of wastewater and effluents. This was not judged as practicable due to the great diversity of water bodies receiving effluents in the Kozani Prefecture. To issue disposal permits, the Kozani Prefecture makes use of the overall procedure of the 1965 national Sanitary Regulation as well as more recent and more specific national regulations (pers. comm. Kozani Pref., 1/12/2005).

Since 2001, also the allowed concentrations of hazardous substances in surface waters of the basin are defined in a CMD for a Special Pollution Reduction Programme for the Lake Vegoritida, Lake Petron and stream Soulou. The quality objectives of this Special Programme aim at maintaining the uses of bathing and fishing as well as restoring the aesthetic value of lakes Vegoritida and Petron. The use of the stream Soulou is restricted to the disposal of urban and industrial effluents into Lake Vegoritida. For reasons of ecology and aesthetics, a minimum level of good water quality should be maintained aiming at reducing the uncontrollable disposal of effluents into the stream.

The Vegoritida basin has not been characterized as a vulnerable zone according to the Nitrates Directive 91/676/EC. Therefore, only the Codes of Good Agricultural Practice should be implemented by farmers on a voluntary basis concerning agricultural diffuse pollution.

Lake Petron and stream Soulou, which mouth into Lake Vegoritida, were characterized in 1999 as a sensitive area of the EU Urban Wastewater Treatment Directive (91/271/EC) due to eutrophication from urban wastewater and industrial effluents (Ministerial Decision 19681/99). In a study of the EC DGXI reexamining the designation of sensitive areas across Europe, Lake Vegoritida was also proposed to be characterized as a sensitive area (Planet Northern Greece S.A & AN.FLO. S.A., 2000).[76]

7.7.2 Chronology of events

First reports on water pollution and interprefectural water quality policies

As early as 1971, local authorities around Lake Vegoritida (community of Amyndeo in the Florina Prefecture and the fisheries station of the city of Edessa in the Pella Prefecture) reported incidents of mass fish kills in the lake.

Alarmed by the situation, an Advisory Committee was set up in 1975 by representatives of all three Prefectures sharing the basin to deal with the sanitary disposal of effluents from industries and the wastewater from the city of Ptolemaida (a main city in the Prefecture of Kozani). This interprefectural Committee published a report, whereby it proposed that the highest intended water use for Lake Vegoritida and stream Soulou should be "drinking water supply and other uses". It also recognised that effluents and urban wastewater needed to be urgently treated. The Committee requested from the main polluters (industries and the city of Ptolemaida) to submit appropriate studies on the treatment of their effluents to the authorities in order to initiate the procedure of issuing effluent disposal permits according to the national 1965 Sanitary Regulation. At that stage, none of the industries held effluent disposal permits yet according to law.

As mentioned in the previous section, an interprefectural decision (2667/76) was issued in 1976 to regulate the disposal of all effluents and wastewater into stream Soulou and Lake Vegoritida, whose water was intended for "drinking water supply and other uses". The issuing

---

[76] The Master Plan of the MDEV (2003) also proposed to consider the Lake Vegoritida as sensitive recipient of urban wastewater (UWWTD 91/271/EC) and to designate its water basin as vulnerable zone (see Nitrates directive 91/676/EC).

of the decision, however, caused the reaction of local associations. They protested that such decisions did not lead to any immediate measures taken in practice. Instead, industries could further dispose effluents with the permission of the State, if they reassured authorities that they would carry out the required effluent treatment studies.

In 1978, a report by the Ministry of Social Services (Department of Hygiene) characterised the water of Lake Vegoritida as unsuitable for drinking. In the same time, a report of the Aristotle University of Thessaloniki stated that water in several parts of the lake might be unsuitable even for bathing and fishing. As a result, in 1979, the interprefectural Advisory Committee issued a revised interprefectural decision (1990/79) defining that Lake Vegoritida was no longer suitable for drinking but only for bathing and lower uses.

Start of more intensive protests on a local level

In view of the increasing water pollution in Lake Vegoritida, the lakeshore communities started more intensive protests in the late 1970s.

In 1979, the MESNA association of the lakeshore community of Arnissa started exercising pressure on public authorities. Its interest was to protect both fish populations and the health of the human population. In cooperation with the community of Arnissa, MESNA organised the first scientific conference on the Protection of Lake Vegoritida from pollution. Complaints on the lake pollution even reached national ministries (especially the Ministry of Development and the Coordination Ministry) but the central government remained rather passive. Although it acknowledged the pollution problem, it pled for more credit of time for large industries, such as the AEVAL nitrogen-fertiliser industry, to find the best (and financially sustainable) solution for combating pollution.

Recognising the need for more scientific evidence

In the early 1980s, the first legal complaints were filed by lakeshore communities against certain polluters in the basin. However, local courts found the polluters innocent due to the lack of convicting evidence (pers. comm. Association MESNA of Arnissa, 27/1/2005). Thus, lakeshore communities realised that more scientific evidence was needed on the pollution levels and invited members of the scientific community and authorities to analyse the water and fish of the lake (Interview Community Agios Panteleimon, 22/1/2005).

Additionally, in 1982, after a joint meeting of the three Prefects of the basin, a Coordination Committee was set up consisting of scientists, the Prefecture and Municipality of Pella in order to gather proposals on how to deal with the lake pollution. This Coordination Committee concluded that the most important polluters were the Municipalities of Ptolemaida and Amyndeo as well as the AEVAL nitrogen fertiliser-industry (all in the Kozani and Florina Prefectures). Moreover, there was no inspection or competent authority to deal with the lake pollution and the lack of data was critical. The Coordination Committee submitted its conclusions to an interprefectural administrative meeting in 6/1982. Therein, it was decided that a common interprefectural decision should be issued on the maximum allowed limits for effluents and competent authorities should inspect the installation of necessary treatment works.

In parallel, a Committee for the Control and Protection of the Lakes of Western Macedonia was set up, whose role among others was to determine the degree of pollution of Lake Vegoritida. This Committee examined the lake water (with funding from the Pella Prefecture) and defined the maximum allowed limits of important quality parameters for fish survival. The Committee concluded that it was possible to return Lake Vegoritida to its initial status, if pollution were limited. In this context, it was important to support the self-purification process of the lake. This Committee further recognised that pollution was also linked to the issue of the lake drop, therefore solutions should be found to abstract less water. The proposals formulated at that stage were to reduce water abstractions from the lake, reduce effluents disposal, analyse water quality on an annual basis and also investigate the quality status of Lake Petron which mouthed into Lake Vegoritida.

Florina Prefecture issues specific regulations for pollution control

After the alarming findings on the interprefectural level and delays in issuing an interprefectural decision on maximum allowed emissions, the Florina Prefecture decided to take concrete action to deal with point sources of pollution within its territory. The 1965 national Sanitary Regulation was considered too abstract to guide inspections on the local level. In 1987, the Florina Prefecture issued its own specific decision on uses of waters and conditions for wastewater and effluent disposal (decision 292/1987) (Interview State Chemical Laboratory, 14/2/2005). Later, in 1990, the Florina Prefecture issued a 2nd more elaborate decision (555/1990) on effluents and wastewater disposal (described above in section 7.7.1).

Lakeshore communities turn to EU justice

Mobilisations of the lakeshore communities on the local level continued until the late 1980s and early 1990s. When it was realized that not much was being achieved on that level, the decision was taken to start mobilizations on a national and European level (Interview Community Agios Panteleimon, 22/1/2005). In 1987, a letter of complaint was sent to the EU denouncing non-compliance of Greece with EU directives concerning the quality of Lake Vegoritida and stream Soulou. According to EU requirements, Greece should have characterised waters receiving effluents, drafted a list of substances disposed, and defined values for the parameters included in the relevant annexes of EU directives. The Greek authorities also ought to have drafted plans for the disposal of effluents and the supervision of the disposal procedure (Iosifidis & Avgitidis, 2000).

In 3/1990, the press also published information on the alleged leakage of toxic substances (clophen, PCB) from the PPC industrial installations around the lake (Newspaper "Edessaiki", 31/3/1990). This news fired off a new wave of local protests and new denunciations were filed both at local courts and the EU.[77] The PPC publicly denied accusations on causing any PCB pollution (Newspaper „Proini", 12/6/90). Shortly afterwards, the authorities announced that no PCB was found in Lake Vegoritida but traces were found in Lake Petron (Newspapers "Proini", 9/5/90, "I Proti", 14/7/1990) leading to great unrest of the local population. At this stage, under enormous local pressure for State action, the Minister of Environment announced that the construction of wastewater treatment plants had been initiated for the cities of Ptolemaida and Amyndeo.

As a result of local complaints, the EC sent Greece a reasoned opinion in 1992 for non-compliance with EU requirements for the quality of Lake Vegoritida and stream Soulou.

In 1994, the Lake Committee of the Region West Macedonia (see section 7.4.2) confirmed that water pollution impacted the ecology, tourism and fisheries of Lake Vegoritida. Whitefish decreased, crayfish disappeared and birds such as pelicans decreased or disappeared. In the same time, fish species resistant to bad water quality increased. There was need for treatment plants for all main settlements and industries in the basin as well as for water quality inspection and penalty mechanisms.

In the ensuing correspondence between the EC and the Greek authorities, the EC judged that the measures taken to improve water quality by the Greek authorities were inadequate (Iosifidis & Avgitidis, 2000) and decided in 1995 to take Greece to the European Court of Justice (ECJ).

Closure of the AEVAL nitrogen-fertiliser industry

In the meantime, in 1995/96, the AEVAL nitrogen-fertiliser industry (one of the key polluters in the basin) closed down, after a period of malfunctioning since 1992/93. However, its close-down was not due to environmental protection measures imposed but due to the high production costs resulting from its out-of-date production technology and loss of competitiveness against foreign products. A part of AEVAL belonged to the Greek Bank for Industrial Development but EU competition rules did not allow anymore State subsidies to support the industry. After 1985, there were some efforts to modernize the industry and treat its effluents, but those efforts were never completed (Interview RWM, 17/2/2005). The closure of the AEVAL industry, however, resulted in a significant reduction of pollution input into the stream Soulou and Lake Vegoritida.

Programme on the stability of water levels and purification of Lake Vegoritida

As already mentioned, a scientific programme on the "Stability and purification of Lake Vegoritida" was initiated by the Region West Macedonia in 1998 under pressure from the local society and the EU infringement procedures. The main outcomes of this programme relevant to water quality are summarized in the following.

---

[77] The PPC was denounced by the president of the community of Arnissa, the president of the agricultural cooperative of Arnissa and the association MESNA of Arnissa for the leakage of PCB at the courts of Florina and Kozani as well as to the EC (Newspaper "Proini Pellas", 30/3/1990).

Wastewater treatment (point sources)

Two pilot treatment works with the low-cost natural method of artificial anaerobic-aerobic lagoons were constructed to serve two lakeshore communities of the Florina Prefecture (Newspaper "Aggelioforos", 29/9/2001).

Additionally, a study on the treatment and disposal of urban wastewater and industrial effluents in the basin was carried out. This study was partly a reply to the request of the Environment Ministry from the three Prefectures sharing the basin, already since 8/1995, to take necessary action for the management, treatment and disposal of effluents in the Vegoritida basin.

In the context of this basin-wide management study, it was proposed to construct 22 treatment works with the method of artificial wetlands and 5 treatment works with the natural method of artificial anaerobic-aerobic lagoons for human settlements. Proposals were also made for the treatment of industrial effluents, which should be executed by the industries themselves.

The study identified three ways of possible financing of the proposed urban wastewater treatment works: i) The EU Cohesion Fund, ii) Self-financing by the municipalities concerned, and iii) The Programme of Public Investments, which is equivalent to 100% State financing. The study authors assessed that financing from the Programme of Public Investments was the most realistic option, pointing out that the Cohesion Fund would probably not be the right option due to the disperse nature of the proposed works and lack of compatibility with the Fund's technical terms. Nevertheless, an effort was made to fund some of the proposed treatment works by the Cohesion Fund and a proposal for ca. 17.6 million Euro was submitted to the EU. As foreseen, the proposal was not approved because the individual settlements in question were under the population limit set in order to fund treatment plants (Interview LVPS, 22/1/2005).[78]

The lake stability and purification programme however assisted in bringing the treatment works for the Municipality of Amyndeo forward involving the construction of a sewage network for three surrounding settlements and the completion of the network connection to a pre-existing, non-operational treatment plant. To this aim, a study on the operational completion of the Amyndeo treatment works was carried out. After the rejection of the funding application to the Cohesion Fund, the project on the Amyndeo (tertiary) treatment works was promoted for funding to the Operational Programme of the Region West Macedonia. To this aim, cooperation between the Municipality of Amyndeo and the Region developed (Interview RWM, 15/2/2005).

In summary, the lake programme had a few positive outcomes in terms of combating point sources of pollution in the basin. Nevertheless, the problem of point pollution still remains

---

[78] According to a letter of complaint of the Vegoritida Society to the EC, the Hellenic Government classified environmental construction projects in the area (sewage systems and wastewater treatment plant) as least urgent (category C) for funding from the Cohesion Fund.

partly unresolved. Except for the still untreated wastewater of most lakeshore settlements, there is also wastewater generated in many small settlements around the city of Ptolemaida. The first step taken, in this respect, so far was to complete a study on the connection of these small settlements to the tertiary treatment plant of Ptolemaida (Interview RWM, 15/2/2005). Additionally, until the end of the case study, it still remained a challenge to adequately treat and control the effluents of industries, including the industrial installations of the PPC. Industrial treatment plants, in operation or planned, still needed to be adequately controlled, monitored and maintained (Interview RWM, 15/2/2005).

Agricultural non-point pollution

A 2-year pilot programme was implemented indicating methods for the reduction of nitrogen inputs from agriculture. However, there was lack of finance to continue this effort beyond the pilot phase. To make further progress in the future, except for funding, it would also be necessary to gain the agreement of individual farmers and of agricultural cooperatives to cooperate with the support of the Prefectural Divisions for Land Reclamation (Interview RWM, 15/2/2005).

Conviction by the ECJ and policy reaction by the Greek State

On 11/6/1998, the ECJ convicted the Hellenic Republic for failing to fulfil its obligations under the EU Directive 76/464/EEC on pollution caused by hazardous substances in Lake Vegoritida and stream Soulou. In more detail, the Greek State had failed to establish programmes, including quality objectives and deadlines for their implementation, in order to reduce the pollution of Lake Vegoritida and stream Soulou by dangerous substances of List II of the Directive.

In 12/1998, the EC sent a letter of formal notice to the Hellenic Republic for failure to comply with the ECJ judgment. In 1/1999 and 3/1999, the Greek State communicated to the EC information concerning measures taken in order to comply with the judgment. The reply of Greece was based on a study by the Aegean University on pollution of Lake Vegoritida with toxic substances of List II of Directive 76/464/EEC, the completion of a monitoring network in 1998 and the lake stability and purification programme. The EC, however, did not judge that the measures reported by the Greek State (limited to studies and monitoring) constituted "programmes including quality objectives and setting deadlines for their implementation to reduce pollution from hazardous substances". Thus, in 9/2000, the EC sent a reasoned opinion to Greece for non-compliance with the 1998 ECJ decision, warning Greece that a second verdict would be translated into financial penalty.

The reasoned opinion of the EC coincided with the finalisation of the lake stability and purification programme. Therefore, the Environment Ministry replied to the EC including elaborate scientific data collected during the programme in order to avoid the financial penalty. In parallel, a Common Ministerial Decision (CMD) was issued in 6/2001 (CMD 15782/1849) adopting a Special Programme to reduce pollution of Lake Vegoritida, Lake Petron and the stream Soulou from hazardous substances. The CMD defined water quality objectives and set deadlines for their implementation to reduce pollution. Concrete aims of this special programme were to determine pollutants, set quality goals, describe a monitoring programme, approve and reexamine terms of permits for effluent and wastewater disposal in the basin and issue special regulations for pollution from non-point agricultural sources (esp. pesticides).

Indeed, after the adoption of the CMD on a Special Pollution Reduction Programme, the EC decided that the Hellenic Republic had taken the necessary measures in order to comply with the ECJ judgment. As a result, in 12/2001, the EC closed the ECJ case on Vegoritida and Soulou.

As required by the CMD, the Environment Ministry set up in 2002 a Supervisory Committee to monitor progress in implementing the Special Programme for lakes Vegoritida, Petron and stream Soulou. The Committee included members from ministries (of Environment, of Development, of Agriculture and of Health), from the Environment Directorates of the Region West Macedonia and the Region Central Macedonia, from relevant divisions of the three Prefectures of the basin, from the municipal enterprises for water supply and wastewater and municipal authorities. This Committee was actually activated, when the non-governmental Vegoritida Society sent a letter of renewed complaints to the EC for non-compliance of Greece with the 1998 decision of the ECJ. Pressure was thus created to activate the Committee in order to deal with this letter of complaint and to draft a report to the EC on the steps that Greece had taken after its conviction.

Chronology summary

From the 1970s until the end of the 1980s, the rivalry over water pollution in the Vegoritida basin was characterised by local action (legal complaints or social pressure) against polluters. In the end of the 1980s, the lakeshore communities resorted to the EU justice. The resulting case before the ECJ in combination with strong social pressure for action mobilised the authorities. In 1994, the first urban wastewater treatment plant in the basin started to operate for the city of Ptolemaida. In 1998, the programme on the stability and purification of Lake Vegoritida was funded by the Region West Macedonia resulting also in the construction of pilot treatment works for two lakeshore communities. In 2001, as a result of the conviction by the ECJ, a CMD was issued by the Greek State establishing a special programme for the reduction of pollution in Lake Vegoritida, Lake Petron and stream Soulou from hazardous substances.

7.7.3   Actor analysis

Table 7.11 gives an overview of the roles played by the main actors in the rivalry over declined water quality, their problem perception, their objectives and resources. Table 7.12 illustrates key relationships between the main rivalry actors, focusing thereby on the development of more or less cooperation observed over the studied process. In many cases, existing relationships (of good or of poor cooperation) developed to a status where no interaction took place anymore. This was, in most cases, due to the partial improvement of the water quality of Lake Vegoritida, which reduced the need for interaction between several actors. The main actor relationships, where conflict remained or developed anew, concerned non-compliant polluters (mainly industrial polluters including the PPC and other industries) and the authorities implementing water quality regulations (see also emphasis in framed cells in Table 7.12). Explanations for these conflictual relationship are given in later sections, especially in section 7.9.2 dealing with the effectiveness of implementation processes of measures and policies. Suggestions on how the actor cooperation potential could improve are made in section 7.9.2 but also indirectly in the synthesis chapter 9.

Table 7.11 Vegoritida rivalry on declined water quality: Problem perceptions, roles, objectives and resources of actors

Actors	Problem perception & roles played in the process *(detailed institutional competence in* Table 7.4*)*	Objectives & resources
EU	**Problem:** The non-compliance of Greece with the requirements of EU water quality directives for Lake Vegoritida and stream Soulou **Role:** It exercised pressure on the Greek State to take action through the ECJ case It rejected the funding application of Greek authorities to the Cohesion Fund for wastewater treatment works in lakeshore communities	**Objectives:** To force Greece to adequately implement EU Directives and establish a special programme for the reduction of pollution in the basin **Resources:** Legal, economic
Ministry of Environment, Physical Planning and Public Works (MEPPW)	**Role:** Main national authority responsible for the correspondence of the Greek State with the EC in the ECJ case	**Resources:** Political/policy-making
Ministry of Development (MDEV)	**Role:** It received complaints concerning the pollution loads of local industries	**Objectives:** To protect interests of industrial users
Region West Macedonia Managing Authority (RWM/MA)	**Problem:** Lack of coordination between different administrative levels around the lake Water quality and quantity are interconnected and must be managed together The Florina Prefecture should become more active and use the lake stability and purification programme results to impose measures on users **Role:** It activated and coordinated the first 1994 participatory committee on Lake Vegoritida (motivation of political leadership) It initiated the lake stability and purification programme It cooperated with local authorities to achieve progress on local treatment works (e.g. with the Municipality of Amyndeo)	**Objectives:** To relax pressure from the ECJ and local society To progress the designation of the basin as natural protected area To progress local treatment of wastewater in human settlements of the basin **Resources:** Economic, political/planning
Regional Directorate for Environment and Physical Planning of West Macedonia (RWM/RDEPP)	**Problem:** Lack of adequate control of the environmental terms approved for water polluting activities was partly due to the overlap of inspection competence of different authorities The PPC started making an effort to reduce the environmental impact of its activities in the basin, especially after receiving approval of the	**Objectives:** To proceed with the designation of the basin as nature protected area To achieve better inspection of environmental terms approved in EIAs **Resources:** Political/regulatory, technical/EIA approval, but inadequate

Actors	Problem perception & roles played in the process *(detailed institutional competence in Table 7.4)*	Objectives & resources
	environmental terms for its operation in the basin. Nevertheless, it will take considerable time for the PPC behaviour to change in practice **Role:** Participant in Supervisory Committee of the 2001 CMD on a pollution reduction programme	human resources
Florina Prefecture (PoF) - especially the State Chemical Laboratory and its Division of Health	**Problem:** Competent Prefectural authorities have incomplete sanctioning power to control polluters; final approval of sanctions should be done by competent authorities and not by elected politicians **Role:** Participated in interprefectural common action to control pollution in the basin since the 1970s It promoted its own water pollution control policies	**Resources:** Political/regulatory, technical but lacking economic resources
Prefecture of Kozani (PoK) - especially its Division for Environmental Protection and Division of Health	**Problem:** It is difficult to control the polluting activities of the PPC, which take place to their greatest part on the territory of the Kozani Prefecture **Role:** The 1975 interprefectural Advisory Committee on the pollution of the basin was set up by decision of the Prefect of Kozani The Kozani Prefecture intensified pressure on the PPC to reduce its environmental impacts after setting up a Unit for the Control of Environmental Quality in 2004	**Resources:** Political/regulatory, technical but lacking economic resources
Prefecture of Pella (PoP)	**Problem:** Pollution of Lake Vegoritida originates mainly from sources outside its territory (mainly in the other two Prefectures of the basin) except for wastewater pollution which also comes from its own lakeshore communities **Role:** It participated in interprefectural common action to control pollution in the basin since the 1970s and funded initial studies to collect data and propose anti-pollution measures	**Objectives:** To improve the water quality of Lake Vegoritida, which used to have a high amenity and fisheries value for its population **Resources:** Political/regulatory
Municipalities and communities (Mun/comm)	**Problem:** Industrial development was the focus in the area for too long; future focus should be on tourism and environment values Problem of urban wastewater pollution caused by municipalities and communities is acknowledged Most problems of the lake cannot be resolved unless a management body is set up Environmental terms in approved EIAs for the PPC activities are not adequately implemented **Role:** Local authorities were strongly mobilised to protect the lake, initially on an individual basis and then as coalition Together with local associations and private individuals, they complained to the EC	**Objectives:** To get external funding for urban wastewater treatment works To improve water quality for the development of (eco)tourism and the return of fisheries in the lake **Resources:** Lacking economic resources. Use of legal resources (locally, regionally and EU level), political pressure (interaction with higher administrative levels, pressure of local MPs to bring issues to the upper policy arena) and of technical resources (external scientific evidence)

Actors	Problem perception & roles played in the process *(detailed institutional competence in Table 7.4)*	Objectives & resources
	culminating to the ECJ case	
	The Municipality of Amyndeo was active in promoting the ecological and touristic value of the lakes Vegoritida and Petron	
Industry (including the PPC, AEVAL and others)	**Problem:** Not all industries acknowledged their responsibility for the pollution impact on water resources. The PPC acknowledged the water pollution problem from its steam-electricity plants, but rejected accusations of PCB leakage incident in 1990 **Role:** Industrial polluters (especially AEVAL but then also the PPC and other industries) were a key target of campaigns of the lakeshore population to protect the lake since the 1970s	**Objectives:** To continue with business as usual with regard to effluents disposal (and keep production costs low). The PPC especially aimed at securing energy production at low cost **Resources:** Political pressure through the financial power of industries and their labour offer
Local associations: Lake Vegoritida Preservation Society (LVPS) and the MESNA association of Arnissa	**Problem:** Failure of public authorities to enforce water quality regulations for industries, the PPC and human settlements **Role:** MESNA played a role in the initiation of the complaint to the EC on the pollution of Lake Vegoritida and stream Soulou The LVPS complained again to the EC in 2002 on the non-compliance of Greece with the 1998 ECJ decision	**Objectives:** To pressurise authorities for the enforcement of water quality regulations To achieve wastewater treatment for all lakeshore communities **Resources:** Legal (local courts, EU), social support, technical/cognitive (public information campaigns) but LVPS lacks knowledge of policies and network of partner organisations
External experts	**Role:** Conducted studies and research projects on the water resources of the basin	**Resources:** Technical and scientific expertise

Table 7.12 Vegoritida rivalry on declined water quality: Relationships among key actors in the process

Actors	Administration								Users		NGOs	Experts
	EU	MEPPW	MDEV	RWM/MA	RWM/RDEPP	PoF	PoK	PoP	Mun/comm	Industry	LVPS/MESNA	experts
EU	▨								📖→/0		📖→/(📖→)	
MEPPW	→/0	▨							→/0		→/(↓)	
MDEV		0/+	▨						→/0	+/+	→/0	
RWM/MA			0/+	▨					0/+			📖/0
RWM/RDEPP		0/+		+/+	▨							📖/0
PoF		0/+		++/+	+/+	▨				−/−	→/0	
PoK		0/+				+/+	▨			−/−	→/0	
PoP		0/+				+/+	+/+	▨	+/0		→/0	
Mun/comm		0/+		€/€				+/0	▨	−/0	++/+	📖/0
Industry			+/+			−/−	−/−		−/0	▨	−/0	
LVPS/MESNA									++/+		▨	
Experts									📖/0			▨

Note (a): The table should be read as the relationship of the actor in the vertical column to each of the actors in the horizontal rows. Blank cells indicate no particular relation observed. Actor abbreviations are explained in Table 7.11.

Note (b): Symbols in the table are explained in the legend below. The presence of symbols left and right of the slash (/) aim at indicating noteworthy changes in actor relationships over the observed process. If left and right symbols are identical, no significant change in the actor relationship was observed. If symbols are different, a certain change took place. For instance, symbols (++/+) indicate close collaboration at first, which was reduced over time. Symbols (–/+) indicate an initial conflict which was followed by certain improvement (formal interaction or sporadic collaboration).

LEGEND OF SYMBOLS:

++	close collaboration (problem-solving interaction)	📖	information supply or exchange
+	only formal relationship or sporadic collaboration	€	financial contribution
0	no noteworthy interaction observed	→	One-sided pressure
–	conflict		

7.8 Process of water basin regime change

This section discusses change in the institutional water regime of the Vegoritida basin, based on the analytical framework presented in chapter 4/4.1.2. The main relevant hypothesis (Hypothesis 2) was that:

> "The less favourable the contextual conditions and the characteristics of the interacting actors for integration attempts, the more likely that Greek water basin regimes fail to reach a more integrated phase."

7.8.1 Dimensions of regime change

7.8.1.1 Extent

An overview of the water uses considered by the Vegoritida water basin regime is given in Table 7.13.

As far as uses which affect water quantity are concerned, the regime extent increased since the 1980s. Water use for irrigation started being regulated in all Prefectures at the latest with the introduction of the 1987 Water Law. Water use for industry also started being regulated with the introduction of the 1987 Water Law but the PPC was exempt from the requirement of permits for the construction of new water works. The regime potential to regulate new PPC abstractions substantially increased after the semi-privatisation of the PPC in 2001. Fisheries and the aesthetic value of raised water levels were not explicitly recognised by the water basin regime from a quantitative perspective. Nonetheless, the aesthetic value of Lake Vegoritida appeared to be an issue of concern, at least in the context of initial discussions to officially define a lake level. However, nature and the water needs of the environment remained undervalued on the policy level due to the lack of the formal designation of the basin as a nature protected area (according to the 1986 national Environment Law). The requirement of the 1987 national Water Law for minimum lake levels and river flows was not translated into a regional regulation either. Towards the end of the case study process, the living environment and nature started being partly considered by the regime. Ecological values came onto the agenda of the lake level definition in the context of a proposed study on the ecological sustainability of the lake ecosystem. This turn in the process came about to fulfill requirements of nature protection policy (EU Natura 2000 network) and not requirements of water policy.

As concerns uses affecting water quality, in the 1970s and 1980s, first prefectural decisions regulated the disposal of urban and industrial wastewater in the basin. Fishing, aesthetic value and bathing gained increased attention in water quality policy aims through the objectives set by the 2001 CMD on hazardous substances pollution in Lake Vegoritida, Lake Petron and stream Soulou. This CMD of 2001 also gave some hints for environmental protection by recognising the aquatic environment as beneficiary of anti-pollution measures. All in all, although reference to the environment is made in several relevant regulations applying to the basin e.g. EU directives for hazardous substances and urban wastewater treatment, no explicit water quality standards were set to safeguard nature. The regime also remained for most of the process incomplete with regard to regulations on diffuse pollution from agriculture. The CMD of 2001 on pollution reduction considered the sole voluntary application of codes of good agricultural practice as sufficient in this respect. However, later in the process (in 2006), the agrienvironmental policy measure adopted for the Vegoritida basin gave incentives for the reduction of water pollution from agriculture.

Table 7.13   Extent of uses in the Vegoritida basin regime

Water uses	Presence in the water quantity domain	Presence in the water quality domain
Agriculture/ irrigation	Regulated through Prefectural restrictive measures since 1989 (the Pella Prefecture imposed restrictions on irrigation wells since 1982). The agrienvironmental policy measure of 2006 for the area around the lakes of the Vegoritida basin also offered voluntary instruments to reduce water consumption for irrigation Prior to 1989, agricultural use was regulated only through authorisations for the construction of wells and connection to power supply	Eutrophication problem from use of pesticides and fertilisers was considered in the quality objectives of the CMD of 2001 on pollution reduction in Lake Vegoritida, Lake Petron and stream Soulou, but it was not judged necessary to set specific policy regulations and measures for the reduction of agricultural non-point pollution The agrienvironmental policy measure of 2006 for the area around the lakes of the Vegoritida basin offered voluntary instruments to reduce water pollution from agriculture
Urban use	Regulated through Prefectural restrictive measures since 1989 Prior to 1989, regional offices for land reclamation and the Local Organisations for Land Reclamation gave authorisations on drilling or extending wells for non-agricultural use	Drinking water supply was prohibited from Lake Vegoritida in 1978 A series of interprefectural and prefectural decisions regulate the disposal of urban wastewater Lake Petron and stream Soulou were characterized as sensitive area of the EU UWWTD 91/271/EC
Industry	Regulated through Prefectural restrictive measures since 1989 Prior to 1989, regional offices for land reclamation and the Local Organisations for Land Reclamation gave authorisations on drilling or extending wells for non-agricultural use as well as concessions to use surface water for industrial uses	A series of interprefectural and prefectural decisions regulate the disposal of industrial effluents The CMD of 2001 on pollution reduction in Lake Vegoritida, Lake Petron and stream Soulou set new quality standards for industries
Fisheries	Not included in the water quantity domain	Included in the water quality domain by national water quality standards set for fishing
Environment	Ecological values of water resources not yet protected by the regime domain since the basin was not designated as natural protected area	
	Nature demands were raised in the discussions on the lake level definition towards the end of the case study process	Good environmental quality standards set in national legislation to meet the uses of the lake, but no explicit quality standards set for safeguarding nature
Recreation	Not included in the water quantity domain	Included in the water quality domain by water quality standards set for bathing. Also the CMD of 2001 aimed at safeguarding (among others) bathing and the lake aesthetic value

7.8.1.2 Coherence of property rights

In terms of the rivalry on the dropping water levels in the basin, internal coherence of property rights remained low over the entire case study process.

Use rights to the water of Lake Vegoritida were conceded to the PPC early on in the process by the central government (initially for the production of hydropower and later for the operation of the steam-electricity plants). In the same time, farmers also had rights to water for irrigation. In detail, farmers of the Skydra plains (neighbouring basin) were entitled to water from Lake Vegoritida, which was abstracted by the PPC and transported to their irrigation networks (with the permission of the Ministry of Agriculture). In the same time, farmers on the lakeshore of Vegoritida had rights to groundwater (through their land ownership). Since groundwater is linked to the level of Lake Vegoritida, the water discharge capacity in wells of lakeshore farmers was adversely affected by the lake abstractions for the PPC and for the Skydra plains. After the PPC stopped abstractions for its own hydropower purposes and only abstracted water to irrigate the Skydra plains, the rivalry between farmers of the two basins escalated in 1989/1990 during extremely dry climatic conditions. The rivalry was resolved with the intervention of the Ministry of Agriculture which cancelled the use rights of the Skydra farmers to the water of the lake (in the same time, providing them with alternative groundwater sources within their own basin). At that stage, this State intervention improved the internal coherence of the property rights system among irrigation users.

Nonetheless, the lingering rivalry between the PPC and other water users (farmers and local communities) after 1990 reveals that the abstraction rights of heterogeneous users continued to contradict each other (incoherent property rights), due to the lack of coordinated cross-sectoral planning. The overarching right of the central government to allow the activity and water abstractions of the PPC in the basin conflicted with:

a) The drinking water rights of local communities and municipalities (authorisation to use water initially awarded by the government and then by the Local Union of Municipalities and Communities).

b) The use rights given to farmers by the Prefectures (water permit issuing authority for use in irrigation).

Secondly, a clear allocation of water rights was impeded by the fact that many illegal irrigation wells have been drilled in the basin, especially in the Florina Prefecture (see also section 7.9.2.1 for the insufficient implementation of relevant regulations). In this context, the result is conflict between farmers in the Pella Prefecture, who have drilled few illegal wells, and farmers in the Florina Prefecture who have drilled the majority of illegal wells.

As far as the rivalry on water pollution in the basin is concerned, the internal coherence of rights to pollute water improved on formal policy grounds since the 1970s thanks to several pollution regulations and the establishment of the effluent disposal permit system. Thus, in principle, the effluent disposal permit system offered the potential for a clear allocation of pollution rights to the different users. Pollution rights were allocated in the form of effluent disposal permits to most industrial users in the basin and to the key urban polluter of the city of Ptolemaida. Towards the end of the case study, also the uses of fishing, bathing and aesthetic value were given "good water quality rights" by the 2001 CMD on pollution reduction. However, in practice internal coherence of rights to pollute water remained low, which is obvious from the lingering rivalry over decreased water quality. Contrary to the requirements of regulations, many users still have no formal use rights to pollute (especially urban users and animal farms which have not applied for effluent disposal permits). Even polluters holding permits to pollute water (especially industries) often exceed the pollution limits set in the permits, at the expense of the "good water quality rights" of other uses (see section 7.9.2 for a discussion of inadequate implementation of water protection regulations in the Vegoritida basin).

7.8.1.3 Coherence of public governance

Governance levels and scales

In general terms, the levels of water management administration in the Vegoritida basin do not fit with the natural basin scale. Competence for water management in the basin is shared between the central government, two Regions and three Prefectures which often followed in the past different prefectural water policies. Although this administrative split-up indicates a need for continuous communication and high coordination among the Regional, prefectural and central government services, the water basin management process was characterised by administrative confusion (Interview Pella Pref., 1/2/2005) and lack of clarity in decision-making and implementation (Interview Municipality Amyndeo, 16/2/2005). Several actors involved considered the absence of an administrative-unifying management body as the main obstacle for a holistic approach to the management of the basin. The administrative split-up also resulted in insufficient water data exchange between authorities and the partial character of several research projects. For instance, the lake stability and purification programme achieved thematic but not spatial integration on the level of the basin, since most actions were taken only on the part of the basin in the Florina Prefecture.

As far as water quantity policy is concerned, competence was in the hands of the three individual Prefectures sharing the basin. Some improvement in the fit between administrative levels and the basin scale took place after the adoption of the 1987 Water Law. The setting up of the first RWMDs marked the start of attempts to coordinate administrative levels for quantitative management on the basin scale. In 1997, the RWMD of West Macedonia was set up with competence over the entire Vegoritida basin for issues of water quantity (before that, the RWMD of Central Macedonia had taken up this role). However, over the case study process, the RWMD had a more bureaucratic rather than management role and it was mainly active in making proposals for Prefectural abstraction restrictive regulations and in the approval of water permits. In the past, there was also difficulty in the communication and understanding between the Prefectures and the RWMD, as well as between the Prefectures themselves concerning restrictive measures (lack of intrapolicy coordination of actors). Over time, the RWMD became more widely accepted by the Prefectures and, by the end of the case study, coordination only remained poor with the Kozani Prefecture. There was also improvement in the coordination between Prefectures for harmonising restrictive measures on water quantity and for reaching an agreement on a lake level definition.

As far as water quality policy is concerned, competence was also in the hands of the three individual Prefectures. There were early attempts for interprefectural coordination in water quality protection (much earlier than for the water levels issue). Namely, an interprefectural committee was set up in the 1970s to coordinate the three Prefectures on pollution issues. Several interprefectural decisions in the 1970s were the result of this attempt. Since the late 1970s, no significant interprefectural interaction and coordination followed, until this was re-animated from a more top-down level. Namely, the implementation of the centrally-adopted CMD of 2001 on hazardous substances pollution is supervised by a Committee involving all three Prefectures, central ministries and the Regions.

Despite problems in the degree of "fit" between administrative and natural basin scales, it is worth noting that both discussed rivalries in the Vegoritida basin eventually became multi-level processes involving actors from different levels.

At the start of the rivalry on dropping water levels, the Prefectures defined their own water policy independent of each other. After the involvement of the Region, the RWMDs became the main policy determinants for Prefectural policies on the basin level. During the rivalry process, there was further official interaction between the central (Ministry of Development) and regional level (RWMD) and between the Region (RWMD) and the Prefectures. However, the local level (Prefectures and municipalities) mainly had one-way interactions with higher levels, whereby local actors exercised pressure on ministries to limit PPC abstractions in the basin.

In the rivalry on water pollution, the levels mainly involved were initially municipalities and Prefectures and later the central level (Environment Ministry), the Region and the EU as far as the EU infringement procedure is concerned. Before the involvement of the Region in pollution issues, the central State interacted directly with Prefectures on the water quality issue. The EU infringement procedure was actually the only case of a (formal) multi-level process involving all levels. The EU, the Environment Ministry and also the Region entered as new levels of governance in the process following the complaint of local actors to the EU. Local actors (local authorities and local associations) were only involved in as far as they used the EU as a legal resource.

Actors in the policy network

The policy arena on water management issues in the case study was only open to actors of the public administration. There were no official participatory arrangements allowing the participation and consultation with actors outside the administration (at least no functional participatory arrangements). Even the interactive discussions, towards the end of the case study, achieving informal consensus between the Region West Macedonia, the Pella and Florina Prefectures over the proposed Lake Vegoritida level at 515.5 masl was limited to actors of the administration excluding other stakeholders.

Although some attempts for participatory arrangements took place (most of which involved only the administration), these were short-lived, thereby hindering the creation of strong actor networks. Participation efforts were partly hindered by the lack of an appropriate legal framework to ensure continuity.

- The 1994 Lake Committee set up by the Region West Macedonia, which was the only participatory approach including water users, had very limited duration (1994-95).

- The Supervisory Committee of the lake stability and purification programme was again of limited lifetime (1998-2000) and only involved officials of the Florina Prefecture and the Region West Macedonia.

- The extended administrative Committee with advisory role on the lake stability and purification programme, which was set up in 2000, was an attempt to coordinate all administrative levels in the Vegoritida basin on issues of research and infrastructure. This extended Committee, however, did not become operational. Progress was hindered by the fact that there were no practical ways to implement coordinated action. There was no administrative authority with competence over the entire Vegoritida basin to take the initiative of coordination (Interview Planet, 8/2/2005).

- Some additional participatory tools (again only on the level of the administration) were used on the issue of water pollution, namely the Interprefectural Committee on pollution in the basin active in the 1970s and the Supervisory Committee for the implementation of the CMD on pollution reduction set up in 2002. The Supervisory Committee of the CMD of 2001 enables increasing interaction between public bodies to comply with the CMD requirements (which were originally imposed by EU requirements).

- The Regional Water Committee required by the 1987 Water Law on the level of the water district of West Macedonia was set up only on paper in 1999[79]. By the end of the case study process, this Committee was still not activated.

Actors outside the administration were involved in the process mainly on an informal basis and on their own initiative to try and influence the regime change process. In the rivalry on the dropping water levels, the main non-administrative actors involved were local associations (the MESNA association of Arnissa and the Vegoritida Society with informal links to public actors and local politicians), the Arnissa agricultural cooperative (with links to its farmer-members, MESNA and Vegoritida Society), the PPC (with strong links to the State) and the national e-NGO Ornithological Society (with links to the Environment Ministry, the Florina Prefecture and the Municipality of Amyndeo). In the rivalry on water quality degradation, the main non-administrative actors involved were local associations (MESNA and Vegoritida Society) and the main industrial polluters (PPC and the nitrogen-fertiliser industry AEVAL).

---

[79] With the following participants: General Secretaries of four Regions sharing the water district (West Macedonia, Central Macedonia, Thessalia and Epiros), Regional Water Resources Directorates of the four Regions, Regional Divisions for Land Reclamation of the four regions, nine Prefects, representatives of Prefectural Councils, unions of agricultural cooperatives, Local Union of Municipalities and Communities, Technical Chamber of Greece, Geotechnical Chamber of Greece, municipal enterprises for water supply and sewage and (upon invitation) the PPC or the National Institute of Geological Research.

Over the case study process, an increase of the number of actors involved was observed. The actor constellation especially grew to include actors engaged in the preservation of Lake Vegoritida:

- The local association of the Vegoritida Society was set up in 2001. Although it had no access to formal water decision-making, its involvement was considered by public actors (especially the Region West Macedonia) as a positive development in the process. The Vegoritida Society was, however, denied member status in the Supervisory Committee of the CMD of 2001 on pollution reduction, although the CMD allows membership to non-governmental organisations with environmental activity in the basin.

- The MESNA association of Arnissa, which was active already since the 1970s, was involved formally only in the 1994 Lake Committee of the Region West Macedonia.

- The Ornithological Society and the Environment Ministry became involved towards the end of the process on the issue of the lake level definition and the cultivation of exposed land. The Ornithological Society has no access to formal decision-making in the basin either but it is influential via its good links to the Environment Ministry, the Municipality of Amyndeo and the Florina Prefecture.

All in all, substantial communication networks between actors could not be established in the basin. There was also lack of interaction and networking between local and national NGOs engaged in the preservation of the lake, namely between the local Vegoritida Society and the e-NGO Ornithological Society.

The PPC, especially, was a rather isolated actor with links only to the central State for most of the case study process. Towards the end of the process, a formal link was created between the Region West Macedonia and the PPC after the semi-privatisation of the PPC in 2001. Since 2001, the PPC must request authorisation for all water abstraction works from the RWMD.

Perspectives and objectives

By the end of the case study, no official management plan was produced for the Vegoritida basin. The only progress in this direction was achieved on the level of research studies, which were however not rooted in policy. The programme for the stability and purification of Lake Vegoritida, which was financed by the Region West Macedonia in 1998, was the only development resembling a water vision for the Vegoritida basin by the end of the case study. For issues of pollution, in particular, a comprehensive study on the treatment and disposal of urban wastewater and industrial effluents in the entire basin was carried out in the context of the programme. A drawback of the programme was that it only involved the Region West Macedonia and the Florina Prefecture, failing therefore to incorporate the perspectives of multiple actors from the whole basin. The CMD of 2001 on pollution reduction, which was partly based on the lake programme results, was also a type of water quality policy vision for the whole basin.

Towards the end of the case study, the preparation of an official water management plan for the entire water district of Western Macedonia began. Its preparation was assigned to consultants by the Ministry of Development and was meant to follow the policy principles of the 1987 national Water Law and the EU WFD. From within the Vegoritida basin, only the RWMD formally participated in the Supervisory Committee of this management plan. Although coordination with Prefectural authorities in the basin was announced, officials in the Prefectures of Florina and Pella complained that there was very sparse coordination with the appointed consultants on this issue (Interview Florina Pref., 26/1/2005; Interview Pella Pref., 1/2/2005).

Reduced coherence in the public governance element of "perspectives and objectives" also resulted from the incomplete recognition of all important rivalries in policy visions and objectives for the Vegoritida basin. Namely, the Prefectural restrictive measures to protect water quantity mainly targeted the use of irrigation. They targeted also uses of industry and water supply but could not impose restrictions on the abstractions of the PPC, which were a source of important rivalries with other users. By the end of the case study, also the conflict of mining and energy policy with environmental and water protection policies was still not recognised in policy (lack of interpolicy coordination). Only a framework policy document for spatial planning and sustainable development of the Region West Macedonia (Environment Ministry Decision 26295, Official Government Journal B 1472/09.10.2003) recognised the rivalry between mining, energy production and agricultural development (due to impacts of aquifer destruction on irrigation). More progress could be observed in the future on the coordination between agricultural and water protection policies, through the implementation of an agrienvironmental policy measure issued for the basin in 2006.

Strategies and instruments

In the context of the rivalry on dropping water levels, the main instrument applied to influence use rights were the water permits issued by the Prefectures and the Region West Macedonia (command-and-control instrument). This type of instrument was not used to redistribute use rights for conflict resolution but was rather used to hinder new abstractions, especially for the use of irrigation. For most of the case study process, this instrument provided no opportunities to authorities for restricting the abstractions of the PPC. Overall, the effectiveness of this instrument was reduced by its poor implementation. Towards the end of the case study, a more flexible instrument was adopted, namely an agrienvironmental policy measure providing incentives for the reduction of water consumption for irrigation (by changing water-demanding crops to non-water-demanding types).

In the context of the rivalry on water quality degradation, the main instruments applied to restrict pollution were effluent disposal permits, also of command-and-control nature. Some other more indirect instruments used included public financing for urban wastewater treatment plants, EIAs on industrial activities (e.g. used as an instrument to require in 2003 the construction of a wastewater treatment plant in the PPC South Mine) and judicial mechanisms in local and EU courts.

In general, very few instruments based on official policies were made available to resolve water problems in the basin. Instead, strategies followed involved mainly investment in research, such as the lake stability and purification programme.

Responsibilities and resources for policy implementation

Most public authorities competent in water management issues in the Prefectures and the Region West Macedonia in the Vegoritida basin, lacked qualified and/or sufficient in number personnel. Additionally, the authority given to public actors for policy implementation was in some cases inadequate.

Firstly, the Prefectural authorities issuing water abstraction permits and applying the Prefectural restrictive measures to protect water quantity had no complete sanctioning authority in case of policy violation. In case of sanctions, water users have the right to appeal to courts. In court, the implementing authorities could then only make limited use of their knowledge and expertise on the water problems of the basin, in as far as relevant information was requested by the judge. Secondly, the Prefectural authorities issuing effluent disposal permits also had no complete sanctioning authority. They had to submit sanctioning proposals concerning non-compliant polluters for approval by the elected Prefect.

Thirdly, the RWMD of West Macedonia, which served as a policy-harmonising authority for water quantity issues on a basin level, was given no similar competence in water quality issues. In fact, there was a more general problem of responsibility separation on water quality and water quantity between different authorities on all levels of the administration (the Prefectures, the Region and the central government). Thus, in the Vegoritida basin, although some integration of quality and quantity aspects was achieved on the level of scientific studies (lake stability and purification programme), these two aspects of water management remained separate in terms of policy implementation competence. Some improvement was noticed towards the end of the case study, whereby the RWMD of the Region was bound to acquire competence both in water quantity and in quality issues by the 2003 Water Law.

Furthermore, for most of the case study process, responsibility for water quantitative use control was split up between different authorities according to water use. The central State maintained for a long time competence over abstractions of the PPC, while the Region's respective competence was not really activated until the semi-privatisation of the PPC in 2001. The Prefectures had competence over abstractions for irrigation. The municipal level (union of municipalities) had competence for water supply abstractions. Issues of ecosystem protection and nature (in terms of water levels) were in the hands of the central State (Ministries of Development and of Environment). The RWMD of West Macedonia had mostly coordination competence, while local authorities were mainly responsible for compliance-control. According to the new Water Law of 2003, management competence for all water uses and responsibility to issue all water abstraction permits is transferred to the RWMDs.

Concerning responsibilities for water pollution control, overlap of competence and insufficient coordination was reported between the Region West Macedonia and the Prefectures (especially the Kozani Prefecture) concerning the inspection of the environmental terms set in EIAs for polluting activities in the basin. Specifically, inspection competence was fragmented between relevant Prefectural divisions and the Regional Directorate for Environment and Spatial Planning. After the set up of a Unit for the Control of Environmental Quality in the Kozani Prefecture in 2001, this competence fragmentation decreased when this Unit took over the task of environmental inspections.

7.8.1.4 *External coherence between property rights and public governance*

External coherence between public governance and property rights in the context of the rivalry on dropping water levels remained low over the case study process. On the one hand, prefectural restrictive measures to protect water quantity restricted the use and ownership rights of farmers to irrigate (mainly by restricting the drilling of new wells and prohibiting direct abstraction from Lake Vegoritida). On the other hand, during most of the case study process, the policy system was not effective in limiting the PPC "rights" in the basin. This ineffectiveness is linked to the basin being an area of national priority for energy issues and the PPC being part of the public sector. Although the PPC was obliged to apply for water use permits for new abstractions after 1989, there was in practice absence of specific permits for this water user (see also implementation issues in section 7.8.3). Instead, halting the direct abstraction of the PPC from Lake Vegoritida in 1997 was the result of the PPC own strategic planning to overcome the impact of the dropping lake on its energy production and not of any water regime (policy) changes. Since the semi-privatisation of the PPC in 2001, the PPC is obliged to apply for both water use and water work permits, which can be translated as an improvement in the coherence between policy and the rights of the PPC in the basin.

Additionally, holders of pre-1989 established abstraction rights, especially for irrigation wells constructed prior to 1989, were not targeted by policy at all, which further reduced the external coherence between policy and the property rights system. Illegal use of water in the basin was also commonplace and users' behaviour was not in agreement with the policy principle that water is a natural good for society (1987 Water Law). Therefore, despite national and prefectural policies, there was a time lag between policy change and limitation of use rights in practice. Therefore, lack of external coherence in this case is also closely related to inadequate policy implementation (see section 7.8.3).

Furthermore, although the environment is somehow included as a user in the public policy system through a clause for the definition of minimum lake levels in the 1987 Water Law, Lake Vegoritida still has no established right to a defined water level.

Finally, the regime external coherence is reduced by the missing land use and spatial planning policy to target the property rights structure for use of the exposed land on the lakeshores. The right to use land exposed from the dropping lake was still out of the control of authorities. Low external coherence, therefore, resulted from the lack of a policy framework to control land use rights for the benefit of water protection.

As concerns the rivalry on water pollution, the regime seemed quite coherent between the policy and property rights elements as changes in policy seemed to cover most relevant uses and users and were closely interconnected to changes in the rights system. Only farmers, who indirectly used water as recipient of their non-point pollution, were by the end of the case study not recognised yet as explicit target groups of water quality policy. All other uses were considered by policy and even indirect users of fisheries, bathing and aesthetics were given formal rights to good quality (standards for good water quality to maintain these uses via the CMD of 2001 on pollution reduction). However, the external coherence between policy and rights to pollute was often reduced in practice by deficient policy implementation. Namely, several polluters who are targeted by policy (especially urban users and animal farms), still have not applied for a formal use right (permit) to pollute (see section 7.8.3 for a discussion of insufficient implementation of regulations).

7.8.1.5 General assessment of regime change

Table 7.14 summarises key developments of the water basin regime elements over the case study process, bearing in mind relevant indicators for a more coherent (and more integrated) water basin regime (see chapter 4/4.1.2 for the list of indicators).

Table 7.14   Change variables of the Vegoritida basin regime

Extent	Indicators for complete extent	Rivalry on dropping levels	Rivalry on water pollution
	All water uses and users are considered by the regime elements (including the recognition of nature as use/user)	Extent increased especially in the late 1980s but is still incomplete concerning recreation, aesthetic values and nature protection	Extent increased since the 1970s and became almost complete except for nature values
Public governance	**Internal coherence indicators**		
Levels of governance	Administrative levels of management fit with the natural scale of water basins	Lack of fit of water management administration levels with natural basin scale	
		No basin administrative-unifying management body	
		Fit increased somehow via the 1987 Water Law	No increased fit
	Process includes actors from different levels	Multi-level process at place	Multi-level process at place (especially due to the ECJ case)
	Coordination and cooperation within each administrative level and between different administrative levels (within the river basin)	Coordination improved both within and between administrative levels	Coordination attempt in the 1970s and then again after the issuing of the CMD of 2001 on pollution reduction
Actors in the policy network	Participatory arrangements for actors, especially non-public actors, with an interest in water resources (formal arrangements such as planning and/or informal consultations)	No official participatory arrangements for actors outside the administration	
		Actors outside the administration took part only informally in the process	
	Establishment and involvement of local associations	The MESNA association of Arnissa was involved formally only in the 1994 Lake Committee of the Region West Macedonia	
		The Vegoritida Society is so far involved only on an informal basis	

	Increasing cooperation, interaction and consensus among actors in the relevant networks	There was increasing actor interaction during the case study process but only weak consensus and cooperation	
	Open networks instead of closed policy communities	Substantial, open communication networks were not established in the basin	
Perspectives and objectives	Development of a (river basin) management plan or a water basin vision (preferably official, but also consider informal initiatives)	No official management plan was produced for the Vegoritida basin, only scientific studies with no obligatory policy implications	
	Incorporation of multiple perspectives in the plan or vision	The lake stability and purification programme did not incorporate the perspectives of all actors in the basin	
	Policy visions and objectives (official visions and visions of stakeholders) recognise new water uses and administer justice to rivalries between different water uses	Incomplete recognition of all new water uses and important rivalries	
	Recognition of relations with other policy fields as coordination topics	Some progress in the coordination between agricultural, environmental and water policy could soon take place	
		No progress in the coordination between energy, mining, environmental and water protection policy	
	Change of water management paradigm, e.g. from hydraulic interventions to "softer" management principles	-	
Strategy and instruments	Availability of water management tools to redistribute use rights and reduce water use conflicts	Mainly command-and-control instruments (permits) to hinder new uses but not to redistribute rights	
		Few tools based on policy were made available and the strategy followed mainly involved investment in research	
	Incorporation of flexible instruments in policy; increasing preference for indirect procedural instruments (e.g. planning, participatory EIA procedures) instead of command-and-control tools and technical works	Adoption of an agrienvironmental policy measure in the end of the observed process	Some indirect, but not necessarily more flexible, instruments included financing for treatment works, EIAs and judicial mechanisms
Responsibilities and resources for implementation	Policy implementation process is sufficiently concerted and equipped with authority, legitimacy, finances, time, human resources, information etc	Insufficient support of policy implementation with adequate authority, financial and human resources	
		Separation of competence on water quantity and quality between authorities	
	Clear, non-fragmented institutional arrangement for implementation	Fragmented framework especially for the issuing of abstraction permits (improvement is expected via the 2003 Water Law)	Fragmented framework, somehow improving towards the end of the case study
Property rights	**Internal coherence indicators**		
	Coordination, not contradiction, of use and	Incoherent property rights due to the lack of	Rights to pollute formally quite coherent after the

	ownership rights when competing uses evolve	coordinated planning of all uses	introduction of effluent disposal permits. In practice, coherence was low due to the lack of adequate policy implementation
Public governance/ property rights	**External coherence indicators**		
	Changes in policy and public governance are reflected in changes of the property rights (e.g. via restrictive regulations such as restrictions on wells, restriction of private rights, establishment of permits for abstractions or pollution) Match between actors targeted by policies and actors with use and ownership rights	Water quantity policy did not match all relevant property rights and relevant actors (low coherence is also closely linked to lack of policy implementation)	Water quality policy covered all relevant uses and users External coherence was often reduced by deficient policy implementation

All in all, the institutional water regime of the Vegoritida basin became quite complex since the 1970s and 1980s. The regime extent increased by considering several uses with claims on water but this was not matched by an increase of the regime coherence. In fact, the regime coherence increased only in some minor aspects such as in the convergence of administrative levels concerning the adoption of common restrictive measures to protect water quantity in the basin and the water level of Lake Vegoritida. Progress was slower in other aspects, such as the establishment of an administrative-unifying management body for nature and environmental protection issues in the basin. All in all, it is concluded that the observed changes in the institutional water regime of the Vegoritida basin did not qualify for a regime shift to a more integrated phase.

7.8.2 Factors affecting regime change

The following discusses, first, factors acting as an external impetus for change in the regime of the Vegoritida basin and, secondly, the contextual conditions and the actor characteristics and interactions which influenced the outcome of the regime change process. Specifically, it is shown that despite the presence of some external change impetus pushing for more integration, the coherence of the regime could not increase substantially due to the absence of favourable contextual conditions and favourable actor characteristics and interactions.

7.8.2.1 External change impetus

EU impetus

Pressure from the European Court of Justice (EU infringement procedure) was a very significant external stimulating force for the generation of water anti-pollution policy by the central government (CMD of 2001 on pollution reduction in the Vegoritida basin) and for the initiation of research activities on the water resource degradation problem (especially the lake stability and purification programme). Particularly the threat of a potential financial penalty from the EU forced the State to take action for the protection of Lake Vegoritida and stream Soulou from pollution. In this context, also water quantity protection issues were examined because of their indirect influence on water quality.

The EU continued acting as an indirect driver for regime change even after closing the infringement procedure. More specifically, EU requirements to conserve areas, which are part of the proposed Natura 2000 network, triggered the Environment Ministry to enter the process on the lake level definition. Thereby, the consideration of nature was demanded, next to irrigation needs and aesthetical values, in fear of new EU infringement procedures for non-compliance with EU nature protection policy.

Problem pressure

Pressure from increasing water degradation was also a key external impetus for changes within distinct elements of the water basin regime. Especially pressure from the PPC abstractions and irrigation water overuse combined with water scarce years around 1990 stimulated the issuing of specific regulations to protect water quantity in the basin. The rising pollution of Lake Vegoritida and stream Soulou (linked to the dropping water levels) was the main impetus for Prefectural regulations to protect water quality since the 1970s. The overall water degradation eventually led to the EU infringement procedure and the Region West Macedonia initiative to fund scientific studies.

Impetus from national policy

In general, national policy does not appear to act as a significant impetus for change in the regime of the Vegoritida basin. In some cases, however, national policy played a supportive role to certain developments (see section 7.8.2.2), acting thereby more as a favourable condition rather than external impetus for change.

New political paradigms

European liberalisation trends in the energy sector led to the semi-privatisation of the PPC in 2001, which also acted as a minor impetus for regime change as far as the role and behaviour of the PPC towards the end of the case study process is concerned.

7.8.2.2 Contextual conditions

In the following, the contextual conditions (relevant to selected characteristics of the actor network and institutional interfaces), which influenced the outcome of the regime change process in the Vegoritida basin, are discussed one by one.

Characteristics of the actor network

Tradition of cooperation

There was no pre-existing tradition of cooperation in the water management sector of the Vegoritida basin, neither were there former appropriate structures to allow and support coordination efforts between actors involved. Instead, the process was characterised by the absence of mutual trust, especially between the PPC and farmers, between the PPC and local associations, between the RWMD and Prefectures as well as among the three Prefectures.

Joint problem

As far as knowledge on water resource degradation is concerned, scientific evidence was available early in the process, especially with regard to the degradation of water quality in the 1970s and the 1980s. Lakeshore authorities and associations approached scientific experts since the late 1970s to start building up knowledge on the status of Lake Vegoritida. Such early scientific evidence was used by local associations and lakeshore authorities to support their first claims to protect the lake. However, over the case study process, the local level was not kept up-to-date with new scientific knowledge gained through research. Especially the scientific results of the lake stability and purification programme were not disseminated to the local population.

As concerns the development of common understanding of the water resource problems in the Vegoritida basin, the drop of Lake Vegoritida was very early recognised as a problem by most actors (authorities, farmers, NGOs and even by the PPC in some aspects). Additionally, the political salience of the water levels problem was recognised, especially because of social unrest linked to farmers' claims on exposed land. The aquifer drop problem, however, was less obvious and less urgent for several actors.

All in all, despite the recognition of the water levels problem by most actors, there was disagreement over the problem causes mainly between water users. From the PPC perspective, natural underground water losses and farmers were co-responsible for the drop of water levels. Vice versa, farmers and the local population blamed the PPC for the problem. The shift of responsibility for the water levels problem between the PPC and the farmers also reveals the lack of common sense of responsibility for the future. There was also little information exchange between sectoral actors such as the PPC and farmers.

Additionally, the lakeshore society remained divided in its perspective of the problem and its objectives in terms of economic development. On the one hand, the majority of the lakeshore population was concerned about the state of the lake and was unhappy about the lack of action on behalf of the authorities. On the other hand, many inhabitants were keen on cultivating exposed land, even illegally. For instance, the population of the community of Agios Panteleimon was divided between those who preferred not to cultivate exposed land and favoured the return of the lake to its earlier status (for fishing and tourism) and those who preferred to cultivate exposed land (mainly large-scale farmers) (Interview Community Agios Panteleimon, 22/1/2005).

There was also limited joint problem perception as concerns the environment and nature. By the end of the case study, very few actors perceived nature protection as an important issue. Some lakeshore authorities (especially the municipality of Amyndeo) were only supported by the e-NGO Ornithological Society in their belief that the environmental value of the area was its most important capital, instead of energy production and industrial development (Interview Municipality Amyndeo, 16/2/2005).

Joint chances

A notion of "win-win" situations from more regime coherence in this case study was mainly reflected in the fact that many actors perceived the establishment of a management body for the Vegoritida basin as a joint opportunity. Joint gains were also seen by most actors (except for industrial polluters) in better water quality which would ensure uses such as fisheries and tourism. Furthermore, for some actors (especially lakeshore authorities, non-governmental associations and certain farmers), joint gains from halting the PPC direct abstractions from Lake Vegoritida motivated them to act and protest in common. The lakeshore population and farmers affected by scarcity were sensitive against any plans of the PPC to further abstract water.

Alternative threat

There was no clear-cut dominant actor in the case study, following unilateral action with severe consequences for other actors in governance terms. Although the PPC could be characterised as dominant due to its economic power, political influence and autonomous presence in the basin, it could not impose an alternative governance model to its benefit.

Institutional interfaces

In total, there were no well-functioning institutional interfaces to provide fertile ground for the success of integration attempts in the water basin regime. A characteristic interview citation was: "The main obstacle is of institutional nature. No institutions and regulations function for water management and land management around the lake (Lake Vegoritida)" (Interview LVPS, 1/2/2005).

Alert mass media

The local press (and less the national press) often covered the issue of water levels and pollution in the basin, especially of Lake Vegoritida. The interest of the press, however, concentrated on marked events such as the massive local protests in 1990 due to the alleged PCB leakage into Lake Vegoritida and into Lake Petron. Press coverage also declined after the PPC discontinued its direct lake abstractions (in 1997) and after a violent conflict between neighbouring lakeshore communities over exposed land was settled (in 1998).

Institutional position of brokers

The RWMD of West Macedonia played a kind of broker role by facilitating the basin-wide harmonisation of abstraction restrictive measures and the discussions on the lake level definition among Prefectures. The technical consultants supervising the lake stability and purification programme also played, for limited time, the role of broker between basin actors in the context of strictly technical discussions.

Institutional representation of rival interests

Most stakeholders were not organised in strong representative organisations to act in a uniform way. Especially farmers had no collective representative in the process, except for some local agricultural cooperatives representing individual lakeshore farmer groups.

Although polluters were in general a much smaller group than farmers, they had no organised representative in the process either. All in all, the number of actors that could potentially be included in a basin participatory process remained large due also to the administrative split-up of the basin.

Alert public

An overall favourable condition in the process was the presence of an alert local public, which exercised social pressure escalating in the late 1980s/early 1990s. In general, there was increased awareness of the population concerning the water problems of the basin, especially of Lake Vegoritida. It was, in fact, the population of lakeshore communities (in the Pella and Florina Prefectures), which was more sensibilised and alert, rather than the population of the non-lake-sharing Prefecture of Kozani. At least in terms of pollution, the Kozani Prefecture was an upstream pollution source, while the Pella and Florina Prefectures were mainly downstream recipients of effluents into the lake. In general, the intensity of local protests could also be attributed to the fact that actions could most of the time be targeted against the water-degrading activities of few sources (especially, of the PPC).

Legal leeway for more integrative approaches

Concerning the existence of legal leeway facilitating more integration, several changes in the water basin regime found necessary policy support in EU policies which defend good water quality such as the Urban Wastewater Treatment Directive, the Hazardous Substances Directive and EU quality standards of water for bathing, fisheries and drinking water.

The national 1987 Water Law also gave some impulse for integration by setting up the RWMD of the Region, by introducing the water permit procedure and by promoting abstraction restrictive measures in the Prefectures. However, the Law was (nation-wide) not fully implemented and only facilitated integration in a very limited way.

The 1986 Environment Law could also have acted as a potential favourable condition but its requirement for the designation of the basin as protected area and the establishment of a management body were not implemented by the end of this case study. It is also worth keeping in mind, that actors' efforts to set up a management body for the Vegoritida basin (in specific, for the area of Lakes Vegoritida and Lakes Petron) were based on environmental legislation (1986 Environment Law) and not on water legislation (1987 Water Law). In fact, only the environmental legislation provided the possibility for management bodies of a more local character for areas designated as protected ones. Instead, the water legislation did not provide possibilities for locally-based management structures but only structures on the broader Regional level encompassing the entire water district of West Macedonia (such as the RWMDs set up in the Regions and the Regional Water Committees for entire water districts).

An unfavourable condition was also the lack of legal possibilities to institutionalise the validity of proposals from scientific studies (especially from the lake stability and purification programme). For instance, the Florina Prefecture, which supervised the programme, lacked institutional competence to translate the scientific programme results into policy for the implementation of appropriate management measures (Interview Florina Pref., 26/1/2005).

Protection of negotiated compromises

Earlier on in the case study process, the conclusions of the 1994 Lake Committee set up by the Region West Macedonia, which were negotiated and accepted by all Committee participants, were not protected by a legal regime and were thus not policy-binding having little effect in practical water management in the basin regime.

7.8.2.3 Actor characteristics and interactions

The following discusses the influence of actor characteristics (of motivation, information and resources) as well as of actor interactions on the making of attempts to change the Vegoritida basin regime towards a more integrated phase. Given the observed distance of the Vegoritida regime from an integrated phase, the following concentrates mainly on the failure to actively promote such attempts. In order to characterise the outcome of the actor interaction processes, use is made of the actor interaction assumptions formulated in chapter 4/4.1.2.3.

Actor motivation is the first crucial factor in the interaction processes of interest. For a regime change attempt towards more integration to take place, deliberate action of motivated actors is needed. In the studied process, only a few actors were motivated to resolve water use rivalries in the Vegoritida basin in the direction of more integration. However, it is encouraging that their number slightly increased towards the end of the case study.

Local society, including local non-governmental associations, municipalities and some farmers affected by scarcity, was motivated to exercise social pressure for the protection of water resources, escalating in the late 1980s/early 1990s. The motivation of local societal actors to cooperate was enhanced by their joint gains from acting together against the PPC activities in the basin as well as by the presence of an alert local public. However, these local actors had limited resources to steer the water management process. For this reason, they resorted to the EU level and managed to achieve a conviction of the Greek State at the ECJ.

Later on, also the Region West Macedonia became more motivated to resolve rivalries in the basin for the sake of protecting water resources. Its motivation was enhanced by the alert local public exercising pressure on the Regional political level. The Region was also an actor with a substantial financial and political resource basis. Due to its initiative, a first participatory Lake Committee operated in 1994/95 and the lake stability and purification programme was funded in 1998.

Towards the end of the case study, the e-NGO Ornithological Society rose as an additional motivated actor to resolve the rivalry between the cultivation of exposed land and the living environment. Its motivation also triggered the Environment Ministry to bring environmental issues onto the agenda of the lake level definition. Nevertheless, despite being a knowledgeable motivated actor, the Ornithological Society had a limited resource basis except for its informal links to the Environment Ministry.

In the end of the case study, also the motivation of the Prefectures in the basin (mainly of the Florina and Pella Prefectures) rose as far as cooperation on the abstraction restrictive measures and an agreement on the definition of the lake level are concerned. Initially, the Pella Prefecture was often faced with a negatively motivated Florina Prefecture, when it came to the introduction of strict abstraction restrictive measures. This often led to conflict between the two Prefectures (*interaction assumption I-8*). The motivation of the Florina Prefecture was later enhanced with the support of new scientific information from the lake stability and purification programme. Its better understanding and its new perception of the water resource problems in the basin induced the Florina Prefecture to adopt a more cooperative behaviour, favouring thus interprefectural coordination on stricter restrictive measures (*interaction assumption I-5*). This interprefectural coordination was also enhanced by the broker role played by the RWMD of West Macedonia in the process of adopting harmonised basin-wide restrictive measures. Only the Kozani Prefecture remained negatively motivated to the introduction of harmonised abstraction restrictive measures proposed by the RWMD for all three basin Prefectures. An agreement to the common restrictive measures would affect a much larger territory of the Kozani Prefecture than before. Thus, this Prefecture was unwilling to bear the costs of this alternative. However, given that the RWMD is more powerful by law than the Prefectures on the formulation of abstraction restrictive measures, at least forced cooperation is expected in this case in the future (*interaction assumption I-6*). Indeed, when the RWMD makes use of its right to request the intervention of the central government, the Kozani Prefecture will probably be forced to follow an alternative more cooperative behaviour.

Most other actors involved in the case study process were characterised by low motivation to support efforts for more integration in the water basin regime. Motivation remained low also under the influence of a missing tradition of actor cooperation in water management in the basin, the initial lack of joint problem perception and the lingering lack of joint chances perception. Moreover, water management problems in the Vegoritida basin were very linked to political interests, especially the issue of cultivating and irrigating exposed land. Thus, interventions of politicians from all administrative levels in the water management process of the basin were often linked to electoral interests rather than motivation to protect the water resources. In the following, the interactions between positively and negatively or neutrally motivated actors are discussed in some more detail, focusing on attempts to change the entire regime or specific elements of the regime. Therein, the discussion is also extended to cover aspects of information and resources of the interacting actors.

Firstly, some proposed measures of the lake stability and purification programme, which aimed at forming incentives to change farmers' behaviour, are discussed (these proposed measures are relevant to change in the regime element of "strategies and instruments" in the public governance system). The proposed measures in question concern the replacement of spray artificial rain with drip irrigation and the replacement of irrigated crops with non-irrigated ones.

Concerning the replacement of irrigation techniques, the Florina Prefecture and farmers were positively motivated to promote this instrument in the basin. However, the Ministry of Agriculture, which was the actor of the central government with the power to bring such an instrument into existence, showed no motivation in this direction within the timespan of the case study. The Florina Prefecture and farmers did not have the necessary policy-making power and could not motivate the support of the Ministry of Agriculture either. In this context of interaction between positively and negatively motivated actors, the balance of power was important. Given the power dominance of the Ministry of Agriculture, the proposed measure did not proceed to realisation (*interaction assumption I-7*).

Concerning the proposed replacement of crop types, the Florina Prefecture positively viewed the introduction of a relevant agrienvironmental measure, based on the knowledge it gained in the lake stability and purification programme. The Florina Prefecture (with the support of its technical consultant) attempted to convince the Ministry of Agriculture to introduce such a measure, since the Prefecture did not have the appropriate policy-making power itself. At start, the Ministry of Agriculture remained inactive, but later in 2006 it actually issued such an agrienvironmental measure for the basin. Thus, once the Ministry also became positive for the introduction of such a measure, a more cooperative interaction developed introducing a rather integrative instrument for the water basin regime (*interaction assumption I-5*).

Secondly, the resolution of the issue of cultivating exposed land appeared as crucial in the regime change process, being quite related to the degree of external coherence between the system of property rights and the system of public governance. On this issue, a few actors such as the e-NGO Ornithological Society, who were motivated to defend environmental values of the water environment, were faced with negatively motivated actors, especially farmers. The positive actors held sufficient information on the negative impacts of the exposed land cultivation to support their arguments. However, this being a highly political issue, farmers proved to be a more powerful actor in this process. Namely, the land cultivation demands of farmers influenced elected politicians, who refused to bear the political cost of removing illegal farmers from exposed land. In other words, elected politicians avoided adopting farmer-unpopular measures. Thus, attempts to resolve the issue of exposed land were by the end of the case study equivalent to a process of obstruction (*interaction assumption I-7*). With the entrance of the politically powerful Environment Ministry in this process to support environmental interests, towards the end of the case study, there seemed to be a potential turn of the obstruction process towards an oppositional negotiation process (*interaction assumption I-8*). The entrance of the Environment Ministry brought about more balance of power between the two actors' fronts. This development can also be viewed as a break-through in the system of powerful actors dominating the process (farmers, elected politicialns) and the opening-up of the governance system to actors who are more motivated to promote environmental protection.

Thirdly, it is worth discussing the failure to achieve change in the water basin regime, by establishing a management body for a to-be nature protected area around Lake Vegoritida (according to the 1986 Environment Law).

Some actors in the basin (local associations, municipal authorities, Region West Macedonia) viewed the option of such a management body positively. Especially local associations and some municipal authorities considered a management body as the only chance for integrated management in the Vegoritida basin, because such a body could ensure continuity of policies in the basin, independent of the personalised motivation and engagement of specific politicians in power. However, some positively motivated actors apparently had insufficient information on the potential institutional structure of such a management body and thus had divergent expectations from its role. Many local actors expected that the management body could act as manager of water resources on a river basin level. However, a management body based on the 1986 Environment Law could only have an advisory role for the protection of the environment and nature and would not be equivalent to a body responsible for water basin management (Interview Ornithological Society, 25/2/2005). Other actors emphasised the need for a politically-independent body with decision-making, implementation and enforcement powers (Interview Florina Pref., 26/1/2005), which again does not match the profile of a management body under the 1986 Environment Law.

In practice, the creation of a management body (according to the 1986 Environment Law) did not move forward by the end of the case study. Key actors (including actors on the central level), having the necessary policy-making capacity to support the set up of a nature protected area around Lake Vegoritida and of a management body, did not act accordingly. The only step taken was the allocation of funding by the Region West Macedonia to carry out the Specific Environmental Study required by the 1986 Environment Law, prior to designating a nature protected area and its management body.

All in all, political will to provide long-term solutions to the Vegoritida basin water problems appeared to be hindered by short-term political thinking to secure local votes. For elected governments, the alternative of creating a management body was not financially costly but could be politically costly, since it could potentially interfere with, for instance, the interests of farmers cultivating exposed land. Additionally, the presence of the PPC with its energy production objectives in the basin may also have reduced the motivation of the central government to protect the area from an environmental point of view. Therefore, these economic interests most probably contribute to the low political motivation for the promotion of the long-awaited nature protection legal framework for the water ecosystems in the basin. Delays in this process were only partly (and not above all, as argued by some) related to the administrative split-up of the basin.

To conclude on the interaction process related to the creation of a management body, it can be summarised that some (in principle) positive local actors are faced with inactive actors on higher political levels. In this context, the balance of power needs to be discussed. The local supporters of a management body have limited capacity in relevant policy-making. Positive local actors also appeared unable to develop a common coordinated strategy in order to motivate the higher level actors towards the activation of a legal framework for nature protection in the basin. Therefore, the positive actors were not powerful enough, resulting in obstruction or at best an ongoing negotiation process for the creation of a management body (*interaction assumption I-7*), at least until there is improvement in the relevant political motivation.

Finally, it should be noted that the PPC, which was an important actor in the basin in terms of its impact on water resources, remained largely non-cooperative (or simply inactive) in attempts to change the regime towards more integration. The PPC was not motivated to support the promotion of a more integrated water regime in the basin, since such an alternative could imply that it has to account for its water use to other actors. The PPC "disliked" potential restrictions to its water abstractions, which would affect its energy production plan. Until the end of the observed process, the PPC maintained a conflictual relationship with farmers, local municipalities and local associations, which blamed the PPC for the drop of water levels in the basin. In fact, the relationship of the PPC with farmers only became really conflictual, once the majority of farmers were affected by the impacts of scarcity. At the start of the process (in the 1970s), farmers even exercised pressure on the PPC not to reduce abstractions from the lake at times of high rainfall, so that they could avoid the flooding of the exposed land which they already cultivated.

All in all, the PPC "afforded" the alternative of non-cooperation in most of the process, because it was backed by the central State in its mission to produce electricity, even if this was at the cost of local water resources. Once water resources in the Vegoritida basin would be fully exploited, the PPC had the alternative of turning to other water sources from neighbouring basins. Unfortunately, the PPC could not be forced to use these alternative water sources, as soon as the alarming drop of Lake Vegoritida was realised. Instead, the PPC turned to water transfer from a neighbouring basin, only later when continuing abstraction from the lake became too costly due to rising pumping costs from the dropping lake.

In the future, the PPC could become more positively motivated to cooperate in the making of attempts towards a more integrated water regime, if it realises any joint gains from cooperation. For instance, in the context of its semi-privatisation since 2001, the PPC may invest more in gaining public acceptance and in maintaining a good relationship with public authorities through better compliance with abstraction regulations.

7.8.2.4 Summary

Some contextual conditions were favourable but most were unfavourable for water basin regime changes to more integration. The influence of these conditions on the regime change process was exercised also by contributing to the shaping of motivation, information and resources of the interacting actors (see Table 7.15).

Table 7.15   Influence of contextual conditions on actor characteristics and the Vegoritida regime

Conditions	Facts	Influence on actor characteristics	Implications for the water basin regime
Selected characteristics of actor network			
Tradition of cooperation	Absence of cooperation tradition	M (↓)	Low degree of cooperation between actors
	Absence of mutual trust between actors		
Joint problem	Joint problem perception of lake level drop	M, I (↑)	Partial coordination of Prefectural restrictive measures to reduce abstractions on the basin level
			Some progress in the promotion of an agrienvironmental measure (for replacing water-demanding crops)
	Disagreement over problem causes	M, I (↓)	No motivation of key water users (mainly PPC but also farmers) to cooperate for more sustainable water use
	Lack of common sense of responsibility		
			No progress in measures for the replacement of irrigation systems
Joint chances	Joint gains realised only by part of the actors	M, I (-)	Motivation of local authorities, associations and farmers to join forces against the PPC
			No motivation of PPC and farmers to support integration efforts
			Lack of progress in the set up of a management body
			Lack of progress in resolving the exposed land cultivation issue
Alternative threat	No dominant actor present threatening to impose its own governance system	M (-)	-
Institutional interfaces			
Alert mass media	Alert local media at times of intensive protests	I, M (↑)	Rise of social pressure leading to a legal case before the ECJ
Alert public	Alert local public	M, R (↑)	Escalation of social pressure, legal case before the ECJ, sensibilisation of the Region West Macedonia which then set up a participatory Lake Committee (in 1994) and funded the lake stability and purification programme (in 1998)
Actors as brokers	RWMD and technical programme consultant as "brokers"	M, I, R (↑)	RWMD facilitated agreement of Prefectures on common restrictive measures on the basin level and common discussions on the lake level definition

218

Conditions	Facts	Influence on actor characteristics	Implications for the water basin regime
			Technical consultant facilitated exchange between basin actors (Florina Prefecture, Region West Macedonia, municipalities, research institutes) on proposed technical measures
Few actors or strong representatives	Not all actors involved had strong representatives (especially farmers)	R (↓)	Reduced negotiation capacity of many actors in the process
Legal leeway for more integrative approaches	1987 Water Law implementation	R (↑)	RWMD set up Establishment of water permits and Prefectural restrictive measures on water use
	Implementation of EU policies for water quality protection	R (↑)	Investment in wastewater treatment Issuing of CMD on pollution reduction in the basin combined with a formal actor exchange platform to monitor progress
	No institutional validity of the lake stability and purification programme to translate research findings into policy measures	R (↓)	Most proposed measures did not proceed to policy formulation and/or practical realisation
Protection of negotiated compromises	Lack of mechanisms to offer institutional validity to coordination and management attempts in the basin	R, M (↓)	Most proposals of 1994 participatory Lake Committee did not proceed to realisation to policy formulation and/or practical realisation

M= motivation, I= information, R= resources

(↑) = positive influence, (↓) = negative influence, (-) = no significant influence

7.8.3 Conclusions and outlook on the regime change process

The analysis of the institutional regime components of property rights and public governance for the Vegoritida basin shows that the specific regime did not qualify for a shift to a more integrated phase in the research period of this case study (1970s until mid-2005).

Several circumstances, especially an EU infringement procedure, EU regulations and pressure from intensive water resource degradation, acted as external impetus for change in the water basin regime. Under this pressure for change, the water regime became increasingly more complex, as more and more uses became regulated (increase of regime extent). However, the increase of the regime extent was not matched by a sufficient increase of the regime coherence for a shift towards an integrated phase. In fact, the empirical analysis concluded that the regime coherence could not increase significantly, under the influence of unfavourable contextual conditions and an unfavourable motivation- information- resources constellation of the actors involved in the regime changes.

Given that the institutional water regime of the Vegoritida basin has not reached an integrated phase so far, it is worth discussing whether chances for this to happen could be better in the near future. The potential for such development is also of interest in view of the current EU policy requirements for more integrated water basin management (see the WFD).

First, it is observed that external impetus pushing the Vegoritida water basin regime to change will be further present in the future. Although the EU infringement procedure on water pollution was closed in 2001, the EC and e-NGOs remain alert on the compliance of Greek authorities with the ECJ decision. The Greek State is, indeed, reluctant to be convicted twice for the case of pollution of Lake Vegoritida and the stream Soulou, since a second conviction would be translated into a financial penalty. Additionally, EU policy requirements for nature protection (Natura 2000 network) will continue to act as an impetus for change until the establishment of a nature protection regime within the Vegoritida water basin. Also, EU policy requirements for integrated water management according to the WFD will be present as an external change impetus. Problem pressure from water resource degradation will be further present but it may not be as strong as in the past, due to the partial improvement of the water status compared to its status in the 1970s/1980s. As far as the intensity of national impetus is concerned, this will largely depend on the degree of adequate implementation of the new 2003 Water Law on the regional level. All in all, it appears that in the future there will continue to be sufficient external impetus present to set attempts in motion for changing the water basin regime to a more integrated phase. In this context, it is crucial to discuss whether also the influence of contextual conditions and the characteristics of actors involved in the regime change process are bound to become more favourable to integration than in the past.

Looking back at section 7.8.2.2, a few aspects of the contextual conditions actually became somehow more favourable towards the end of the case study process. The degree of cooperation seems to have increased at least within the public administration (more cooperation takes place between Prefectures and between Regional, municipal authorities and the Prefectures). The problem of unsustainable water management became acknowledged in common by most actors in the basin, even though the lack of shared responsibility for the future of the basin was not overcome, especially by several farmers and the PPC. As concerns the presence of legal leeway to support integrative approaches, no improvement could be observed but there was at least intention to use the respective opportunities provided by the 1986 Environment Law and the 2003 Water Law.

Additionally, also the motivation of certain actors rose towards the end of the case study process to support regime change attempts towards more integration. Increased motivation was observed by local associations, the Region West Macedonia, the e-NGO Ornithological Society, partially by the Prefectural authorities sharing the basin as well as by the Ministry of Agriculture which introduced a new more flexible policy instrument to protect water resources in the basin from agricultural pressures (agrienvironmental measure). However, it could not be assessed whether more motivation (especially political motivation) would be present in the near future to resolve two crucial issues in the process, i.e. the issue of cultivating exposed land and the creation of a management body for a to-be nature protected area.

All in all, few aspects of the contextual conditions appear to be somehow more integration-favourable now than in the past and could exercise their future influence on the regime change process via a constellation of some more motivated actors. Nonetheless, it cannot be claimed that, in total, contextual conditions in their present state would be very favourable to the establishment of a more integrated regime. The institutional water regime of the Vegoritida basin would have a greater adaptive potential towards a more integrated phase, if its contextual conditions are further improved as recommended in the synthesis chapter 9.

7.9 *Process of implementation of measures and policies adopted*

This section discusses, first, any water use sustainability implications observed from the implementation of measures and policies adopted during regime change and, secondly, the effectiveness of the implementation process in itself. The main relevant hypothesis (Hypothesis 3) was:

> "The less new incentives provided from regime change and the less adequate the implementation of any new incentives, the more likely that there are less positive sustainability implications of regime change in water use."

Indeed, the following shows that even the few new measures and policies adopted in the change process of the Vegoritida basin regime had only minor implications for more sustainable water use. This can be explained by the inadequate implementation of the measures and policies, under the influence of an unfavourable motivation-information-resources constellation of the actors involved in implementation (see chapter 4/4.1.2.4 for the theory-based assumptions on the outcome of actor interactions in implementation).

7.9.1 Sustainability implications of measures and policies

Implications on water resources and the environment

Towards the end of the case study, the level of Lake Vegoritida was more or less stabilised which is considered an environmental improvement compared to its continuous decline since the 1950s. The stabilisation of the lake was mainly due to:

- The discontinuation of large abstractions directly from the lake, first for the PPC hydropower plant and the irrigation of the Skydra plains (discontinued in 1991) and then for the PPC steam-electricity plants (discontinued in 1997). The discontinuation of the PPC direct abstractions was not related to changes in the water regime itself but was the result of the PPC strategic energy production planning in the face of water resource scarcity. The discontinuation of direct abstractions for the irrigation of the Skydra plains was a more obvious implication of developments in the regime change process, considering that it was the result of lakeshore social protests and the intervention of the central government.

- The positive inflow of water into the lake from the PPC water transfer scheme from the River Aliakmon basin (since 1997).
- The increased rainfall patterns towards the end of the case study.

It is worth adding that nature and environmental aspects also started being considered in discussions on the lake level definition towards the end of the case study.

As far as the protection of the basin aquifer is concerned, there were no significant sustainability implications of measures or policies taken by the end of the case study. Although the Prefectures issued relevant restrictive measures, these were not strictly implemented in the largest part of the basin. Additionally, the PPC continued to exploit the aquifer for its mining activities and to some extent for its stream-electricity plants.

Regarding water quality, many water bodies in the basin, such as stream Soulou and Lake Petron, remained polluted over the case study process. Only the water quality of Lake Vegoritida somehow improved. Improvement of water quality degradation was due to:

- The closure of the nitrogen-fertiliser industry AEVAL in the 1990s.
- The construction of the urban wastewater treatment plant for the city of Ptolemaida in 1994, as a result of EU directives and local pressure to the central government.
- Policies for effluent disposal permits and EIAs, which forced several industries to install treatment works.

Most of these factors can be considered related to developments (policies and measures) in the regime change process, except for the closure of AEVAL which was the result of global market forces.

All in all, it was observed that changes in the environmental status of water resources in the Vegoritida basin were only partially the result of changes in the water basin regime. External influences played an additional important role in this respect, in specific external natural factors (changing rainfall patterns) and external socioeconomic factors (industry closure due to increase of competitiveness in a more globalised economy).

Economic development implications

By the end of the case study, no positive economic development implications from measures and policies adopted could be noted. The improvement of water quality and the stability of the level of Lake Vegoritida may, however, have positive economic implications in the near future (return of fisheries increase, etc.). Prefectural measures restricting the drilling of new wells and the uncertainty over the issue of cultivation concessions and property rights to exposed land have an adverse economic impact on farmers who are eager to cultivate more exposed land.

Social development implications

The limitations imposed on irrigation and agriculture in the basin in combination with the new EU CAP developments may decrease the importance of farming as an employment source in the long-term. As a result and assisted by the improvement of water quality, fisheries and (eco)tourism may rise as new sources of employment and livelihood.

7.9.2 Influence of actor characteristics and interactions on implementation

This section discusses and explains the extent of adequate implementation of measures and policies adopted in the regime change process. Implementation is discussed as a social interaction process influenced by the motivation, information and resources of the key interacting actors. Where relevant, the influence of public governance elements of the water basin regime on the actors' characteristics and indirectly on implementation is also discussed.

The main policies discussed as to their implementation are the Prefectural restrictive measures to protect water quantity and Prefectural regulations on effluents and wastewater disposal. The Common Ministerial Decision (CMD) of 2006 on an agrienvironmental policy measure for the Vegoritida basin and the CMD of 2001 on a special pollution reduction programme for Lake Vegoritida, Lake Petron and stream Soulou were adopted only towards the end of the case study and their implementation could not yet adequately be discussed.

7.9.2.1 Restrictive measures to protect water quantity

The following discusses the implementation of restrictive policy measures and of the water permits system to control water use, firstly, for irrigation and, secondly, for energy production by the PPC.

The effectiveness of implementation concerning the water use for irrigation was different in the Prefectures sharing the basin. Differences in implementation were especially marked between the Pella and the Florina Prefectures. The Pella Prefecture achieved stricter implementation, while the Florina Prefecture showed reduced implementation effectiveness, and it is interesting to explore why this was the case.

The motivation of farmers (target group) in the Florina Prefecture was, in general, low in terms of compliance with abstraction restrictive measures and permit requirements. Farmers disliked being limited in their ability to irrigate and in general disliked the command-and-control character of these instruments (→influence of the public governance element of "strategies and instruments"). Indeed, during the observed process, farmers in the Florina Prefecture, but also in the whole basin, were not given other alternative instruments in order to reduce their irrigation consumption, such as support to replace their irrigation systems or to replace irrigated crops with non-irrigated ones. For this reason, the following focuses only on farmers' behaviour and actor interactions with respect to those instruments available during the process, i.e. abstraction restrictions and permits.

Over the 1980s and 1990s, farmers in the Florina Prefecture could afford a non-cooperative behaviour with regard to abstraction restrictions, because they were initially not greatly impacted by scarcity due to the drop of the aquifer. The aquifer is by nature shallower on the side of Florina and, despite the aquifer drop, Florina farmers could easily reach and abstract groundwater at low cost (either legally using power but also illegally using gasoline).

223

Furthermore, Lake Vegoritida itself is also shallower on the side of the Florina Prefecture and thus, as it was dropping, it revealed much more exposed land, which could be claimed for farming, there than in the Pella Prefecture. As concerns the Pella Prefecture, the aquifer is deeper there and, due to the dropping water levels, the Pella farmers were early forced to abstract water from even greater depth at increasing cost. As a result, they could afford less than the Florina farmers a non-cooperative attitude concerning water protection measures and the authorization of wells. The option of non-authorisation of wells would imply for them high costs to run wells illegally on gasoline instead of legally on power. It seems therefore that the dropping of water levels of the aquifer and of Lake Vegoritida had somehow unidirectional externalities in favour of farmers in the Florina Prefecture, who proved less motivated than the Pella farmers to comply with abstraction regulations and restrictions.

Given the negative motivation of farmers in the Florina Prefecture, it is important to examine the motivation of the implementing authorities in order to evaluate the outcome of the implementation process.

In general, the links between quantitative water management and political interests in the Vegoritida basin unavoidably influenced the motivation of Prefecture officials with regard to the implementation of regulations. In the Florina Prefecture, it was considered necessary to overcome issues of political cost in order to be able to impose fines for illegal and/or excessive water abstractions. For instance, for considerable time, there was lack of political will to pressurize (voting) farmers with prohibitions on the cultivation and irrigation of exposed land (Interview Florina Pref., 26/1/2005). Additionally, the effectiveness of stricter measures adopted in 1998 was questioned by officials in the implementing authorities themselves. Given the excessive number of wells already drilled by the time strict restrictive measures were finally adopted in the Florina Prefecture after 1998, few chances were seen in the effect of these measures in practice. On the contrary, more chances were seen in subsidy programmes that would promote the installation of drip irrigation techniques to replace very water-consuming irrigation systems (Interview Florina Pref., 26/1/2005).

Therefore, given a constellation of an unmotivated target group and a little motivated or neutral implementing authority in the Florina Prefecture, adequate implementation of the restrictive measures and the water permits system was for considerable time obstructed (*interaction assumption II-11*).

Towards the end of the case study process, the motivation of the Florina Prefecture implementing authority was enhanced through its active involvement in the lake stability and purification scientific programme (see also section 7.8.2.3). New stricter restrictive measures issued by the Florina Prefecture in 2000 were largely based on the conclusions of this scientific programme. Through its involvement in the programme, the Florina Prefecture was convinced that it needed to effectively implement measures. Beforehand, there was little conviction for strict enforcement both on the political and on the administrative level (→influence of change in the public governance element of "perspectives and objectives"). Nevertheless, the Florina Prefecture still encountered difficulties in its effort to impose strict restrictive measures in the Vegoritida basin (Interview Planet, 8/2/2005). In the context of this new constellation of more positive implementers but a still unmotivated target group, the availability of information and resources in the hands of implementers became important. In fact, in the Florina Prefecture as well as in the entire Vegoritida basin, there was lack of information on the actual patterns of water consumption due to the lack of metering devices and limited resources for field inspections to gather the necessary information. Even though water permits for irrigation defined the maximum discharge of water to be abstracted, the volume of water abstracted in reality was not measured. For instance, even though a permit defined that up to 50 cm/h (discharge) could be abstracted, farmers might have been abstracting 60-80 cm/h (Interview LVPS, 22/1/2005). Therefore, despite the increased motivation of the Florina Prefecture authorities, the implementation process was further hindered by insufficient information in the hands of the implementers (*interaction assumption II-17*). Even in cases where implementers were informed by third parties about illegal practices in the field, implementation remained inadequate due to the implementers' reduced power to impose sanctions (*interaction assumption II-15*) (→influence of the public governance element of "resources and responsibilities for policy implementation"). In a specific case of denunciation of illegal wells in the lakeshore community of Agios Panteleimon, the sanctioning process ended up in court (in May 2004). Even though the court did not have sufficient expert knowledge to judge the legality of these abstractions, insufficient use was made of the expert knowledge of the competent Florina Prefecture authority (Land Reclamation Division). As a result, farmers, who claimed that the wells in question were drilled prior to 1990 and should be maintained as pre-1989 established rights, were found innocent. In the same time, however, the competent Florina Prefecture authority knew that the specific location of these wells was flooded by the lake until 1990 (Interview Florina Pref., 26/1/2005)!

All in all, farmers in the Florina Prefecture remain largely uncooperative in the implementation of abstraction regulations even today, because they count that this alternative will not have severe consequences. This is related to the weak enforcement capacity of the implementers. In order to improve this situation, the enforcement capacity of the respective implementing authorities should be enhanced or alternative instruments for the reduction of abstractions should be adopted and implemented, such as agrienvironmental measures which reduce indirectly abstractions through crop change.

In the Pella Prefecture, the implementation of abstraction restrictive measures and of the water permits system started earlier and was more effective than in the Florina Prefecture. The more effective implementation was largely due to the increased sensitivity of both the implementing authorities and the public. For Pella Prefecture officials and citizens, Lake Vegoritida was for a long time a place for recreation and its protection was perceived as an important task to which they were committed to (Interview Pella Pref., 1/2/2005 and 8/3/2005). Therefore, the Pella Prefecture implementing authorities were positively motivated for the implementation of abstraction restrictions, while Pella farmers could be characterised as at least neutral considering also the sensitivity of the Pella Prefecture public (including local farmers) for the protection of Lake Vegoritida and its aquifer. Given this constellation of actors, it is important to further examine the information level in the hands of the implementing authorities.

In the past, effective implementation of restrictive measures in the Pella Prefecture was due to an operational and experienced competent authority, able to carry out frequent field inspections and gather necessary information. This resulted in a situation of constructive cooperation of actors for quite effective implementation (*interaction assumption II-13*). After the Prefecture became 2nd level of local self-administration in 1995, changes took place in the administrative structure of the Pella Prefecture. The structural changes involved a personnel reduction of the implementing authority for restrictive measures, reducing thereby also its capacity for field inspections (Interview Pella Pref., 8/3/2005). This resulted in decrease of the information and resources in the hands of the implementers and thus in some decrease of the effectiveness of implementation (→influence of the public governance element of "resources and responsibilities for policy implementation").

Except for the implementation of restrictions on the use of water for irrigation, it is worth discussing the implementation of restrictions on the (substantial) abstractions of the PPC in the basin. The PPC was little affected by new regulations based on the 1987 Water Law, except for the fact that it was obliged to apply for water use permits for any new abstractions after 1989. Even so, the PPC did not apply for any water use permits for abstractions in the Vegoritida basin until it was semi-privatised in 2001. The main implementing authority in the basin with competence over the PPC abstractions was the RWMD of West Macedonia, since it was competent for issuing water permits for use in energy production. However, given that the PPC was part of the public sector and under the "protection" of the central government, the motivation (and power basis) of the RWMD to control the PPC abstractions was limited, making thus a non-cooperative behaviour a non-costly option for the PPC. The lack of any really positively motivated actors in this process resulted in the lack of the application of the water permits system for the PPC abstractions (*interaction assumption II-1*). The change of status of the PPC to a semi-privatised corporation in 2001 resulted into higher motivation of public authorities to control the PPC activities. Since 2001, the PPC is obliged to apply for both water use and water work permits and, so far, it appears itself more cooperative than in the past to follow abstraction permit regulations.

7.9.2.2 Regulations to combat water pollution

The Florina and Kozani Prefectures are the main Prefectures hosting polluting activities in the Vegoritida basin. In the Florina Prefecture, even though frequent inspection of the implementation of effluent disposal regulations took place in the field, there was frequent breaching of the law. Non-compliance was observed either in terms of breaching the conditions set in existing effluent disposal permits or in terms of not having a disposal permit at all (Interview State Chemical Laboratory, 14/2/2005). In the Kozani Prefecture, it was especially difficult to adequately implement regulations on the polluting activities of the powerful PPC. Towards the end of the case study, the PPC was making efforts to change its behaviour, especially after the approval by the Environment Ministry of the environmental terms for the operation of its steam-electricity plants and mines (Interview RWM, 17/2/2005) (→influence of the public governance element of "strategies and instruments").

The general characteristics of the effluent disposal regulations, including the rise of production costs due to the required effluents treatment, imply that the motivation of the target group (mainly industries) was not positive to implementation (→influence of the public governance element of "strategies and instruments"). Therefore, to explain the outcome of the implementation process of the effluent disposal regulations, the following paragraphs need to concentrate on the characteristics of the implementing authorities.

In the Florina Prefecture, the implementing authorities were diligent about the implementation of the Prefectural regulations to control pollution. After the adoption of the 1986 Environment Law, the Florina Prefecture was one of the first Prefectures nation-wide which set up (in 1986) a Unit for the Control of Environmental Quality[80] involving the Prefectural Division for Industry, the Division for Health and the State Chemical Laboratory (Interview State Chemical Laboratory, 14/2/2005). This way, the Florina Prefecture developed quite early a relatively effective mechanism for field inspections of its water quality regulations. Although there was lack of continuous monitoring of pollution parameters in the effluents of each industry and in the main lakes and rivers, inspections in the Florina Prefecture functioned satisfactorily delivering sufficient information to the implementing authorities.

Given the presence of positively motivated implementers in the Florina Prefecture with sufficient information from a longer tradition of inspections, the effectiveness of the implementation process was more determined by the power balance of the actors involved. As in the case of water quantity protection policies, political interference played a role. The implementing authorities of the Florina Prefecture (similarly to other Prefectures) did not have complete authority for imposing penalties in case of non-compliance with water quality regulations. The political leadership of the Prefecture and courts were also involved. Fines and penalties could be issued only by decision of the Prefect, after proposal of the implementing authorities. Given the involvement of political personalities in this sanctioning procedure, it became difficult to impose restrictions on polluters especially after the Prefectures became locally elected self-administration in 1995. Namely, there was at times pressure by the financial interests of industries on the Prefectural political leadership to avoid penalties. The reduced sanctioning power of the implementing authorities decreased their overall power basis and led to a moderately only adequate implementation of water quality regulations (*interaction assumption II-15*) (→influence of the public governance element of "resources and responsibilities for policy implementation"). Instead, it has been proposed that, if penalties and fines could be imposed directly by the implementing authorities of the

---

[80] According to article 26 of the 1986 Environment Law.

Prefectures, implementing authorities would be more powerful in this respect (Interview State Chemical Laboratory, 14/2/2005).

In the Kozani Prefecture, a Unit for the Control of Environmental Quality was set up late in the case study process (in 8/2004) under pressure by the local society and local authorities to establish a control mechanism for air pollution emitted by the PPC. Until then, the Kozani Prefecture authorities could not control the PPC pollution directly. Since the creation of the Quality Control Unit, environmental inspections began to be carried out more efficiently in the Kozani Prefecture, also in terms of water pollution (Interview RWM, 17/2/2005). Therefore, for the greatest part of the case study process, the implementing authorities of the Kozani Prefecture had only limited inspection and information gathering capacity, especially as far as the main polluter on their territory is concerned, i.e. the PPC (→influence of the public governance element of "resources and responsibilities for policy implementation"). All in all, although the implementing authorities might have been positively motivated for efficient implementation, they had insufficient information and were also faced with a largely unmotivated target group. This constellation only led to inadequate symbolic implementation of water quality regulations for important pollution sources. However, as the Kozani Quality Control Unit becomes more informed over time about water pollution loads and the characteristics of the main water polluters, authorities could gather more adequate information to improve the future implementation (*interaction assumption II-17*).

All in all, as in the case of non-compliant users who abstract water, users who pollute water often choose not to cooperate with the implementing authorities on water quality regulations, counting on the authorities' weak enforcement capacity. Weak enforcement was related either to the lack of real information and capacity to monitor emissions and/or the powerful status of industries themselves which influenced sanctioning decisions. The prospects of cooperation between polluters and authorities could improve first by enhancing the enforcement power of the authorities and secondly by involving the polluters themselves in the selection of anti-pollution measures (see also relevant recommendations in chapter 9 on public governance elements and the regime contextual conditions in the Vegoritida basin).

7.9.3 Conclusions on the implementation process

In summary, only minimal positive implications of regime change on the sustainability in water use could be observed in this case study on the Vegoritida basin. This is, on the one hand, attributed to the few relevant new incentives (policies and measures) given by the non-integrated Vegoritida basin regime. On the other hand, this is attributed to the inadequate implementation of the few new incentives provided. The inadequate implementation of these incentives also partially explains the observed low coherence within the system of property rights of the regime and between the systems of property rights and public governance.

In fact, the few positive changes observed in the environmental status of water resources in the basin were only partially the result of changes in the institutional water regime and partially the result of influences external to the regime (changing rainfall patterns as well as industry closure due to the increase of competitiveness in a more globalised economy).

The inadequate implementation of new incentives provided from regime change was explained by the unfavourable motivation-information-resource constellation of the actors interacting during implementation. For better implementation of the specific measures and policies, more effort is needed especially to raise the motivation and resource basis of the implementing authorities. This largely involves putting aside political costs, which are often related to a stricter implementation of water protection regulations, disregarding pressure from powerful water users. Efforts to enhance the motivation, information and resources of the actors involved in implementation are not only crucial for the water protection regulations examined in the previous sections but also for the effectiveness of recent policy developments, such as the 2001 CMD for the reduction of hazardous substances pollution in the Vegoritida basin and the 2006 CMD for an agrienvironmental policy measure.

8 Institutional water regime of the Mygdonian basin

8.1 Introduction to the case study on the Mygdonian basin

Water basin description

The Mygdonian water basin is located in northern Greece in the administrative Region of Central Macedonia (Prefecture of Thessaloniki, sub-Prefecture of Langadas), approximately 12 km north-east of Thessaloniki, the second largest city in Greece. The basin is 2120 km² large (Koutrakis & Blionis, 1995). It includes Lake Koronia in its western part (upstream) and Lake Volvi and River Rihios in its eastern part (downstream). River Rihios ultimately mouths into the Aegean Sea.

Figure 8.1   The Mygdonian water basin

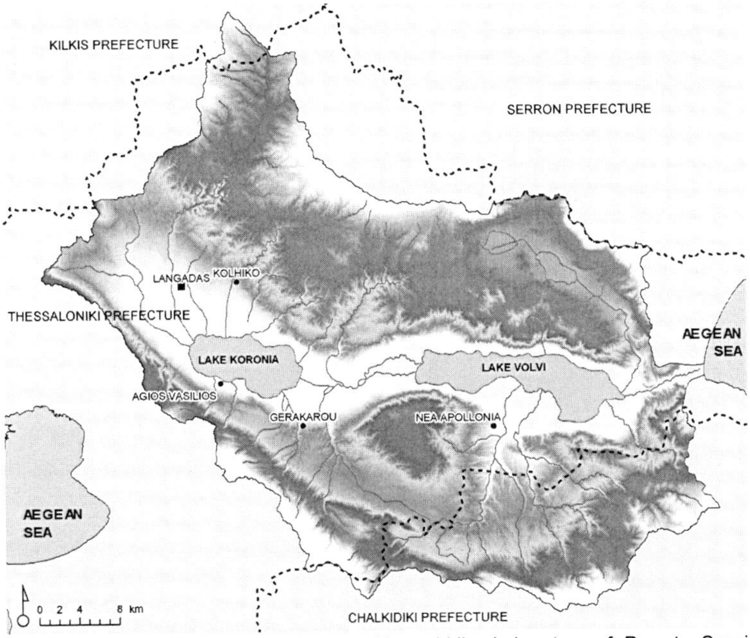

Source: Map produced by Dr. Thomas Alexandridis, Laboratory of Remote Sensing and GIS, Faculty of Agronomy, Aristotle University of Thessaloniki, Greece.

The torrents of the Mygdonian basin drain into lakes Koronia and Volvi. Lake Koronia receives water from three main torrents and a ditch initially constructed for flood protection (Knight Piesold et al., 1998). Lake Koronia could in the past overflow to Lake Volvi through another artificial ditch.

The subbasin of Lake Koronia is 837 km² large (Knight Piesold et al., 1998) and will be largely the focus of attention in this case study, since it has been the main source of water use conflicts in the Mygdonian basin.

In the Koronia subbasin, there is a shallow aquifer (0-50 m) and a deep aquifer (60-500 m). The shallow aquifer and Lake Koronia are hydrologically connected. The degree of exchange

231

between the shallow and deep groundwater aquifer of the basin is unknown due to the lack of appropriate monitoring data (Zalidis et al., 2004a).

Lake Koronia is a relatively shallow lake. Its maximum depth used to be more than 5 m, but by the mid-1990s it declined to less than 1 m (Zalidis et al., 2004a). By 2001, the lake depth declined further to approximately 0.8 m (Papakonstantinou & Katirtzoglou, 1995). The water surface of Lake Koronia also gradually declined from 47 km² in 1970 to 30 km² in 1995. During the same period, its water volume was reduced from $150\text{-}200*10^6$ m² to $30*10^6$ m³ (Zalidis et al., 2004a). The reasons for the quantitative degradation of Lake Koronia are discussed in more detail further below in the case study.

The climate of the area surrounding Lake Koronia is transitional between a Mediterranean and temperate one (Mitraki et al., 2004). The warmest and driest months are July and August. Annual precipitation during the last century ranged from 262-722 mm, while the mean annual value is 455.8 mm (Mitraki et al., 2004).

Demographics and employment

Three municipalities border directly with Lake Koronia: Municipalities of Langadas, of Koronia and of Egnatia. The city of Langadas is the major center of urban development (with 6,500 inhabitants). The local population level is relatively stable thanks to the proximity of the city of Langadas to the large urban centre of Thessaloniki. In 1991, the total population of the three municipalities was 21,846 inhabitants (Zalidis et al., 2004a, based on data of the National Statistical Service of Greece). There is nevertheless a tendency for a population decrease in the settlements on the shores of Lake Koronia. The disappearance of fish due to the lake degradation in the 1980s and 1990s, led to the unemployment of fishermen who either turned to agriculture or abandoned the area.

The local population is mainly employed in the agricultural sector. 5,000 of 12,500 employed inhabitants (in the three municipalities bordering Lake Koronia) work in the primary sector. Especially since the 1990s, farming became more intensive (Zalidis et al., 2004a).

Methodological issues of case study

The reasons for selecting the basin of Mygdonian as a case study were elaborated in chapter 6. The methodology used for data collection in the case study was described in chapter 4/4.3.2. Here, additional information is given on the focus of the specific case study and the sources of data used.

As mentioned above, this case study focuses mainly on the subbasin of Lake Koronia. Lake Volvi, which neighbours the Koronia subbasin, faces similar water problems as Lake Koronia but to a smaller extent. Lake Koronia, being also shallower and smaller than Volvi, faces greater pressure from overabstractions for irrigation, pollution by agrochemicals, industrial effluents and domestic sewage (Koutrakis & Blionis, 1995). This case study deals with Lake Volvi only in as far as it is relevant to the management process of the Koronia subbasin. However, it should be kept in mind that pressure seems to be increasing also on the relatively good status of Lake Volvi, which is the second largest Greek lake with a surface of

68 km². Specifically, pressure grows from the operation of greenhouses on its shores, illegal fishing and water-demanding irrigation techniques in addition to the operation of collective irrigation networks fed by its water (Newspaper "Kathimerini-Oiko", 14/1/2006).

The events of this case study are presented as one chronology and are analysed as one process. This aproach was followed because most changes in the institutional water basin regime, although partly related to different water use rivalries over water quantity and water quality, were highly interdependent. Only the regime component of property rights is described in separate for rights to abstract water and for rights to pollute water.

A wide range of written documents and primary data were collected in the field as well as through personal interviews and communications with actors involved in the case study process (see Annex III and Annex IV for details).

Structure of case study

Section 8.2 gives information on the use of water in the Mygdonian basin, emphasising key pressures on the water resources, while section 7.3 describes the key water use rivalries. In the following, section 8.4 introduces the main actors involved in the water management process of the basin. Section 8.5 describes the property rights of the main water uses in the basin to serve as the basis for assessing the property rights component of the water basin regime. Section 8.6 describes the chronology of events in the management process of the basin, while section 8.7 provides an analysis of the key actors involved and their interrelationships. The actor analysis serves as useful basis for the subsequent evaluation of actor interactions in the regime change and implementation processes. Sections 8.8 and 8.9 discuss the case study findings according to the analytical framework for change in institutional water basin regimes (see chapter 4/4.1.2). In specific, section 8.8 evaluates the process of making changes to the water basin regime and section 8.9 evaluates the process of implementing new incentives (measures and policies) resulting from regime changes.

8.2 Key water uses and their impacts

The uses of water resources in the Mygdonian basin are the following:

- Agricultural use: Irrigation of fields using lake water and the aquifer. Agricultural activities impacting water resources also include animal farms.

- Industrial use: Lake Koronia and streams receive effluents of industries. Industries also abstract water from the basin aquifer.

- Urban use: Water for drinking is abstracted from the basin aquifer. Lake Koronia and streams also receive untreated urban wastewater.

- Fisheries.

- Recreation: Recreation was important in the 1970s when a center for water sports was based on Lake Koronia. Due to the lake drop, however, water sports and rowing organized by lakeshore communities was discontinued (Kazantzidis et al., 1995).

- Lake Koronia together with Lake Volvi and their surrounding area function as an important wetland habitat for flora and fauna.

The following describes key water uses in the Mygdonian basin and their impact on water resources, with a specific focus on the Koronia subbasin.

Agriculture

Agriculture was present early in the basin of Mygdonian due to its fertility. In fact, the plains around lakes Koronia and Volvi have been defined by the Ministry of Agriculture as land of high productivity (Newspaper „Aggelioforos", 13/4/04). Since the 1970s, agriculture in the basin has become more intensive and dependent on irrigation (Knight Piesold et al., 1998). Except for the excessive use of water quantity, agriculture is also source of runoff of nutrients and chemicals from farms (Tsougrakis, 2000).

Impacts on quantity of water resources

The irrigated surface in the subbasin of Koronia increased gradually from 2,730 ha in 1968 (9.6% of the total cultivated area) to 6,300 ha in 2003 (>20% of the total cultivated area). Most irrigated areas are adjacent to Lake Koronia (Mitraki et al., 2004).

According to Table 8.1, water use in agriculture makes up for 85% of the total annual water consumption in the Koronia subbasin. Only in the period 1995-2000, there was a 23% increase in water consumption for irrigation around the lakes Koronia and Volvi, as a result of the increase in irrigated cultivated areas (Newspaper „Thessaloniki", 22/11/2002).

Table 8.1   Annual water consumption of main water uses in the Koronia subbasin

Uses	Agriculture	Industry	Water supply
Water consumption	$59*10^6$ m³/y [a] (data of 2000)	$7*10^6$ m³/y [b]	$3*10^6$ m³/y
Source of water	Shallow and deep aquifer Lake Koronia (until 1995)	Deep aquifer	Deep aquifer (a few users also from surface water)

Source: Zalidis et al. (2004a) ; Verani & Katirtzoglou (2003).

a. Actual withdrawals are $65*106$ m³/y reduced by 10% flow returns to the aquifer.

b. After industrial use, water becomes wastewater discharged into torrents, canals or into Lake Koronia.

As far as the types of crops cultivated are concerned, the area is dominated by crops of maize, wheat and tobacco, which were well subsidized by the EU Common Agricultural Policy (Arabatzi-Karra et al., 1996). Particularly hard wheat increased dramatically since 1993 due to the allocation of high subsidies. In the late 1990s, the cultivation of tobacco decreased because of the rise in labour costs (Knight Piesold et al., 1998).

In terms of irrigated crops, clover is the most extensive cultivation on 56% of the total irrigated area around Lake Koronia, together with other crops used as animal food. Maize covers 20% of the irrigated area while fruit and vegetables ca. 15% (Knight Piesold et al., 1998). Some important trends were obvious for irrigated crops in the period 1971-1996. Maize and clover increased around Lake Koronia, compared to a decrease of these crops in the rest of the Koronia subbasin. Especially cotton and clover are cultivated to a great extent in fields next to the lake. Considering that both these crops are resistant to salt, this trend may indicate a degradation of soil conditions in the proximity of Lake Koronia (Knight Piesold et al., 1998).

Most crops cultivated in the Koronia subbasin are irrigated between May and September. Irrigation water is gained from the aquifer via private wells[81] and to a lesser extent from collective irrigation networks (fed by wells or direct abstraction from Lake Koronia). Farmers use mainly water-intensive irrigation techniques such as spraying guns and artificial rain often during unsuitable (too warm) times of the day (Arabatzi-Karra et al., 1996). Altogether, the increase of irrigated crops combined with water-intensive irrigation practices was one important factor leading to the drop of the aquifer and of Lake Koronia in the basin.

The operation of six collective irrigation networks around Lake Koronia (to their majority managed by local communities) also pressurized water resources and was itself then adversely affected by the drop of the lake level and the drop of the aquifer. Two irrigation networks used to abstract water directly from Lake Koronia with a potential abstraction rate up to $15*10^6$ m³ in one cultivation period. When the lake began to drop, additional canals were opened to facilitate the entrance of water into these irrigation networks. Finally, direct abstraction from the lake for these two irrigation networks was prohibited by the Prefecture of Thessaloniki in 1995 due to the lake drop and the degradation of water quality making the lake water unsuitable for irrigation purposes. The operation of the other collective irrigation networks around the lake which were fed by aquifer water was also adversely affected by the overexploitation of water resources in the basin. For instance, one of two wells serving the irrigation network at the location Kolhiko ceased to operate due to the dropping groundwater level (Knight Piesold et al., 1998).

Impacts on quality of water resources

Agricultural activities around Lake Koronia pollute surface and groundwater with nutrients such as nitrogen, phosphorus and potassium due to increased fertilizer use and pesticides (see CMD 35308/1838 on a Special Programme for the Reduction of Pollution of Lake Koronia from hazardous substances, Official Government Journal B´ 1416/12.10.2005). It is estimated that the quantities of used agrochemicals exceed the required ones by 20% (Arabatzi-Karra et al., 1996).

The extensive use of fertilizers contributed to the eutrophication of Lake Koronia. This was aggravated by the fact that Lake Koronia is a shallow lake whose water volume was dramatically reduced since the 1980s. Additionally, the excessive water amounts used for irrigation could not be absorbed by the soil and drained back into the lake after washing off excessive fertilizers from the soil. The eutrophication of the lake led to the proliferation of microorganisms which consume more oxygen. When oxygen levels drop below a certain level, some organisms such as certain fish cannot survive, as was the case in Lake Koronia in 1995 when mass fish kills were observed (HOS, 2003).

---

[81]   Many irrigation wells draw water from the shallow aquifer with a discharge of less than 100 m³/h. Some irrigation wells draw water also from the deep aquifer.

Effluents from animal farms are a source of agricultural organic pollution and nutrients in the Koronia subbasin (Knight Piesold et al., 1998). There are numerous animal farms near human settlements and in the proximity of Lake Koronia, while large flocks of sheep and goats graze on the hills above the lake.[82] Nutrients and ammonia produced by domestic animals contributed to the eutrophication of the lake. In general, animal farms in the basin do not treat their effluents but discharge them directly into cesspits or septic tanks[83] (Arabatzi-Karra et al., 1996). Effluents from cesspits or septic tanks, which do not end up in the aquifer, are usually emptied on the fields of animal farmers or into torrents and streams of the basin (Tsagarlis, 1998), ultimately ending up into Lake Koronia. The development of animal farming without proper planning in the Koronia subbasin also has indirect impacts on the types of crops cultivated. Namely, the cultivation of irrigated crops such as clover in the basin serves to a large extent as food in local animal farms.

Industry

Several industries and business enterprises operate in the Mygdonian basin, 80% of which are located close to Lake Koronia. Large industries present include mainly textile production, fabric-dye industries, food processing and dairy product industries (Tsagarlis, 1998). The fabric-dye industries settled in the Koronia subbasin in the early 1980s (Tsagarlis, 1998), following the trend of a blooming clothing manufacturing industry all over Greece. More fabric-dye industries settled in the mid-1980s, while existing ones further expanded their operations in the early 1990s (Arabatzi-Karra et al., 1996). Besides large industrial units, there is a significant number of smaller business enterprises dispersed in the wider area around the lake including units of food processing, metal constructions, marble processing, ceramics, furniture and sale of construction material (Grammatikopoulou et al., 1996).

The main reasons for the concentration of industries in the Koronia subbasin are (Arabatzi-Karra et al. (1996); Interview Langadas Sub-Pref., 5/4/2005; Interview Young Farmers Union, 14/4/2005):

- Provision of incentives for setting up industries and business enterprises in this area from the national Development Laws 1262/82 and 1892/90.

- Easily accessible groundwater resources.

- Proximity to the urban centre of Thessaloniki.

- Availability of cheap farmland which could be bought for setting up industries, compared to the high prices for land in the Industrial Zone of Thessaloniki.

- Favourable location for commerce and transport close to the borders with other countries.

---

[82] The area surrounding Lake Koronia is mainly private agricultural land. Livestock grazing takes place on public land (wooded areas, narrow strips of land by the lakes or along torrents) (Tsougrakis, 2000).

[83] Cesspits are in essence sealed tanks which need to be emptied on a very frequent basis. Unlike cesspits, septic tanks are designed to allow seepage into the surrounding land.

Except for their large number, industries settled in the basin in a very disperse manner given the lack of a spatial plan[84] for the development of the area. The location of industries was determined by accidental and profit-oriented criteria such as the offer of cheap land and its proximity to the highway (Arabatzi-Karra et al., 1996). The disperse settlement of industries contributed even more to the unsustainable development model of the region around Lake Koronia. An effort was made to control the location of industries by setting up an industrial park in the area in the 1990s (with considerable delay considering the policy requirement for setting up this industrial park in the 1987 General Municipal Plan of Langadas). Most industries, however, had already settled by then elsewhere and none were re-located to the industrial park, firstly, because it was not obligatory and, secondly, because they had no incentives to do so (hardly any infrastructure existed to host industries in the park) (Interview Consultant to Master Plan, 20/4/2005). Even after the designation of the Mygdonian basin as a National Wetland Park in 2004, the anarchistic model of industrial development continued. It is still not obligatory for small industries to settle in the industrial park, if they meet all necessary conditions to settle elsewhere (Interview Ornithological Society, 28/3/2005).

Since the late 1990s, the presence of industry in the Koronia subbasin began to decline. Several water-demanding industries (especially fabric-dye industries) began abandoning the area in fear of the unavailability of water resources in the near future (Newspaper "Aggelioforos", 13/8/2002). Most importantly, several fabric-dye industries relocated parts of their production to cheaper production countries.[85]

Impacts on quantity of water resources

The most water-demanding industries in the basin are fabric-dye and dairy operations (see total industrial water consumption in Table 8.1). Industries (especially dye industries) abstract water from the deep aquifer of the basin at 250-300m (Knight Piesold et al., 1998). In the beginning of their operation, some dye industries abstracted water also from the shallow aquifer but they had to switch to the deep aquifer because of their increased water needs and in order to avoid local conflicts with farmers who also used the shallow aquifer for irrigation (Verani & Katirtzoglou, 2001). Most water consumed by industry is discharged back into the water system as effluents (Ydromeletitiki E.E. et al., 2002).

Impacts on quality of water resources

Approximately 17,000 m³ of industrial effluents are discharged daily into Lake Koronia (Knight Piesold et al., 1998). Industry is responsible for the pollution of the surface waters of the basin with salts, organic load, heavy metals and toxic substances.

Dye industries are considered as the most significant source of pollution in the Koronia subbasin (Tsagarlis, 1998). Their daily discharge of effluents amounts to 14,680 m³

---

[84] In fact, it is argued that spatial planning has been practically inexistent so far all over Greece (Interview Ornithological Society, 28/3/2005).

[85] In the 1990s, there used to be 11 fabric-dye industrial units in the basin. By 2005, only 2-3 units remained really in operation (Interview Technical consultant to the Master Plan, 20/4/2005).

(Tsagarlis, 1998). Dye industries especially discharge large quantities of salt effluents (Ydromeletitiki E.E. et al., 2002). The concentrations of inorganic salts (of Na, Ca, Mg, Cl) rose in Lake Koronia since the early 1980s. By 1995, the concentration of Cl-compounds increased by five times making the water of Lake Koronia unsuitable for irrigation (Knight Piesold et al., 1998). The specific conductivity of the lake water remained relatively stable at 1,300 µS/cm from 1977-1991, but then increased rapidly to > 6,000 µS/cm by 1993-1994 and reached a maximum of 7,700 µS/cm in 1997 (Mitraki et al., 2004). The control of salt pollution in the basin was especially hindered by the fact that the water quality regulations of the Prefecture of Thessaloniki did not define any limits to the concentration or total load of salts in the effluents discharged to the waters of the basin. Combined with the high cost of treating salt effluents, the lack of appropriate regulations led to the uncontrolled salt pollution of Lake Koronia (Ydromeletitiki E.E. et al., 2002).

As regards other types of pollution, despite the installation of effluent treatment by most dye industries in the 1990s, treated effluents did not meet the required standards for basic parameters, especially Chemical Oxygen Demand (COD). Nonetheless, the observed values of COD indicated that the dye industry was not the only source for the increasing pollution of the lake (Tsagarlis, 1998). Also two other large industrial units in the basin have a significant impact on water quality, namely a dairy product industry and the canning industry of the Union of Agricultural Cooperatives (Arabatzi-Karra et al., 1996) (the latter is now closed).

Urban use

Water use by human settlements mainly has an impact on the quality of water resources in the Koronia subbasin through the discharge of domestic wastewater. Human settlements either discharge their wastewater untreated into torrents of the basin which end up into Lake Koronia or they use cesspits or septic tanks (Ydromeletitiki E.E. et al., 2002).

In total, Lake Koronia receives 548,000 m³/y of urban wastewater. The city of Langadas alone discharges untreated wastewater of 1,500 m³/d (Tsagarlis, 1998). The old sewerage network of Langadas (planned to be replaced) serves 80% of the city population. There are a few other settlements in the basin served partially by old sewerage networks. These networks were constructed without any plans or guidelines and, although in many cases they were initially constructed to collect only rainwater (Prefecture of Thessaloniki, 2000), inhabitants improvised connections of their domestic wastewater collection to the pipeline system.

An urban wastewater treatment plant was constructed for the city of Langadas with funds of the Region Central Macedonia, but was not put into operation by the end of this case study. Before put into operation, the sewerage network of the city needs to be completed and connected to the treatment plant. The remaining population in the water basin is served by septic tanks or cesspits emptied in regular intervals by freight trucks (Prefecture of Thessaloniki, 2000). The impacts of wastewater discharge into cesspits or tanks can be significant due to the potential pollution of groundwater and the pollution of surface water from the illegal discharge of collected wastewater into surface waters by freight trucks.

Fisheries

Lake Koronia used to be one of the most fish productive lakes in Greece, the source of livelihood for hundreds of families as well as a place for recreation (Birtsas, 2005). In the 1960s, ca. 500 fishermen fished in the lake. The number of fishermen decreased dramatically over time (Knight Piesold et al., 1998) as fish production gradually dropped due to the increase of pollution, the disruption of water exchange with Lake Volvi and the degradation of riparian vegetation.

Annual fish production of Lake Koronia in the 1950s reached 1,050 tones, but in 1993 it only reached 67 tones (Grammatikopoulou et al., 1996). The fish population decrease which began in 1960 was linked to the use of new detergents affecting fish reproduction and the intensification of fishing with new technologies and fishing equipment (Liberopoulou, 1994). In the 1990s, commercial fish such as eel, perch and pike disappeared from Lake Koronia (Grammatikopoulou et al., 1996). A mass fish kills' incident in 1995 extirpated the lake commercial fish populations (Grammatikopoulou et al., 1996). Thereafter, fishing as a professional activity was discontinued and some fishermen of Lake Koronia moved to fish in Lake Volvi (Arabatzi-Karra et al., 1996). The social and economic impacts of the fisheries decline were a reduction in the number of professional fishermen and their income, discouragement of young men from undertaking the profession and weakening of the fishermen's lobbying power (Koutrakis & Blionis, 1995).

The ultimate collapse of the fish population was due to the inadequate management of commercial fish species, especially during the drop of fish production after 1990, and due to the degradation of the natural lake ecosystem, which resulted in both qualitative and quantitative alterations in the fish population (Knight Piesold et al., 1998).

The partial recovery of the lake water level after 2002 due to increased rainfall led to the re-appearance of certain fish species. However, fish populations are still far from building an ecologically stable population (Interview Langadas Sub-Pref., 5/4/2005).

Wetland functions

The Mygdonian basin is a rare ecosystem complex of lakes, rivers, riparian forests, wet prairies and farmland. Because of the large extent of shallow lakes and wetlands, Lake Koronia (and the neighbouring Lake Volvi) is an important staging area for many migrating birds and hosts a great number of protected species (Zalidis et al., 2004b).

The wetlands of the Mygdonian basin are protected by several international and national regulations which outline guidelines for the development and management of the area. In detail, the area is designated as:

- Protected area under the RAMSAR Convention for Wetlands of International Importance (designation on 21/08/75). In total, there are 11 Ramsar protected wetlands in Greece.

- Area of Special Protection of "Lakes Volvi, Koronia and Redina Gorge" under the EU Birds Directive and a proposed Site of Community Interest under the EU Habitats

Directive (proposed site in the NATURA 2000 network). The area hosts on a regular basis 20,000 aquatic birds, including the species *Pelecanus crispus*, *Phalacrocorax pygmeus*, *Podiceps cristatus*, *Aythya ferina* and *Aythya fuligula* (EC, 2002). In total, the wetland of lakes Koronia and Volvi hosts 204 bird species, 142 of which use the area to stage during migration (Arabatzi-Karra et al., 1996).

• National Wetland Park of the lakes Koronia, Volvi and the Macedonian Tembi (since 2004), whose limits coincide with the boundaries of the Mygdonian basin. This designation follows the Common Ministerial Decision (CMD) 6919/2004 which established a protection regime for the area according to the national Environment Law 1650/86.

The degradation of both quantity and quality of water resources in the basin has an impact on environmental values and wildlife. After the incident of mass fish kills in 1995, several fish-feeding birds (pelicans, cormorants and herons) as well as flamingos also disappeared from Lake Koronia (Arabatzi-Karra et al., 1996). The bird populations supported by Lake Koronia were again hardly hit at an incident of mass bird kills in 2004.

8.3 Key water resource problems and rivalries

As a result of different pressures on water resources, the current state of the Koronia subbasin is significantly altered from its status in the beginning of the 20th century. The main problems facing its water resources are two-fold based on a combination of water quality and water quantity reduction. Pressure increased during the 1980s, when rapid agricultural and industrial development aimed at creating employment. This development rate was supported by the central government and especially by a system of agricultural subsidy schemes with different objectives than environmental protection. Thus, a land use model was promoted in the basin which was not sustainable in terms of water resource use (Knight Piesold et al., 1998). As a result of increasing environmental degradation, Lake Koronia became famous nation-wide from two ecological disaster incidents of mass fish and bird kills in 1995 and 2004 (more information is given in section 8.6).

This section identifies the main rivalries between water uses and users in the basin, which emerged due to increasing water demand and rising pollution levels.

Rival fronts involved mainly the environment, industry and agriculture. There was often debate whether agriculture or industry was more responsible for the water resource degradation of the Koronia subbasin. According to Tsougrakis (2000), in the past, this "false" dilemma prevented authorities from taking the necessary actions. This debate actually evolved into a type of rivalry between agriculture and industry concerning the problem perception and share of responsibility between these two users. Each user considers that the other should first take action or be the target of restrictive measures by the State (Interview Former expert of the Information Centre of Lake Koronia and Volvi, 23/3/2005).

As far as the use of fisheries is concerned, it acted as ally to the environment in terms of its demands for good water status of Lake Koronia. However, fishermen themselves did not come in direct conflict with industries or agriculture, since they were quite disorganized institutionally (Interview Former expert at Information Centre, 23/3/2005).

Rivalry of agricultural and industrial abstractions vs. the environment and fisheries

The water balance of the Koronia subbasin became deficient since the beginning of the 1980s, resulting in the drop of Lake Koronia (see Figure 8.2) and of the aquifer.

Climatic conditions were one of the factors responsible for this quantitative degradation. After the mid-1970s, a long lasting drought period began drying out large areas of the lake and reed beds (Arabatzi-Karra et al., 1996). Dried-out reedbeds were burned and land exposed by the dropping lake was turned into cultivated land (Liberopoulou, 1994) (see Box 8.1 for more information on the cultivation of exposed land).

Figure 8.2   Absolute height of the level of Lake Koronia in 1982-1999

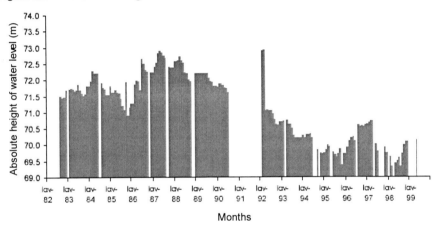

Source: AUTH (Aristotle University Thessaloniki, School of Agriculture, Applied Soil Sciences Laboratory) (2004).

Box 8.1   Cultivation of exposed land around Lake Koronia

In 1984, exposed land from the drop of Lake Koronia was temporarily conceded by the Ministry of Agriculture to landless farmers for cultivation. The integrated administration and control system (IACS) for the management of EU agricultural subsidy payments[86] even subsidized such cultivations (Interview Young Farmers Union, 14/4/2005). In 1991 (15/9/1991), a prefectural decision prohibited the cultivation of exposed land and reed beds. Therefore, after 1991, concessions to cultivate exposed land of Lake Koronia were not renewed and no new ones were issued. Nevertheless, many farmers illegally cultivated exposed land at least until 1995 (Kazantzidis et al., 1995). In winter 1991/92, the first effort took place to penalise illegal cultivation and to evict farmers by administrative decisions (Blionis, 1992). After the administrative eviction decisions, the IACS stopped subsidising crops on exposed land surfaces (Interview Young Farmers Union, 14/4/2005). Nevertheless, the long-lasting legal procedures resulted in lack of penalties (Blionis, 1992). Often, farmers asked for the intervention of the Prefect of Thessaloniki to provide amnesty and cancel the fines imposed.

In 1998, the Environment Ministry funded the on-site demarcation of the boundaries of zone A of the protected area of lakes Koronia and Volvi. Within this zone, agricultural activity was prohibited and this contributed to the reduction of illegal cultivation of exposed land (Interview Ornithological Society, 28/3/2005).

---

[86]   The development of the IACS for the management of agricultural subsidy payments was required in all member states by the EU in the context of the Common Agricultural Policy. The IACS is a system capable of storing information on every land parcel used for agricultural purposes (including details on parcel size, crops grown and ownership records).

Later, in 1989-1994, there was another prolonged drought period. Mean annual rainfall in the area dropped from a high of > 600 mm to 300 mm between 1989 and 1993, but after that it increased again to > 400 mm. Therefore, the progressive drop of the lake from at least 1989 to 1993 and beyond to 2001 cannot be ascribed principally and only to climatic factors (Mitraki et al., 2004). In reality, while natural water resource replenishment in the basin decreased due to climatic conditions, water demand continued to rise due to economic development.

The explosive development of agricultural and industrial activities in the 1980s had a significant quantitative impact on water resources. The level of the shallow aquifer in the western part of the Koronia subbasin dropped progressively during 1969-1981 as irrigated surface increased (Mitraki et al., 2004). The western part of the basin was also the location of extensive aquifer abstractions by fabric-dye industries. In total, water consumption by agriculture exceeded by far that of industry (see Table 8.1). However, industry exerted a stronger water demand locally while agriculture affected the entire basin (Mitraki et al., 2004).

The quantitative degradation of the water resources was further aggravated when, in 1990, large pumps were installed on the lake to supply two collective irrigation networks (Newspaper "Thessaloniki", 3/7/2001). Since the early 1990s, there was also a sudden increase in the drilling of private wells due to a combination of factors. First, there was an increase in agricultural activity supported by EU development programmes which funded infrastructure for the exploitation of farmland. Secondly, in 1990, the public Land Reclamation Service (of the Ministry of Agriculture in the Prefecture of Thessaloniki) stopped drilling wells on behalf of farmers (see similar development in the Vegoritida basin in chapter 7/7.6.1). From then on, the drilling of wells was exclusively in the hands of the private engineering sector and the ability of public authorities to control the drilling activity was reduced (Interview Prefecture, 1/4/2005). Altogether, since 1990, the large pumps installed for the collective irrigation networks together with numerous private wells abstracted large amounts of water from the basin.

All in all, the dramatic decrease of the level of Lake Koronia can be attributed to the systematic groundwater exploitation for irrigation and industry, the abstraction of water directly from the lake for irrigation until 1995 and the decrease of natural inflow due to dry weather. The result of the lake drop was:

- Eutrophication and anoxic conditions on the lake bottom, combined with pollution from urban and industrial effluents and fertilizers.

- Decrease of fish population and of habitats suitable for birds and amphibians.

- Reduction of the lake surface by ca. 15 km² until October 1995. For the (often illegal) cultivation of the exposed land, irrigation water was mainly abstracted directly from the lake.

All in all, the quantitative degradation of water resources in the Koronia subbasin is linked to a rivalry between agriculture and industry versus the living environment and commercial fisheries over water resources. Data on water consumption indicate that quantitative degradation is to its larger extent due to overabstraction for irrigation. In the same time, agricultural users prefer to shift or share the blame with industrial users. Fisheries had no strong presence as a conflictual party but their aims in the rivalry were largely served by the objectives of environmental protection.

Rivalry of discharge of pollutants vs. the environment and fisheries

In the beginning of the 1970s, Lake Koronia still had good water quality and it was even considered as potential source of drinking water. In 1973, the French company BRGM carried out first systematic research on the potential of lakes Koronia and Volvi as water supply source for the city of Thessaloniki (Tsagarlis, 1998).

Since then, agricultural, industrial and urban activities gradually polluted water resources in Lake Koronia, affecting both the environment and fisheries. In the same time, water quality degradation was also a result of the decrease of the water volume of the lake.

Early eutrophication of Lake Koronia was attributed to the untreated wastewater of the city of Langadas and the use of fertilisers in the surrounding fields (Tsagarlis, 1998). The city of Langadas started discharging untreated urban wastewater into the lake approximately in 1978 (Newspaper "Thessaloniki", 5/4/93). After the start of explosive industrial development in the 1980s, also industrial effluents were added to the group of pressures on water quality. During the late 1980s and early 1990s, the lake became hypertrophic (Mitraki et al., 2004). A study based on monitoring data of 1999-2000 concluded that the lake is biologically "dead" and its water is unsuitable for any use (AUTH, 2002).

The main conflict over water quality in the Koronia subbasin involved the uses of industry, urban settlements and agriculture (polluters), the living environment and fisheries. In terms of the polluters, agriculture is given much less the blame than industry and human settlements. Fisheries have been again a less obvious part of the conflict than the environment but more present than in the conflict over water quantity. The rivalry over water quality status in the basin was the source of most legal developments in the process described in the chronology of the case study.

8.4   Actors in the process of the Mygdonian basin

The actors involved in the water management process of the Mygdonian basin, with specific focus on the water use rivalries described above, include actors from the administration, water users, non-governmental groups and external experts. Table 8.2 presents the main relevant actors from the different levels of the administration, while Table 8.3 lists the main water users, non-governmental actors and external experts. More specific information on the actors' roles in the case study process is given in section 8.7.

Table 8.2   Actors of the administration (Mygdonian basin)

Administrative level	Actors	Main tasks and competence in water management	Policy field relevance
International	Ramsar Convention	Monitors progress of the implementation of the Ramsar wetlands treaty in the Greek Ramsar site of	Wetland protection

Administrative level	Actors	Main tasks and competence in water management	Policy field relevance
	(Conference of the Parties)	Koronia	
European Union (EU)		Judicial competence as far as the implementation of EU directives is concerned	All fields
		EU Cohesion Fund was source of funding for the Master Plan to restore Lake Koronia and the UWWTP of Langadas	
		EU was source of agricultural subsidies	
National		Definition of national framework legislation, policy-making	
	Ministry of Environment, Physical Planning and Public Works	Competence for national implementation of the international Ramsar treaty for wetlands	Spatial planning
		National competence for the implementation of EU and national environmental legislation:	Environmental protection
		- EU water quality and environmental directives	Water management (> 2003)
		- 1986 Environment Law	
		- 2003 Water Law	
	Ministry of Development	National competence for the implementation of the 1987 Water Law	Industry
		National competence for the regulation of industries	Water management (1987 – 2003)
	Ministry of Agriculture	Competence in national water legislation relevant to agriculture, prior to the 1987 Water Law, including earlier regulations on authorisations for wells	Agriculture, fisheries
		Management of agricultural subsidies	
		Formulation and funding of agrienvironmental policy measures	
	Ministry of Economy	Involved in the funding application of the Koronia Master Plan to the Cohesion Fund	
Regional			
	Region of Central Macedonia	Application of national legislation and responsibility for planning on the regional level	All fields
	Regional Water Management Department (RWMD)	Main authority responsible for the implementation of the 1987 Water Law and the 2003 Water Law:	Water protection (quantity extended also to quality by 2003 Water Law)
		- 1987 Water Law: Water management competence for the water district including the Mygdonian basin; issued permits for water use and water works for energy production and for multiple use of water; approved all permits for water use and water works, before they could be issued by other authorities	
		- 2003 Water Law: It is now responsible for issuing <u>all</u> water permits	
		- Proposes restrictive measures for quantitative water protection to be issued by the Prefecture	
	Regional Directorate for the Environment and Physical Planning	Approval of certain EIAs (depending on type of project)	Environment, spatial planning
		Competence in issues of environmental and nature protection	

Administrative level	Actors	Main tasks and competence in water management	Policy field relevance
Prefectural			
	Prefecture of Thessaloniki	Regulatory and administrative competence, responsibility for planning on the Prefectural level	
		Elected Prefects give final approval for sanctions in case of non-compliance with Prefectural regulations	
		Prefectural Council is consulted on the approval of EIAs for projects within the Prefecture	
		Prefecture approves certain EIAs (depending on project type)	
	Division of Water Resources and Land Reclamation	Issues Prefectural restrictive regulations on water quantity protection	Water management (quantity) in irrigation, land reclamation
		Under the 1987 Water law, it issued permits for water use and water works in irrigation, animal farms, aquaculture (this competence was transferred to the Region by the 2003 Water Law)	
		It used to be the regional Service for Land Reclamation of the Ministry of Agriculture, responsible for regulating the drilling of wells for all water uses (prior to 1987 Water Law)	
	Division of Environmental Protection	Since 1995, it issues all effluent disposal permits in the Prefecture	Water quality
		It is responsible for environmental control and inspections	Environmental protection
	Division of Industry	Under the 1987 Water Law, it issued water permits for industrial abstractions (competence transferred to the Region by the 2003 Water Law)	Industry
		It issues operation licenses for industries and is therefore competent for the legality of the operation of industrial units in the Koronia subbasin	
	Division of Health	Until 1995, it was competent for issuing disposal permits of wastewater and effluents	Water quality
			Health
	Division of Agricultural Development	Day-to-day management of the agrienvironmental policy measure adopted for the basin in 2003	Agriculture
	Sub-prefecture of Langadas	It is subordinate to the Prefecture and responsible for the administration of a certain geographic area (Langadas) of the Prefecture. This geographic area includes the subbasin of Koronia	
		The competence of its divisions is more restricted than of the respective divisions of the Prefecture	
	Sub-prefectural department of Fisheries	Responsible for fisheries in Lake Koronia (located in the sub-Prefecture of Langadas)	
Municipalities and communities		Competence for the delivery of water supply, sewerage and wastewater treatment services	Water supply
		Under the 1987 Water Law, the Union of municipalities and communities issued water permits for domestic use, mainly drinking water supply (competence transferred to the Region by the 2003 Water Law)	Wastewater treatment
		Local self-administration and rural police forces have responsibility for compliance-check of terms of water	

245

Administrative level	Actors	Main tasks and competence in water management	Policy field relevance
		use permits and wise use of water in their territory (assisted by the permit-issuing authorities) (Presidential Decree 256/89)	
Local organisations for land reclamation		Responsibility for management and distribution of water in collective irrigation networks	Agriculture

Table 8.3   Water users, non-governmental groups and experts (Mygdonian basin)

Actor	Actor information
Users	
Farmers	Farmers were represented mainly by three legally recognised organizations (Interview Young Farmers Union, 14/4/2005):
	1) Union of Agricultural Cooperatives of Langadas
	90% of farmers of the area of Langadas are part of this Union. The individual agricultural cooperatives did not have a role in the process. The Union's role in practice was mainly limited to the management of agricultural subsidy funds. Nevertheless, this Union was also occasionally contacted by public authorities and invited to meetings for issues relevant to Lake Koronia e.g. on the Master Plan for its restoration
	2) Union of Young Farmers of the Thessaloniki Prefecture
	It is a relatively recent Union founded in 1/1996. It represents a small group of farmers aged between 18-40 years old (200 members in the whole Prefecture). This Union is more progressive than other farmer associations regarding farming practices. It keeps track of new policy developments as well as alternative farming techniques such as bio-farming
	3) Farmers' Association of Langadas
	All farmers are members of this association, including also all large-scale farmers who cultivate 80% of land in the area
Industries	Industries can be represented by a number of unions (Commercial and Industrial Chamber, Chamber of Business Enterprises, Union of Industries of Northern Greece)
	However, in the process of the case study, industries participated in discussions and negotiations only on an individual basis and not through representative unions
Fishermen	There was no active association representing fishermen of Koronia in the process. An existing cooperative did not operate properly. Fishermen were in fact not organised at all, since they were not united to one another by profit after the collapse of fisheries (Interview Langadas Sub-Pref., 5/4/2005)
Associations and NGOs	
Union for the environmental protection of the wider region of Langadas	This Union was the most important local NGO
Hellenic Ornithological Society	This is an important NGO on the national level with significant local action in the conservation of the Koronia wetland
Ecological movement of Thessaloniki	Ecological organisation with links to the municipal and prefectural party "Ecology, Solidarity and Coalition of Citizens" which is represented in the Prefectural Council of Thessaloniki
External experts	
Technical Chamber of Greece	The Technical Chamber of Greece represents nation-wide professional engineers. It is a public legal entity with elected administration and has been defined as the official adviser of the State in technical matters
Academic community	
Greek Centre for Biotopes and Wetlands	Centre for research, public awareness and information aiming at the sustainability of natural resources in Greece (based in Thessaloniki)

8.5 Property rights

The following discusses rights related to water quantitative use, rights related to water quality and rights to use water for fisheries, with specific emphasis on the Koronia subbasin.

8.5.1 Rights to water quantitative use

General regulatory framework

The implementation of the 1987 Water Law in the Prefecture of Thessaloniki took off after 1989, with the introduction of the system of permits for water use and water works as well as of Prefectural decisions to restrict excessive water use. In fact, regulations to protect water resources were issued in the Prefecture even earlier than that, but these were limited to groundwater protection only.

Any abstractions which started after 1989 without a permit are considered illegal, while abstractions prior to 1989 based on custom or law were maintained ("pre-1989 established rights"). Prior to 1989, authorisation from public authorities was only needed for the construction of wells and their connection to the power supply system, but not for the use of water itself. In the context of the national agricultural electrification programme, connection to cheap power supply was granted by the regional Land Reclamation Service of the Ministry of Agriculture (which was later transformed into the Division for Water Resources and Land Reclamation of the Prefecture). The electrification programme led to a switch of many water abstraction pumps in the Koronia subbasin from gasoline to power supply (electric pumps increased from 280 in 1971 to 1260 in 1996) (Knight Piesold et al., 1998). Wells, which had been connected to power supply by the Land Reclamation Service prior to 1989, were also considered legal by its successor Prefectural Division for Land Reclamation. The RWMD of the Region, however, which was responsible for implementing the principles of the 1987 Water Law regionally, did not consider this pre-1989 connection of wells to the power supply network as adequate sole reason to grant the status of a "pre-1989 established right based on custom or law which can be maintained". It was argued that such wells connected to power supply prior to 1989 had not been legalized in a strict sense by any specific law of the Ministry of Agriculture. The situation was further complicated by the lack of specific policy guidelines from the national level regarding the definition of such "pre-1989 established rights". In this context, the RWMD emphasised the necessity for a national regulation, defining criteria and evidence needed for the identification and verification of pre-1989 established rights in the sense of the law (Interview Prefecture, 1/4/2005; pers. comm. RCM, 31/3/2005).

The first Prefectural decision specifically aimed at the protection of the Koronia subbasin was issued in 4/1995, after a period of prolonged drought from 1989 to 1993 and under the influence of an awareness project for lakes Koronia and Volvi (see section 8.6 on the process chronology). This Prefectural decision in 1995 prohibited the direct abstraction of water from Lake Koronia except for two collective irrigation networks which had only to reduce abstractions by 20% annually, the drilling of wells within a zone of 300 m around the lake and the cultivation of land exposed by the dropping lake.

Despite these restrictive regulations, the lake continued to drop over the summer of 1995. Following the incident of mass fish kills in 8/1995, the Prefecture revised its Prefectural decision on water use in the Koronia subbasin. In the meantime, Prefecture officials had admitted that the imposed restrictions were not implemented adequately partly due to the lack of personnel of local police stations which were competent for inspections. It was also observed that local farmers continued to abstract water directly from the lake (Koutrakis & Blionis, 1995). The amended Prefectural decision of 9/1995 included:

- Extension of the prohibition on the drilling of wells to 500 m around the lake.

- Changes in the operation of two collective irrigation networks which abstracted water directly from the lake:

    o Discontinuation of the operation of one irrigation network.

    o Investigation of the possibility to stop operating the second irrigation network if alternative water sources were to be found.

- Increase of fines for those who cultivated exposed land.

In 9/1996, the drilling of all new wells was prohibited in the entire basin of Lake Koronia, except for use in drinking water supply, until the end of 1998.

The Prefecture issued even stricter measures to restrict excessive water use around Lake Koronia in 2000 and again in 2004 following specific conditions set in the 2004 CMD establishing a National Wetland Park in the Mygdonian water basin (see Annex VI for an overview of the Prefectural restrictive measures).

Overview of rights

Table 8.4 gives an overview of the main rights to use water quantitatively as well as the relevant policies and regulations.

Table 8.4   Rights to quantitative water use in the Koronia subbasin

Use	Users	Use rights	Relevant policies and regulations
Living environment	Flora and fauna	No water right defined (no minimum lake level defined according to the 1987 Water Law)	1987 Water Law
			2003 Water Law
			1986 Environment Law
			CMD 6919 of 2004 designating the National Wetland Park
Irrigation	Farmers	*Surface water:*	1987 Water Law
		After 1989: rights on the basis of permits for water use and water works are required	2003 Water Law
			L. 1988/1952 on drilling wells
		In 1995, water uptake from Lake Koronia for irrigation (or other purpose) was prohibited. The direct uptake was again prohibited in 2004. Illegal	Prefectural regulations and restrictive

Use	Users	Use rights	Relevant policies and regulations
		private use of lake water has been limited	measures on water protection
		Groundwater:	CMD 6919 of 2004 designating the National Wetland Park
		Before 1989: authorisation to drill private wells and gain electricity was required	
		After 1989: rights on the basis of permits for water use and works are required; uses prior to 1989 based on law or custom are maintained	
Industry	Industries	*Groundwater:*	1987 Water Law
		Before 1989: authorisation to drill wells was required	2003 Water Law
		After 1989: rights on the basis of permits for water use and works are required; uses prior to 1989 based on law or custom are maintained	Decision of the Ministry of Agriculture of 1968 (26278/3128/701) on drilling wells for several uses where restrictive measures apply
		There are only a few industrial wells in the basin with a permit according to the 1987 Water Law. Most industrial wells were drilled prior to 1989, while there were also some cases of illegal drilling of industrial wells for industrial use after 1989 (pers. comm. RCM, 31/3/2005)	Prefectural regulations and restrictive measures on water protection
Recreation		No right defined	
Fisheries	Fishermen	No minimum lake level defined	
Water supply	Municipalities communities	*Groundwater:*	1987 Water Law
		After 1989: permits for water use are required; uses prior to 1989 based on law or custom are maintained	2003 Water Law
		Illegal wells for water supply mainly concerned wells drilled on private property to provide water to houses where no public water supply network was available (pers. comm. RCM, 31/3/2005)	Prefectural regulations and restrictive measures on water protection
		After 2003 (2003 Water Law): permits for water uses and water works for water supply are required	CMD 6919 of 2004 designating the National Wetland Park

Water use rights in the subbasin of Koronia are mainly linked to the use of wells for water abstraction from the aquifer. Most wells around Lake Koronia are private. First attempts to count wells in the subbasin of Koronia took place in the early 1990s. By 1992, 2,117 wells were registered with the Land Reclamation Division of the Prefecture but the (potentially high) number of illegal wells remained unknown to the authorities (Grammatikopoulou et al., 1996). As early as 1995, experts of the Prefecture admitted that there was no official control on wells, especially for irrigation (Koutrakis & Blionis, 1995). In 1998, Knight Piesold et al. (1998) made a rough estimation of 2,000 wells in the basin, including illegal and abandoned ones. Later on, a detailed survey of wells recorded 1,975 wells in the basin; 237 of these held a water permit (for irrigation) following the 1987 Water Law, while 1,738 wells were drilled prior to the 1987 Water Law and to their greatest extent were connected to the power supply network following earlier regulations.[87] Information from other experts and the press

---

[87] Prefecture of Thessaloniki, Division for Water Resources and Land Reclamation, Document Nr. 09/2005, Subject: Control of illegal wells in the Prefecture of Thessaloniki, 7/2/2005.

indicates that the number of existing wells (by 2004) was even higher reaching 2,300, including legal, illegal and abandoned wells for several uses (irrigation, industry, drinking water) (Interview Langadas Sub-Pref., 5/4/2005; Interview University of Thessaloniki, 29/3/2005; Newspaper "Macedonia", 10/12/04).

8.5.2 Rights to water as medium for pollutants

Table 8.5 gives an overview of rights to use water as a medium for the transport and absorption of pollutants. In the following, key issues of the relevant regulatory framework affecting water pollution rights are described. Rights of industrial and urban polluters are then discussed in more detail, additionally to the information given in Table 8.5.

Table 8.5   Rights to water as medium for pollutants in the Koronia subbasin

Use	Users	Rights to use water	Relevant policies and regulations
Living environment	Flora and fauna	Right to water of good quality set by policy	CMD of 2004 on the designation of the National Wetland Park of Lakes Koronia and Volvi
			CMD of 2005 on a "Special pollution reduction programme for Lake Koronia" (it referred to ecosystem aspects but did not set specific ecosystem aims)
Water as medium for discharge of pollutants	Industries	Rights defined by effluent disposal permits	Prefectural regulations
		Most but not all industrial units in the Koronia subbasin hold effluent disposal permits (see Table 8.6 for details)	EU Urban Wastewater Treatment Directive
			CMD of 2005 on a "Special pollution reduction programme for Lake Koronia"
	Municipalities, communities	Rights defined by effluent disposal permits	Prefectural regulations
		No urban polluters in the basin hold relevant disposal permits	EU Urban Wastewater Treatment Directive
	Farmers	Rights for animal farms to pollute water are defined by effluent disposal permits	Codes of Good Agricultural Practice
		Most farms in the basin do not hold permits (illegal effluents disposal). A survey in 1997 indicated that, in the basins of Koronia and Volvi, only 24 of 95 animal farms held a disposal permit.[88] Most animal farms do not even have operation licenses (Arabatzi-Karra et al., 1996)	Prefectural regulations
			CMD of 2005 on a "Special pollution reduction programme for Lake Koronia" sets standards for pesticides and nutrients in the water of Lake Koronia
		Rights of crop farms to pollute water were defined indirectly by posing limitations on the application of fertilisers in the CMD of 2006 on nitrate pollution reduction	CMD of 2006 on an Action Programme for the reduction of nitrate pollution of agricultural origin in the plain of Thessaloniki-Pella-Imathia
Recreation	Broader public	No specific right defined	
Fisheries	Fishermen	Rights to water of good quality set by policy	EU Water Directive for Fish

---

[88] Calculated by the author based on Tsagarlis (1998).

Use	Users	Rights to use water	Relevant policies and regulations
		standards	CMD of 2005 on a "Special pollution reduction programme for Lake Koronia" (it set aim to safeguard fisheries as a use)

Regulatory framework

Except for national and EU legislation which defines a general legal framework for the protection of water quality, specific Prefectural regulations aimed at the restriction of water pollution in the subbasin of Koronia. The earliest Prefectural Decisions of this kind were issued in 1968 and 1972.[89] These were based on the principles of the 1965 national Sanitary Regulation and they defined Lake Koronia as a potential recipient of wastewater whose quality should remain good enough to serve drinking water supply. In fact, despite the great quality degradation of Lake Koronia in the coming decades, the intended use of Lake Koronia remained that of drinking water supply until 2002, when a Prefectural Decision re-defined the intended use of the lake to water suitable for fish survival and irrigation.[90]

The first applications of polluters for effluent disposal permits were submitted to the Prefecture of Thessaloniki in 1973. At that stage, the Technical Service of Municipalities and Communities and the Health Division of the Prefecture were defined as competent authorities for the issuing of effluent disposal permits according to the 1965 Sanitary Regulation. However, the Technical Service of Municipalities and Communities could not play any significant role in this process nation-wide due to lack of scientific personnel. Therefore, the implementation task was left in the hands of the Health Divisions of the Prefectures. Since the Health Divisions were not equipped with the appropriate scientific personnel either, committees of experts from various Prefectural Divisions were set up to control effluent disposal and treatment studies submitted by polluters and to control the construction of effluent treatment units (Tsagarlis, 1998). In 1995, in the context of reforming Prefectures to a 2nd level of local self-administration, the Prefecture of Thessaloniki finally set up a separate Division for Environmental Protection which took over competence on the Sanitary Regulation and effluent disposal permits. Permits in the Prefecture of Thessaloniki were from the on issued by a Joint Committee of the Prefectural Divisions for Environmental Protection, for Health and for Industry (Interview Prefecture, 12/4/2005).

The first Prefectural Decision which defined specific conditions and maximum limits on the concentration of pollutants in wastewater and effluents disposed into Lake Koronia was issued in 1983 and amended in 1994. The Prefectural Decisions of 1983 and 1994 did not define specific conditions for effluents disposal into other waters of the Koronia subbasin except for the lake itself. In practice, the same conditions defined for emissions into Lake Koronia also applied to the entire drainage network of its subbasin (pers. comm. Prefecture,

---

[89] Prefectural Decisions Nr. 103349/5-9-68 and Nr. KY/56860/30-12-72.

[90] Prefectural Decision Nr. 30/1585/28.3.2002, Official Government Journal B 525/2002.

12/4/2005). However, industrial activities located further than 3 km from Lake Koronia, which was in fact the case for most industries in the area, were allowed to exceed by 30% the pollutants' limits set for Lake Koronia. Additionally, the 1994 Prefectural Decision prohibited any increase of the discharge of wastewater and effluents around Lake Koronia from the quantities defined in the effluents treatment studies submitted to the Prefecture. However, in practice, no attention was ever really paid to this prohibition (Tsagarlis, 1998).

In 2004, the Prefectural Division for Environmental Protection invited other divisions to make proposals for amending the quality objectives for effluents discharged into surface waters of the Prefecture. In this context, the department for fisheries of the Sub-prefecture of Langadas proposed to characterise some of the streams and torrents draining into Lake Koronia as waters suitable for cyprinoids with their own minimum water quality objectives for fish life. Such characterization would consider the need for good water quality also in the drainage network of the lake. The approach of the 3 km limit around the lake above which pollution parameters in effluents could increase by 30% was characterised as an insufficient technical definition which often did not respond to the ecosystem needs of the entire Koronia subbasin (Interview Langadas Sub-Pref., 5/4/2005).

There was also critique that the method of issuing effluent disposal permits did not consider the self-purification capacity of the environment. Receiving waters were considered as deposits with unlimited absorbing capacity. In this context, there were considerations to establish a new method for issuing disposal permits in the Prefecture based on the total pollution load in kg/d that each industry could discharge, instead of setting limits on the concentration of each pollutant in the discharged effluent. This alternative method emphasized the total pollution load discharged and the absorbing capacity of the receiving water (Tsagarlis, 1998).

By the end of this case study, no relevant amendments had been made to the Prefectural Decision on wastewater and effluents.

In 2005, water quality objectives for Lake Koronia were defined in the CMD for the adoption of a Special Pollution Reduction Programme for the Lake, which set maximum allowed concentrations of organic and toxic substances in the water of the lake. With this development, the approach for the water pollution control of Lake Koronia moved towards a combination of specific water quality objectives and emission limit values.

Additional policy requirements for the water quality of Lake Koronia were also set in the context of important EU Directives. In 8/1999, Lake Koronia was designated by the Environment Ministry as sensitive area according to the EU Urban Wastewater Treatment Directive (Official Government Journal B΄ 1811/99). Furthermore, the wider region of Lake Koronia was designated as a nitrates vulnerable zone according to the EU Nitrates Directive 91/676/EEC and in 2006 a Common Ministerial Decision was adopted for an Action Programme to reduce nitrate pollution from agriculture in a broader area including lakes Koronia and Volvi.

Rights of industry

According to Table 8.6, the number of industries operating in the Koronia subbasin with a final permit to dispose effluents increased since the mid-1990s (see Annex VIII for a detailed list of industries and the status of their effluent disposal permits). This progress in the official issuing of final permits was partly due to political pressure exercised to protect the international Ramsar wetland site of lakes Koronia and Volvi (Interview Prefecture, 12/4/2005).

Table 8.6   Effluent disposal permits of industries in the Koronia subbasin

Year	Number of industries	Industries with final permit	Industries with temporary permit	Industries with expired temporary permit	Industries with no permit
1996	20	7	7	3	3
1998	22	11	3	4	4
2005	16	15	-	-	1

Source: Numbers compiled from a variety of reports and personal communications.

In the past, most industrial units in the Koronia subbasin did not operate legally to the full extent. Most had no operation license or only a temporary one for several years. The main reason for the lack of permanent operation licenses was the inadequate treatment of wastewater, since effluent treatment and final effluent disposal permits were necessary conditions for the concession of a permanent industrial operation license. By the mid-1990s, only few industrial units in the basin had carried out all works necessary for their legal operation. Most industries discharged their effluents untreated to nearby torrents making industry incompatible with the environmental use of the Koronia wetland (Arabatzi-Karra et al., 1996). Even industries with installed effluent treatment did not meet the required quality standards or operated the treatment unit only occasionally (Grammatikopoulou et al., 1996). At times, fines were imposed on some industrial units (Knight Piesold et al., 1998) either for not having an operation license or for inadequate operation of their effluents treatment facilities.

By the end of this case study, even after issuing final effluent disposal permits for most industries, there was still an issue of non-compliance of industries with the terms set in disposal permits (Interview Prefecture, 12/4/2005).

Rights of urban users

In the first decades after issuing the national 1965 Sanitary Regulation, unregulated disposal of urban wastewater was not alarming for the water status in the subbasin of Koronia, because there was no intense wastewater pollution problem. The situation changed following the large increase of the population of the city of Langadas, which brought the need for a change in the regulation of wastewater (Interview Prefecture, 12/4/2005).

Still, even after the issuing of specific Prefectural regulations on wastewater disposal, municipalities and communities in the Koronia subbasin were unable to meet up to the challenge of treating wastewater on the basis of approved wastewater disposal permits. Not even the city of Langadas, the largest urban centre in the basin, held a disposal permit for its wastewater disposed through an old sewerage network (Tsagarlis, 1998; Prefectural Division for Environmental Protection, pers.comm., 2005).

Most municipalities and communities in the basin are served either by incomplete sewerage networks or cesspits and tanks. The few settlements with sewerage networks do not hold permits for wastewater disposal (Tsagarlis, 1998). Wastewater from cesspits and tanks is also discharged without any permit or control into surface waters, although disposal is only allowed at the existing treatment plants in the Prefecture of Thessaloniki. It is difficult to estimate the amount of effluents (industrial and urban) from cesspits and tanks which are illegally discharged into surface waters. A rough estimate for the whole Prefecture of Thessaloniki is 1,000 m³/d (Tsagarlis, 1998). The transport of wastewater to treatment plants, which is required by legislation, raises costs for the transporting freight trucks which consumers usually deny to pay (this was also the case in the city of Langadas). Therefore, the illegal practice of disposal in remote locations is preferred, despite the potential health impacts (Interview Prefecture, 12/4/2005).

8.5.3 Rights to water as medium for fisheries

For the use of fisheries, good environmental status of Lake Koronia is necessary to support fish habitats and fish populations.

Fishing rights for inland waters in Greece are usually conceded by the State as a priority to fishing cooperatives or to private individuals for a certain period of time based on legislation and fishing codes. The concession holder is then responsible for controlling fishing activity, charging fishermen with ca. 10% of their earning from fish catches and paying an agreed rent to the State. Lake Koronia, however, was the only lake in the country whose fishing reserves were controlled directly by the central State. The reason was that Koronia used to be the most productive lake in the country and the State made profit from this form of direct taxation of fish catches (Interview Langadas Sub-Pref., 5/4/2005).

After the extirpation of Koronia's fish populations in 1995, fishing rights were practically eliminated via fishing prohibitions by the Prefecture. Active management of fish resources of the lake was also undermined after the administrative reform of the Prefectures to 2nd level of local self-administration in 1995. During the reform, the regional Fisheries Department of the Ministry of Agriculture, formerly competent for fisheries in Lake Koronia, became part of the Prefectural administration and its staff was drastically reduced. After its attachment to the Prefecture, the Fisheries Department was no longer directly competent for Lake Koronia since competence on fish resource management remained in the hands of the Ministry of Agriculture. However, the Ministry of Agriculture remained rather inactive in its role of fisheries manager, leaving the commercial fish populations of Koronia practically unmanaged (Interview Langadas Sub-Pref., 5/4/2005).

In 2002, the lake started to rise again and, despite bad environmental conditions, fishermen illegally introduced some fish species into Lake Koronia (Birtsas, 2005). The fish reproduced quickly and fishermen started to fish again, although the Prefectural fishing prohibition was still valid. Fishing prohibitions by the Prefecture were reinstated in 2003, in 2004 (following the incident of mass bird kills) and again in 2005. Illegal fishing became very extensive in Lake Koronia also due to lack of resources to inspect the lake continuously. Given the bad environmental conditions of the lake, illegal fishing could pose a health threat due to the low quality of fish (Newspaper „Aggelioforos", 20/4/2004).

8.6 Chronology of case study events

This section presents the common chronology of events related to the two key water use rivalries presented in section 7.3. The management process of the Mygdonian basin, with specific focus on the Koronia subbasin, evolved around a series of water protection regulations issued occasionally and two core efforts to establish a management scheme in the area, the one steered by the Prefecture (Master Plan for the restoration of Lake Koronia) and the other steered by the Environment Ministry (legal protection framework for the designation of a National Wetland Park).

Gradual water degradation and sporadic administrative (re)action

Although warnings on the continuous degradation of the status of Lake Koronia were given by scientists and academic experts since the 1970s,[91] in the 1980s there were still no effective regulations and environmental controls applied around Lake Koronia (Knight Piesold et al., 1998).

On behalf of the public administration, the central government (specifically the Environment Ministry) was the first who started taking some sporadic action by carrying out studies in the Koronia subbasin in line with the national commitment of Greece to the international Ramsar treaty for wetlands. These studies were mainly concerned with the delimitation of boundaries and zones of protection of the Ramsar wetland area.[92] The studies assigned by the Environment Ministry aimed also at serving as the basis for a Common Ministerial Decision (CMD) which would establish a legal protection regime for the wetland according to the 1986 national Environment Law.

In the 1990s, the emerging severe environmental problems of Lake Koronia also sensibilised to some extent the Prefecture of Thessaloniki which issued several decisions on:

- The prohibition of drilling new wells around Lake Koronia and of any type of surface water abstraction from the lake (1995 first Prefectural Decision to protect water resources of Lake Koronia).

---

[91] Greek Biotopes/Wetlands Centre, Document Nr. 7950, Addressed to the Prefecture of Thessaloniki, Subject: Comments on the revised plan for the restoration of Koronia, 13/7/2004.

[92] In 1986, the Environment Ministry completed a programme for the definition of boundaries of wetlands of the Ramsar Convention, which included the area of Lakes Koronia and Volvi.

In 1992, the Department for Environmental Planning of the Environment Ministry co-funded with the EU a study on the boundary definition and management of the wetland of lakes Koronia and Volvi aiming at defining three zones of protection for the wetland.

- The prohibition of cultivation of exposed areas of the dropping lake (relevant Prefectural decisions since 1992).

In 1995, an informal committee of the Prefecture (consisting of the Divisions of Industry, of Environmental Protection and of Health) drafted a technical report on polluting activities in the basin and proposed to impose fines on several industries (Grammatikopoulou et al., 1995). Although, at times, fines were imposed on some industries for illegal effluent disposal by the Prefectural Division for Environmental Protection (Kazantzidis et al., 1995), implementation of regulations remained inadequate and water quality in the 1990s continued to deterioriate.

First awareness efforts

In 1994-1995, the Greek Biotopes and Wetlands Centre, in cooperation with the Environment Ministry and the Ministry of Agriculture, carried out the first public information and awareness campaign on the environmental status of lakes Koronia and Volvi within the framework of the MedWet initiative[93]. This campaign was targeted at decision-makers, fishermen, farmers as well as teachers and schoolchildren.

Fishermen involved in the campaign activities were also quite positive to protection measures. On the contrary, farmers were very sparsely represented in the discussions. Moreover, although the campaign had a positive influence on the public administration, officials of the administration expressed disappointment about the lack of practical action and the lack of interest of the local society (Interview Former expert at Information Centre, 23/3/2005).

The campaign received a high degree of response from mass media, making the pressures on the ecosystem of Koronia more widely known to decision-makers and the public. Press publicity also contributed to bringing the problems of Koronia to the National and European Parliament by Greek MPs (Koutrakis & Blionis, 1995). In the beginning of 1995, the issue was brought to the European Parliament by a Greek MP asking whether the EC could force Greece to take the necessary measures for implementing the Ramsar convention and two relevant EU directives on birds and habitats in the wetland of Koronia.

Environmental collapse and start of problem definition

Through 1995, Lake Koronia continued to drop. In August 1995, an ecosystem catastrophe took place whereby mass fish kills extirpated the entire fish population of Lake Koronia and professional fishing was discontinued. A month later, large populations of birds died on the lake (Knight Piesold et al., 1998; Arabatzi-Karra et al., 1996).

The incident of mass fish kills rang the alarm bell on the environmental status of Lake Koronia. The severity of the situation steered up the public authorities. Based on the initiative

---

[93] MedWet stands for the Initiative for the Conservation and Sustainable Use of Mediterranean Wetlands. MedWet is a forum where twenty-five Mediterranean countries, specialized wetland centers and international environmental organizations meet as equals to discuss, identify key issues and take positive action to protect wetlands, for man and for biodiversity (source: www.medwet.org).

of the Sub-Prefecture of Langadas, a Rescuing Plan for direct action to save Lake Koronia was then prepared (Grammatikopoulou et al., 1996). The Rescuing Plan was the first holistic effort to document the environmental problems of the Koronia subbasin and to propose measures. Based on data already available, it was drafted on a voluntary basis by scientists of the Sub-Prefecture of Langadas and the Regional Division for the Environment and Spatial Planning.

The Rescuing Plan concluded that the lake was hypertrophic and anoxic conditions prohibited the survival of living organisms. It suggested that the water balance of the system was steadily negative, since annual water consumption exceeded the renewable water resources of the basin (Grammatikopoulou et al., 1995). At that stage, only 3 out of 20 industries operating in the basin held an effluent disposal permit and many did not even have an operation license (Kazantzidis et al., 1995). Most fabric-dye industries discharged their effluents untreated into torrents to avoid the costs of effluents treatment. Control of industries by the authorities was almost non-existent and took place only after alarming events such as the mass fish kills of 1995. The Rescuing Plan proposed measures such as restrictions on the drilling of wells, water recycling in industry, construction of small dams in the upper catchment and on-site delimitation of the protected wetland area (Grammatikopoulou et al., 1995). For certain environmental pressures, such as salt pollution from dye industries, no proposals were made because these pressures were not considered severe yet at that stage due to lack of appropriate knowledge and data (Interview Langadas Sub-Pref., 5/4/2005).

In terms of the practical implementation of the Rescuing Plan, progress was only made in the on-site delimitation of the protected wetland area. Most other proposed measures of the Plan were not taken forward because their promotion was not in the competence of the sensibilised sub-Prefecture of Langadas. There was need for solutions from higher administrative levels including the central government (Interview Langadas Sub-Pref., 5/4/2005).

Master Plan for the restoration of Lake Koronia

In 1997, under the pressure of the environmental problems of Lake Koronia, the Prefecture of Thessaloniki took the initiative to approach the Cohesion Fund of the EU DG-XVI and request funds for the restoration of Lake Koronia. As a first step, the EU assigned a preliminary study on the status of the lake to a Danish consultancy. The results of the Danish report (Rambold, 1997) convinced the EU of the bad environmental status of Koronia and the Cohesion Fund leadership committed itself to fund studies and works for the restoration of the lake. In 12/1997, the Cohesion Fund assigned the preparation of a Master Plan for the Environmental Restoration of Lake Koronia to a British-Greek consortium co-funded by the EU and the Greek government. The Prefecture took over responsibility for the development of the Master Plan and for the coordination of its drafting process.

Drafting process of the Master Plan

Phase A of the Master Plan consisted of the description of the causes of the environmental problems of Lake Koronia and of proposals for main technical solutions.

An intra-administrative Working Committee was set up by the Prefecture to supervise the Master Plan, including representatives of the Regional Division for Environment and Spatial Planning, of several Prefectural divisions[94] and the Sub-Prefecture of Langadas. The Prefectural Division of Environmental Protection was leading the process. The purpose of this Committee was to monitor progress of the technical consultants on the Master Plan and to inform and consult with actors relevant to the execution of certain projects, such as the Municipality of Langadas regarding the construction of its wastewater treatment plant and sewage network (Interview Langadas Sub-Pref., 5/4/2005; Interview Consultant to Master Plan, 7/4/2005).

The preliminary proposals of the draft Master Plan were presented at a public meeting in 5/1998 attended by the Minister of Environment, the deputy Minister of Agriculture, the Director of the EU Cohesion Fund, the Prefect of Thessaloniki, divisions of the Prefecture as well as representatives of the local administration of the Koronia communities (Knight Piesold et al., 1998). Environmental groups were not invited, although they had on several occasions requested in written to be informed on the Master Plan by the Prefecture (Interview Ornithological Society, 28/3/2005).

The draft Plan proposed several projects of a budget of 10 billion drachmas to be funded to 85% by the Cohesion Fund. The Director of the Cohesion Fund and the Minister of Environment committed themselves to secure the funds needed. In the meantime, in 1998, the construction of a wastewater treatment plant for the city of Langadas very near to Lake Koronia had already started (Newspaper "Aggelioforos", 10/5/98) financed by the EU and national sources (Interview Consultant to Master Plan, 20/4/2005).

In 8/1998, the Master Plan for the Restoration of Koronia was finalised after a process of intra-administrative consultation.

Main proposals of the Master Plan

The main objective of the Master Plan was to restore the lake back to its status of the 1970s in terms of water quality, water quantity and ecological aspects. Two concrete objectives were set:

- To reach the lake level of the 1970s (approximate lake volume of $200*10^6$ m³ and lake depth of 5m). To this aim, the Plan suggested that $45*10^6$ m³ of water should be added to the system annually. Water deficiency in the basin was calculated at $30*10^6$ m³ annually and was mainly attributed to overuse for irrigation. The Plan recognized that the assessment of the basin water balance was only average calculation based on available data and did not indicate the real water quantities consumed.

- To achieve water quality suitable for the survival of cyprinoids.

Table 8.7 presents the main measures proposed by the Master Plan.

---

[94] Divisions of Industry, of Land Reclamation, of Technical Services, of Agricultural Development and of Environmental Protection.

Table 8.7   Proposed measures of the Master Plan for the restoration of Lake Koronia

Proposed action	
I. Restoration of the level of Lake Koronia	**II. Restoration of good water quality**
Water transfer from River Aliakmonas	Execution of a planning study for the treatment of industrial effluents and wastewater, including effluents from animal farms, in the area of Lake Koronia.
Diversion of two torrents from Lake Volvi to Lake Koronia	Construction of UWWTP for the city of Langadas and extension for industries
Hydrological restoration by using water from the deep aquifer (short-term measure)	Construction of a sewage network for the town of Langadas
Restructuring existing irrigation networks	Langadas wastewater treatment with maturation lagoon
Reduction of water loss from irrigation (survey of wells in the basin, water meter installation, restrictive measures for water use, introduction of water pricing, replace spraying with drip irrigation)	Improvement of industrial effluents treatment
Agrienvironmental programme to reduce water use for irrigation	Treatment of effluents from animal farms
III. Achieving ecological balance of the lake	**IV. Supportive administrative structures**
Management of wetland vegetation	Issue the CMD for the National Wetland Park and create the respective Management Body
Enrichment of fish fauna	Set up a body competent for the implementation of the Master Plan and environmental inspections (reporting directly to the Environment Ministry)

Source: Knight Piesold et al. (1998).

Lack of public consultation

The drafting process of the Master Plan was criticized for lack of public consultation and participation since the Plan was finalised only with the involvement of the public administration. On few occasions, legally recognized representatives of water users only, such as the Union of Agricultural Cooperatives of Langadas, were invited to relevant meetings. Environmental groups were not involved at all.

Only after the finalization of the Master Plan, a public information and awareness campaign was initiated aiming at informing the broad public about the already finalised Master Plan. In this context, several seminars took place, also inviting water users such as industries and farmers. Nevertheless, considering the finalised state of the Master Plan, this information campaign was not equivalent to a consultation process on the Plan itself (Interview Ornithological Society, 28/3/2005).

As a result of the lack of public consultation during the drafting process, objections were expressed on certain proposals of the final Master Plan when a more participatory meeting[95] took place in 11/1998 to discuss the submission of the funding application to the Cohesion Fund (Newspaper "Macedonia", 17/11/98).

From the perspective of the NGOs, the final Master Plan avoided putting pressure on dominant interests in the area and did not bring about radical changes in key sectors of production (Ecological Movement of Thessaloniki, 2002). Proposed projects relied heavily on

---

[95]   Representatives of the Environment Ministry, the Prefecture, the Region, mayors of local municipalities, environmental groups and the Information Centre of Koronia and Volvi were present.

technical works and put off action in terms of policy measures. There was critique that large technical works (especially a proposed transfer of water from the basin of the River Aliakmon) were proposed under political pressure, although, initially, the Master Plan included several worthwhile proposals which could change the unsustainable development model of the basin. Large technical works enjoyed political support to show to the local population that action was being taken on the ground and to secure large engineering contracts (Interview Ornithological Society, 28/3/2005).

The scientific community also doubted whether certain proposals of the Master Plan were technically feasible, economically and environmentally sound.[96] A report of the Technical Chamber of Greece argued that water management solutions in the basin could not be achieved only through technical works. The problem was much related to the development model of the wider area, having strong economic and social dimensions and was therefore deeply political. The scientific community called for policy measures and actions to bring about changes in the development trend of the basin. Critique targeted the omission of the Master Plan to deal with the inability of the State and the public authorities to adequately implement regulations, licensing procedures and inspections. The Master Plan was specifically criticized for the lack of an environmental and socioeconomic scoping study for ecosystem restoration, the choice of inappropriate measures for hydrological restoration (e.g. the water transfer from River Aliakmon), the lack of adequate data to substantiate proposals and the lack of assessment of the ecological effectiveness of the proposed measures. Overall, a more modest and more realistic restoration target should have been adopted with slower but safer results (Delimbasis et al., 1998).

Also the green coalition participating in the Prefectural Council of Thessaloniki (the party "Ecology, Solidarity and Coalition of Citizens") denounced the Master Plan as technocratic and short of alternatives. It argued that external scientists and technical experts should have been involved (Newspaper "Macedonia", 19/11/98).

Even the technical consultants of the Master Plan had reservations concerning the proposed water transfer from River Aliakmon both on environmental and financial grounds (high operational cost)[97]. However, under pressure by local institutions and interest groups and after the consent of the Cohesion Fund leadership, the proposed transfer was finally put forward in the final Master Plan for further examination through cost-benefit analysis and EIA studies (Interview Consultant to Master Plan, 7/4/2005).

---

[96] Expression of opinion with regard to conservation actions for Greek Ramsar wetlands and to the applicability for removal from the Montreux Record (Annex IV), 7th Meeting of the Conference of the Contracting Parties to the Convention on Wetlands (Ramsar, Iran, 1971), "People and Wetlands: The Vital Link", an José, Costa Rica, 10-18 May 1999.

[97] The Master Plan's annual operational costs reached 2 billion drachmas (Newspaper "Macedonia", 29/7/04), especially due to the proposed water transfer.

Lack of coordination between central and local administration

The Master Plan drafting process was also short of cooperation between different levels of the administration, especially between the central and the Prefectural level. Although the protection of Koronia, being a Ramsar wetland of international importance, is officially in the hands of the Environment Ministry, the Ministry was not consulted during the drafting process of the Master Plan by the Prefecture. The Prefecture was therefore leading the drafting process of a potential management strategy for Lake Koronia without consulting with the Environment Ministry. The Master Plan was in fact developed in parallel but not in coordination with the preparation of a legal act to designate the wider basin of Koronia as a nature protection area by the Environment Ministry.

As a result of the lack of cooperation between the central and the Prefectural level, experts of the Environment Ministry (and other competent actors of the central administration) expressed their disagreement with several proposed projects of the finalised Master Plan, such as the water transfer from Aliakmon (Interview Ornithological Society, 28/3/2005; Environment Ministry phone comm., 18/11/2005). The fact that the Master Plan did not proceed to full realization (as described in the following) is partly due to the fact that its preparation was kept at the local level, while the restoration of Lake Koronia is of national importance (Interview University of Thessaloniki, 29/3/2005).

Master Plan stops half-way (formal rejection of proposed measures)

By 12/1999, funding from the Cohesion Fund was approved only for part of the Master Plan (phase A), in specific for: the purchase of monitoring equipment and water meters, studies and EIAs on exploiting the deep aquifer, on the transfer of water from the River Aliakmon and on the partial diversion of two torrents as well as studies on ecotourism development and on the treatment of industrial effluents (Statement of G. Kougiamis, Prefectural Councilor and President of the Prefectural Corporation of Thessaloniki, Newspaper "Chortiatis", Oct-Nov/2001). However, after the completion of all relevant studies and EIAs (phase A), most proposed measures and works of the Master Plan did not proceed to practical realisation (phase B). Several projects (mainly those proposed to restore the water level of Lake Koronia) did not receive the necessary approvals of environmental terms through the process of EIAs.

More specifically, the proposed diversion of two torrents from Lake Volvi to Lake Koronia was rejected by the Environment Ministry due to adverse effects on Lake Volvi. The proposal was also faced with protests of inhabitants around Lake Volvi (Statement of M. Tremopoulos, Prefectural Councilor from the "Ecology, Solidarity and Coalition of Citizens", Newspaper "Epochi", 8/12/2002).

The proposed hydrological restoration by using water from the deep aquifer was rejected by the Ministry of Development. The main argument was against spending energy to abstract water of good quality from the deep aquifer (at 300 m depth) in order to simply add it to a lake of very degraded water quality, without previously examining the lake ecological balance and food web (Interview University of Thessaloniki, 29/3/2005).

The proposed water transfer from the River Aliakmon proved to be the most controversial project of the Master Plan. From the start, there were serious doubts by several actors whether it could go ahead on environmental grounds.

On the one hand, there was intra-administrative disagreement within the Working Committee of the Prefecture supervising the Master Plan, whereby the sub-prefecture of Langadas (especially its Department of Fisheries) and the Regional Division for Environment and Spatial Planning disapproved this line of action. The Sub-prefecture characterized the water transfer as unsustainable and too costly in operation. It argued that an ecological approach should involve managing the water within the basin itself prior to any water transfer from another basin. It also objected the potential adverse effects of the modified discharge of the River Aliakmon on its delta (which is also a Ramsar site) (Interview Langadas Sub-Pref., 5/4/2005).

Objections to the water transfer were also expressed by the Environment Ministry, the Technical Chamber of Greece, some representatives of the local administration, environmental NGOs and the green party represented in the Prefectural Council.

On the other hand, the water transfer scheme was warmly supported by the local public of Langadas, especially farmers, because it would give them enough water to continue with business-as-usual in irrigation. For this reason, the proposed water transfer is considered to have hindered awareness raising of farmers on the environmental problems of the basin (Interview Langadas Sub-Pref., 5/4/2005).

The proposed water transfer also contradicted relevant international and EU principles for sustainable water management. On the EU level, the WFD had already been issued and its principles for integrated water management influenced national discussions on the suitability of the water transfer solution (Interview Langadas Sub-Pref., 5/4/2005). Even stronger pressure was exercised by the Ramsar Convention for wetlands of international importance. In 5/1999, in COP 7 (Conference of the Contracting Parties) of the Ramsar Convention, Greece did not manage to remove the Ramsar site of lakes Koronia and Volvi from the Montreux Record of sites (black list of Ramsar sites). In its Resolution VII.12, COP 7 "expressed the hope that development and implementation of additional measures for Lake Koronia will enable the removal of this site from the Montreux Record in due course". In preparing for COP 8, the Environment Ministry assigned the preparation of a report on the state of the Greek Ramsar sites. The report concluded that efforts of the Prefecture on the planning level for restoring Lake Koronia (Master Plan) were not yet implemented in an effective manner. Lakes Koronia-Volvi were categorised as a site still in "clear deterioration" (Red Alert), while the restoration of Koronia remained a controversial issue both politically and scientifically (Greek Biotopes/Wetlands Centre et al., 2002). During COP 8 of the Ramsar Convention (in 11/2002), a Resolution was prepared against Greece for not taking restoration measures for Lake Koronia. The Resolution was finally worded as follows: "(the COP) congratulated the Government of Greece for its stated intention to take appropriate action in line with Resolution VIII.16 on the *Principles and guidelines for wetland restoration* for the Lake Koronia Ramsar Site, taking into consideration environmental constraints, based on the availability of natural resources, socio-economic characteristics and other peculiarities of the catchment" (Resolution VIII.10). The wording of this Resolution indicated the commitment of the Greek State at COP 8 to take necessary action. According to the recommended restoration principles of Ramsar, it is not acceptable to transfer water from one water basin to the other (unless there are very good reasons). Therefore, although the COP 8 Resolution gave more credit of time, it pressurised Greece to reconsider the proposed water transfer from River Aliakmon (Interview University of Thessaloniki, 29/3/2005).

Finally, the proposed water transfer was formally rejected by the Environment Ministry which did not approve the relevant EIA study. The water transfer also lacked legality in terms of the 1987 Water Law, which required approval of interbasin water transfers by the respective Regional Water Committees and the Interministerial Joint Water Committee. The proposed transfer never reached the Interministerial Joint Water Committee, since it was rejected already on the regional level (Interview University of Thessaloniki, 29/3/2005). The Regional Water Committee of Central Macedonia referred the proposed water transfer to the Management Body of the newly set up National Wetland Park of Lakes Koronia and Volvi, which in turn rejected the transfer (pers. comm. RCM, 31/3/2005 and 21/4/2005).

After the formal rejection of several projects of the Master Plan on the national level, it did not come as a surprise when the application submitted to the Cohesion Fund for funding phase B of the Master Plan was rejected in 12/2002. The rejection was linked to the absence of approved EIAs for certain technical works as well as the lack of a Programme Contract which should define the management responsibilities of different actors for the operation of the technical restoration works after their completion (Interview Head expert of revised Master Plan, 29/3/2005).

Progress of non-rejected Master Plan measures

Some of the actions proposed by the Master Plan were not rejected; therefore, some attempts were made to bring their application forward.

The proposed measures of the Master Plan affecting industries, and especially the issue of effluents treatment plants, were discussed directly with industries either in the context of the public information seminars funded through phase A of the Master Plan or in separate meetings with individual industries. Industries did not agree to most of the proposals of the Master Plan which would affect their operation including the installation of water meters to measure their water consumption and their effluents discharge volume (Interview Young Farmers Union, 14/4/2005; Interview Consultant to Master Plan, 20/4/2005).

Funding from phase A of the Master Plan was, however, used to purchase water meters to measure water consumption in agriculture. The aim was to install meters for the management and control of water use and, in the same time, collect data on real water consumption patterns. Nevertheless, these water meters could not be installed in practice by the Prefectural authorities due to objections and protests from farmers (HOS, 2003). Farmers argued that this was the result of failed communication and a top-down approach on behalf of the Prefectural authorities which caused the opposition of farmers. Very few water meters could finally be installed with the intervention of the National Agricultural Research Institute for strictly scientific purposes, namely to compare water consumption rates of spraying and drip irrigation (Interview Young Farmers Union, 14/4/2005).

Concerning the agrienvironmental measure proposed by the Master Plan to reduce agricultural pressure in the protected area, it was finally approved via a relevant CMD issued in 5/2003 under the leadership of the Ministry of Agriculture.[98] The aim of the measure was to support farmers to use sustainable agricultural production practices in lakeshore areas without reducing farmers' income. Farmers' participation would be voluntary.

Applications of farmers to enter the agrienvironmental measure could be made for irrigated fields around Lake Koronia and Volvi which were cultivated at least once in the 2 years prior to the programme (2002-2003) with maize, alfalfa or clover. When entering the programme, farmers were committed to reduce their consumption of irrigation water by 25-30% and their use of nitrogen fertiliser according to the Codes of Good Agricultural Practice by one of the following methods:

- Combination of alternating crops with uncultivated farmland:

    o Create a permanent uncultivated surface of 3% of their land which is registered with the programme.

    o Alternate crops on a surface of 22-27% of their land which is registered with the programme with one of the following non-water-depending crops: soft wheat and dry clover.

- Turning 30% of their land registered with the programme to fallow land.

The implementation of the agrienvironmental measure kicked off in 2004. However, only very few applications were made for the first call and the effectiveness of the measure was minimal on the ground (see also section 8.9.2.4). The agrienvironmental policy measure was later amended by issuing a new CMD in 12/2005, following requests of farmers and the competent Prefectural authority for the simplification and optimisation of the measure.

Drafting a legal framework for nature protection

As already mentioned, the Master Plan of the Prefecture was developed in parallel to a legal act for the designation of the wider basin of Koronia as a nature protection area by the Environment Ministry. The purpose of this legal act, which was in preparation since 1996, was to be a powerful legal protection tool defining zones of protection and the activities allowed within the wetland protected area. Through this act, also the establishment of a Management Body and of an integrated Management Plan for the wetland would be promoted.

---

[98] The measure is co-funded by the EU and the Greek State.

In 11/1996, the first part of a study assigned by the Environment Ministry in preparation for the legal act (Arabatzi-Karra et al., 1996) was completed. This study was the basis for the Specific Environmental Study required by the 1986 Environment Law prior to designating any area as protected.

In 9/1998, the preparatory study was finalised under the supervision of the Ministries of Environment, of Agriculture and the Directorate for Environment and Spatial Planning of the Region Central Macedonia. In 1998, the on-site delimitation of zone A of the protected area of Koronia took place according to the draft Specific Environmental Study (SES) (Interview University of Thessaloniki, 29/3/2005). The SES also proposed measures specifically relevant to the sustainable management of water resources in the Koronia subbasin, indicating the need to consider also the conclusions of the Master Plan of the Prefecture. The proposed measures on water resources mainly aimed at giving general directions for later concrete action (pers. comm. RCM, 21/4/2005). In practice, mainly some secondary activities were supported, such as fish fauna studies in Lake Volvi, depending on the availability of funds (Environment Ministry phone comm., 18/11/2005).

In the meantime, from 1997-2000, a Programme Agreement between the Ministries of Environment, of Agriculture and local authorities around Lake Koronia was in operation to implement several projects for the enhancement of the protected area prior to its designation. The priority projects in this context did not really involve any water management activities but were limited to the operation of an Information Centre for lakes Koronia and Volvi, visitor management and management works for riparian forests in the basin. Responsible for the implementation of projects was a Joint Committee of the contracting parties of the Agreement. The Joint Committee was technically supported by a well-operational Advisory Committee comprising NGOs, local interest groups (cooperatives of farmers, fishermen and hunters' association), civil services and scientific institutions (Tsougrakis, 2000). The Programme Agreement was a significant coordination step for the area and achieved progress in raising awareness of local civil services and the public (less of water users).[99] It also served as good preparation for the later creation of a management body for the protected area. The disadvantage of the Agreement was that there was no feedback evaluation system in order to use the results of this activity in next activities in an organized way (Interview Ornithological Society, 28/3/2005).

In the drafting process of the legal act for the protected wetland area of Koronia, local authorities (especially the Prefectural Council of Thessaloniki and the Sub-Prefecture of Langadas) were consulted. There was also some degree of public participation combined with presentations of the SES on the local level. During these public presentations, several comments, positive and negative, were expressed and even tension was created. Observers

---

[99] Expression of opinion with regard to conservation actions for Greek Ramsar wetlands and to the applicability for removal from the Montreux Record (Annex IV), 7th Meeting of the Conference of the Contracting Parties to the Convention on Wetlands (Ramsar, Iran, 1971), *"People and Wetlands: The Vital Link"*, an José, Costa Rica, 10-18 May 1999.

of the process, however, characterised the design of the public participation process "as quite experimental and immature, being in some aspects even negative" (Interview Former expert at Information Centre, 23/3/2005).

Overall, the local population was mainly concerned about potential obstacles to local development resulting from the legal act (Interview Former expert at Information Centre, 23/3/2005). The Union of Agricultural Cooperatives of Langadas submitted written comments to the Environment Ministry. The main comments of farmers were relevant to the limitations imposed on the size of new farming buildings and to the restrictions imposed on the replacement of existing wells. More specifically, farmers considered the size of farming buildings allowed as too small, especially for animal farms. Secondly, farmers requested the possibility to replace existing wells in the Koronia subbasin that were not operational anymore (due to technical failure, clogging, earthquakes etc). In the end, farmers were disappointed that their comments were not considered by the Environment Ministry in the final legal act issued (Interview Young Farmers Union, 14/4/2005).

EU infringement procedure and legal battles in local and national courts

The European Commission initiated itself an infringement procedure against Greece in 6/1999 (letter of formal notice) for environmental damage caused to Lake Koronia. The procedure was interrupted in 7/2001, after the Greek authorities had submitted evidence that industrial plants around the lake were compliant with conditions set in effluent disposal permits and that the lake pollution was caused by sources other than the local industry. The Greek authorities further argued that they had started implementing a programme designed to improve the ecological status of the lake. However, new data and controls showed that the management of the lake was very inadequate. DG Environment of the EC received information from DG Regional Policy according to which the reduction of the surface covered by the lake was due to overabstraction for irrigation purposes.[100] In the meantime, Lake Koronia completely dried out in August 2002. The EC re-opened the infringement procedure and sent a new letter of formal notice in 12/2002 asking from Greek authorities further clarification of their compliance with 3 EU directives in Lake Koronia (Birds Directive, Habitats Directive and Directive on water pollution from hazardous substances). The EC noted that no specific protection programme had been adopted in order to control pollution and restore the lake. The industries around the lake continued to discharge effluents with substances prohibited by the EU Directive on water pollution from hazardous substances. Secondly, despite the characterization of the wetland area as an Area of Special Protection (SPA) under the EU Birds Directive, there was no coherent legal framework in place for sustainable management and the effective protection of the lake and its birds (EC, 2002). In their reply, the Greek authorities argued that the bad status of Lake Koronia was partly due to extreme dry weather conditions, without however denying that water consumption had

---

[100] Summary of question and answer session between EP Mihail Papayannakis and the EC on 4/11/2002. Accessed online on 2/11/2005: www.europarl.europa.eu.

increased abruptly and significantly to satisfy agricultural and industrial activities. The Greek authorities also argued that two legal acts to set up a legal framework for nature protection around the lake and for reducing pollution from hazardous substances were under preparation. Without noting any significant progress towards the adoption of these two legal acts to protect Lake Koronia by 12/2003, the EC issued a reasoned opinion against Greece.

In the meantime, legal action had started being taken against the lake polluters (mainly industries but also the city of Langadas) also within Greece.

In 10/2000, 25 industries as well as the ex-Mayor of Langadas were taken to court (one-member district court of Thessaloniki) for causing the pollution and degradation of Lake Koronia (Newspaper "Kathimerini", 14/10/2000). In the trial, approximately 25 industries were convicted and fined. However, according to a local environmental group, convictions by such local level courts had no effect in practice on the behaviour of industries. Some of the convicted industries were very strong financially and invested further in appeals to postpone payments of the fines, which in time were reduced until they became insignificant for the industry to pay (Interview Local Union for the Environment of Langadas, 6/4/2005).

In 2/2003, the Council of State (the superior court of Greece) examined a case brought forward by three ecological organisations from Thessaloniki and the region of Langadas[101] which denounced the Environment Ministry for not protecting Lake Koronia from industrial pollution (Newspaper "Kyriakatiki Avgi", 23/2/2003). The ecological organisations denounced the fact that industries persistently breached the law and were granted temporary effluent disposal permits by the public administration without solving the problem of pollution. In its decision 2680/2003, the Council of State ruled in favour of the ecological organisations and obliged the Thessaloniki Prefecture and its competent authorities (Divisions for Environmental Protection, for Municipal Planning and for Industry) to carry out immediate strict inspections on the operation of industries in question and their treatment installations. According to the decision of the Council of State, adequate time should be given to these industries to comply with regulations. In case of non-compliance, the Prefecture was obliged to seal-off all industries not holding operation licenses and effluent disposal permits.[102] By autumn of 2005, however, the decision of the Council of State had still not been implemented in the full by the Prefecture.[103]

Creation of a National Wetland Park and its Management Body

In 8/2002, a Management Body for the protected area of lakes Koronia and Volvi was set up by Law 3044/2002 (this Law set up in total 27 management bodies of protected areas all

---

[101] Ecological Movement of Thessaloniki and Environmental Lawyers (both are ecological organisations from the city of Thessaloniki) as well as the Union for the Environmental Protection of the region of Langadas.

[102] Press release of ecological and environmental organisations of Thessaloniki, 3/6/2004, Subject: Lake of Agios Vasilios – Shocking decision of the Council of State.

[103] Information from a statement of the prefectural party of Ecology, Solidarity and Coalition of Citizens. Accessed online on 10/03/2006: http://www.e-ecology.gr/DiscView.asp?mid=737&forum_id=6&.

over Greece). The Management Body was set up as a legal entity of private law supervised by the Environment Ministry. Its competence is described in Box 8.2.

The Management Body, similarly to all management bodies set up by Law 3044/2002, assumed mainly a consultative role in the process of issuing permits for certain activities and a supportive role to public law-enforcement authorities. Its main responsibilities lie in monitoring and evaluating the application of management instruments in the protected area of its competence, collecting and elaborating scientific data as well as carrying out information and public awareness campaigns.

Box 8.2   Competence of the Management Body for the protected area of lakes Koronia and Volvi

According to L.2742/99 and the CMD 6919/2004, the Management Body competences include:

- Drafting and responsibility for the implementation of regulations for the administration and operation of the protected area and of respective management plans

- Monitoring and assessing the implementation of regulations and limitations of the CMD 6919/2004. The Management Body takes care of the collection and processing of environmental data for the protected area of its competence

- Delivering its expert opinion prior to the environmental approval of any works and activities in the protected area and for any issue where its opinion is requested by the competent authorities

- Assisting competent administrative and legal authorities in the control of implementation of environmental legislation for works and activities in the protected area

- Drafting studies and executing any necessary technical or other works included in the Management Plan of the protected area. Construction, repair and maintenance of necessary infrastructure works and having the necessary scientific staff and equipment to exercise management functions

- Executing national or European programmes and actions relevant to the protected area

- Information, education and awareness of the public on the issues of the Management Body responsibility and objectives. To this aim, Management Bodies may set up information centres within the protected area of their responsibility, organise education programmes with other institutions, organise relevant seminars and workshops and publish relevant material

- Promoting ecotouristic programmes, issuing of licenses for guides and for scientific research within the protected area.

- Managing public land handed over to the Management Body or rented by the Management Body.

In 7/2003, the administrative board of the Management Body was set up by decision of the Environment Ministry with 11 participants from different levels of the administration (local, Prefectural, Regional and central), experts of established technical professional chambers, environmental groups and the academic community (see Table 8.8). The administrative board is responsible for exercising the competences of the Management Body as described above.

In the initial period of its operation, the Management Body was active on an irregular basis. At first, it was active at least until 11/2003, when its president resigned to take part in the national preelection campaign. After a period of inactivity, the Management Body met again in 5/2004. At start, the Management Body had in general no capacity to deal with substantial management issues (Interview Ornithological Society, 28/3/2005) also due to the lack of financial and institutional support (Newspaper "Macedonia", 7/12/04).

Table 8.8   Administrative board of the Park Management Body

Participants of the board	Representation
Regional administration	2 regular members (Head of Region; Regional Division for Environment and Spatial Planning)
Prefectural self-administration	3 regular members (Prefecture of Thessaloniki: Forestry Division and Assistant of Prefect; Sub-Prefect of Langadas)
	2 substitute members (Prefectural Councilor of the Prefecture of Thessaloniki and Fisheries Division of the Prefecture of Chalkidiki)
Local self-administration	2 regular members (Mayor of Langadas; Administrative Council of Zervohori)
	2 substitute members (Union of the Local Self-Administration of Thessaloniki; Mayor of Arnea/Chalkidiki)
Central administration	1 regular member (Ministry of Development)
	2 substitute members (Environment Ministry /Department of Natural Environment; Ministry of Agriculture /Forest inspector of location Stavros)
Experts	2 regular members (one representative of the Geotechnical Chamber of Greece; one scientific expert (usually of the academic community))
NGOs	1 regular member

A legal act for the official designation of the area of lakes Koronia and Volvi as a National Park was issued in 3/2004 as a Common Ministerial Decision (248/5.3.2004) on the "*Characterisation of the wetland system of lakes Koronia-Volvi and Macedonian Tembi as a National Wetland Park, the delimitation of protection zones A, B and C and definition of uses, conditions and building restrictions therein*". This CMD mainly defined the boundaries of the protected area, divided it in three zones and defined measures for protection in each of the three zones.

The CMD for the designation of the National Wetland Park was issued with considerable delay, considering that its preparation by the Environment Ministry had started already in 1996. The lack of political will at a national and local level played an important role in this respect. In fact, delay in signing such legal protection acts was commonplace and linked to local reactions in most affected protected areas and wetlands all over Greece out of fear to slow down development (Interview Ornithological Society, 28/3/2005). Political pressure to issue the CMD for the Koronia wetland increased under the threat of a legal case before the ECJ. In this context, the CMD was issued 2 days before the 2004 national elections (Interview Former expert at Information Centre, 23/3/2005). Another reason for signing the CMD at that point was to receive a loan from the European Bank of Investment for a major highway construction project next to the Koronia wetland (the Egnatia Highway Project). Compliance with European legislation on nature protection was set as a condition for the approval of the loan (Interview Prefecture, 12/4/2005).

Revision of the Master Plan and change of the restoration paradigm

Revision and consultation process

In autumn 2002, a new political party took over the leadership of the Prefectural administration of Thessaloniki and also took over the implementation of the Master Plan. In the beginning of 2004, the Prefecture's new political leadership set up a Committee of Guidance and Coordination for the Restoration of Lake Koronia with the aim to submit a new funding application to the Cohesion Fund for the Master Plan. Participants to this Committee were mainly divisions and experts of the Prefecture but also the main technical professional chambers of Greece (Technical Chamber and Geotechnical Chamber), representatives of the agricultural sector of the area concerned as well as representatives of the local population (elected officials from the local municipalities and communities) (Newspaper "Typos Thessalonikis", 8/4/2004; Interview Prefecture, 11/7/2006).

The Guidance and Coordination Committee was also supported by a Committee of external academic experts and experts of the National Agricultural Research Institute, who examined the proposed measures for restoring Lake Koronia and gave a negative evaluation of the original Master Plan (Newspaper "Avriani Makedonias-Thrakis", 15/1/2004; Newspaper "Macedonia", 10/1/2004). On that basis, the Prefectural Guidance and Coordination Committee decided to assign certain studies to revise the Master Plan. It was agreed that a group of experts of relevant authorities and external experts would revise the Master Plan, which would then be additionally evaluated by an external international expert as required by the EU rules for funding applications (Anonymous, 2004). The Committee of Guidance and Coordination assisted by the Committee of external experts monitored the progress of the revision (Interview Head Expert of the revised Master Plan, 29/3/2005). Also, the Management Body of the protected area of lakes Koronia and Volvi supported the revision of the Master Plan (Newspaper "Egnatia", 17/1/2004).

The revision of the Plan was carried out bearing in mind the following (Zalidis et al., 2004a):

- The resolution of COP 8 of the Ramsar Convention requesting amendment of the restoration philosophy for Lake Koronia.

- The EU Water Framework Directive 2000/60, the EU Habitats Directive 92/43 and the EU Birds Directive 79/409.

- The official rejection of EIAs for several proposed technical projects of the original Master Plan.

- The CMD 6919/2004 defining lakes Koronia and Volvi as a National Wetland Park.

Against this background, the two main objectives of the revised Master Plan were (Zalidis et al., 2004a):

- To define the best possible restoration scenario which, based on the availability of the water basin resources, would provide the best conditions for long-term functional and structural rehabilitation of the wetland/lake system, for maximum habitat diversity and conservation of the fauna, especially fish and bird fauna.

- To identify the appropriate measures at both lake and water basin scale, which would contribute to the restoration of Lake Koronia and would address the sources of degradation.

The approach of the revised Plan was to achieve sustainable restoration of the entire Mygdonian water basin (Newspaper "Polis News", 4/3/2004). The original Master Plan aimed at hydrological restoration primarily by transferring water from another water basin instead of managing the resources of the Mygdonian basin itself. In the revised Master Plan, restoration aimed at a self-sustainable water basin, in accordance with the Ramsar restoration principles, even during "difficult" hydrological periods encountered in the Mediterranean region. The new aim was to restore the lake so that it could deal with critical events in the future and sustain its food web. To this aim, a self-sustainable reference level was defined, which was different than the reference of the lake status in the 1970s as adopted in the original Master Plan. The new approach was also based on the principle that the Mygdonian water basin is one operational system with links between lakes Koronia and Volvi and between the basin and the two lakes (Interview Head Expert of the revised Master Plan, 29/3/2005).

The revised Master Plan was finalized in 9/2004 after being approved by the Prefectural Council and following a short consultation process with actors outside the administration. The Prefectural Committee of Guidance and Coordination for the Restoration of Koronia invited for a short time (of 3-4 months) comments on the revised Plan from selected stakeholders including environmental NGOs. Stakeholder consultation was also required by the rules of the Cohesion Fund for funding the specific project (Interview Head Expert of the revised Master Plan, 29/3/2005).

Overall, the revised Plan was evaluated as more modest and the consultation process was judged as improved compared to the original Master Plan. Although the Plan was revised under great time pressure allowing only one-off and not phased consultation, even the environmental NGOs had more faith in a better result (Interview Ornithological Society, 28/3/2005). Despite some comments on individual proposed projects, the NGO Ornithological Society agreed on the basic principles of the revised Plan welcoming the emphasis on wetland functions and ecological characteristics (HOS, 2004). The green party represented in the Prefectural Council of Thessaloniki also characterised the revised Plan as clearly improved with only few points in need of reconsideration (Tremopoulos, 2004). Farmers however complained that the revised Plan was not open for comments by agricultural water users on the local level. The only meetings which took place locally concerned the sewerage network and treatment plant for the city of Langadas (Interview Young Farmers Union, 14/4/2005).

As far as the interaction between the central administration and the Prefecture is concerned, the process revising the Master Plan may also be assessed as improved. There was written communication with the relevant central authorities, who were asked for their expert opinion, early in the revision process. Especially, the Department for the Natural Environment of the Environment Ministry gave a positive evaluation of the general principles of the revised Plan and submitted positive expert opinions for individual projects which required its approval of environmental conditions. The EIAs of newly proposed projects were more easily approved by the Environment Ministry since the proposed interventions were "softer" than those of the original Plan (Environment Ministry phone comm., 18/11/2005). Proposed projects and works were also submitted for approval to several other public authorities and to the Management Body of the National Wetland Park (pers. comm. RCM, 21/4/2005).

Main proposals of the revised Master Plan

The revised Master Plan proposed a scenario for the restoration of Lake Koronia according to which the surface of the lake should increase to 3,525.7 ha (compared to 3,438.9 ha in 2003) and its water volume to $83.8*10^6$ m³ (compared to $57.2*10^6$ m³ in 2003). This scenario aimed at the creation of a wetland which should combine a system of open water embedded within a highly vegetated marsh. The maximum depth of the lake should be 4 m.

The specific measures proposed by the revised Master Plan with the aim to support the food web and biodiversity, to reestablish a positive water budget and to improve water quality are outlined in Table 8.9.

Table 8.9   Proposed measures of the revised Master Plan

Proposed restoration measures	Proposed works and action
Establishment of wetland and creation of deep water habitats	- Establishment of wetland of a surface of 376 ha with controlled periodic flooding, by constructing a dike on the western side of the lake. This will re-create degraded wetland habitats and contribute to non-point pollution control - Creation of deep water habitats by dredging part of the lake bottom (117 ha) for the reproduction, over-wintering and protection of fish, which in their turn will benefit fish-feeding bird species
Improvement of hydraulic characteristics and reversible operation of the ditch connecting lakes Koronia and Volvi	- Partial transfer to Koronia of 25% of the yearly water flow of two torrents, Scholariou and Lagadikion, normally mouthing into Lake Volvi – this way, the level of Koronia could reach its restoration goal within 6-7 years and the measure is not expected to significantly affect the water balance of Lake Volvi - Reversible operation and hydraulic modification of the ditch connecting Koronia and Volvi – this will strengthen the water potential of the lake and contribute to the progressive rise of the water level. It is not expected that the connection of the two lakes will decrease the water quality of Lake Volvi
Treatment and disposal of urban wastewater and industrial effluents to minimise the inflow of nutrients and toxics into Lake Koronia	- Extension of the UWWTP of Langadas with two additional units to receive and pre-treat effluents from domestic cesspits of small settlements and effluents from cesspits of small industries. Pre-treated effluents will then be further treated in the UWWTP of Langadas
	- System of maturation ponds, where the treated urban wastewater from the UWWTP of Langadas will be mixed with industrial effluents already treated by individual industries. Ponds serve for further treatment to reduce remaining COD and colour (mainly through sedimentation) before the discharge of wastewater and effluents into Lake Koronia
	- Construction of a pumping station for the UWTTP of Langadas and a separate sewerage and stormwater drainage system within the city of Langadas
	- Treatment of saltwater effluents of the dye industries in a central treatment unit and disposal into the sea
	- Construction of sewerage networks and wastewater treatment units (artificial wetlands) for settlements of Kolhiko, Drakontio, Lagyna, Kavalari (proposed measure of secondary priority)
Measures of horizontal support to the revised Master Plan	- Monitoring of management interventions and development of decision support system
	- Actions for public information, awareness and volunteer work
	- Contracting a project management consultant
Application of sustainable	- Agrienvironmetal programme of the Ministry of Agriculture

Proposed restoration measures	Proposed works and action
agricultural practices	
	- Regulations for the improvement of farming practices
	- Irrigation area stabilization
	- Farmer support measures (Centre for agrienvironmental information and development including office for informing farmers)
Rehabilitation and management of the groundwater aquifer	- Collective irrigation networks and shallow aquifer technical enrichment
Upper-watershed stream management measures to reduce sediment transport to the lake and improve water infiltration from streams into the shallow groundwater aquifer	- Upper-watershed stream management measures (torrents of Bogdana, Kolhiko, Kavalari) (proposed measure of secondary priority)
Support of private sector environmental investments	- Private sector investments in dye industry for saltwater treatment and reuse
	- Support for private investments in drip irrigation

Source: Zalidis et al. (2004a).

Especially, measures to restore the shallow aquifer in combination with the agrienvironmental measure of the Ministry of Agriculture, the proposed regulations to stabilize the surface of irrigated areas and the proposed private investments to increase irrigation efficiency (investments in drip irrigation) are expected to gradually restore the shallow aquifer of the basin by saving $20*10^6$ m³ annually. Particularly the measure of private investments to increase irrigation efficiency was expected to improve significantly the hydrological status of the aquifer. In fact, some drip irrigation systems for maize were already installed in the basin based on private investments of some large-scale farmers (Interview Young Farmers Union, 14/4/2005).

Towards the end of the case study, the proposed support for private investments in drip irrigation started to be actively promoted through the cooperation of the Prefecture, the Region and the National Agricultural Research Institute. In specific, after a proposal of an expert working group of the National Agricultural Research Institute for radical irrigation reductions in the basin by replacing water-demanding techniques with drip irrigation, the Prefecture and the Region cooperated for the promotion of a relevant pilot measure using the mechanisms of the national System of Certification and Supervision of Agricultural Products. In the context of this initiative, the replacement of water-demanding techniques with drip irrigation could be funded in a pilot area of 1,400 ha, if it could be certified that farmers follow the principles of sustainable farming and the standards of biofarming (fulfillment of agrienvironmental principles during production). To this aim, the Region funded a study for the development of a model of Integrated Production Management (Agro 2.2) which was submitted to the national Certification System mentioned above (Interview Prefecture, 11/7/2006). In the context of this activity, emphasis will be given to the installation of drip irrigation for the cultivations of clover around Lake Koronia (Oiko Kathimerini, 12/3/2005). However, considering that certifying the fulfillment of sustainable farming principles on the ground is still a difficult task in the Greek context, the revised Master Plan also proposed the establishment of a local consultation and support centre aiming at advising farmers on the use of fertilizers and more sustainable farming practices (Interview Head Expert of the revised Master Plan, 29/3/2005).

Concerning the measures proposed for the treatment and disposal of wastewater and effluents, these were largely based on the conclusions of a planning study for the treatment of effluents and wastewater in the Koronia subbasin (Ydromeletitiki E. E. et al., 2002) which was carried out in phase A of the Master Plan. The proposal of the original Master Plan to fund treatment works for animal farm effluents was dropped in the revised Plan because it was concluded that pollution from animal farms should be best treated on a local, private basis at individual farms following current regulations. In case of remaining animal farm cesspits and tanks, effluents would be pre-treated in an extension unit of the Langadas urban wastewater treatment plant (UWWTP) for cesspit effluents of small industries (Interview Consultant to Master Plan, 20/4/2005). It was also concluded that the most important pollution factor of Lake Koronia was saltwater effluents of fabric-dye industries (Ydromeletitiki E. E. et al., 2002). Therefore, the revised Master Plan proposed a specific measure to combat saltwater pollution. As first step, it was proposed to separate saltwater effluents from other effluents within the dye industries. Secondly, in order to avoid high construction and operational costs for a saltwater treatment facility, the planning study proposed to collect and transport saltwater effluents to the sea. Transport to the sea would take place after treating effluents for COD and colour in a central treatment unit attached to the Langadas UWWTP (Interview Consultant to Master Plan, 20/4/2005).

Reaction to technical measures proposed by the revised Master Plan

Despite enjoying wider public acceptance than the original Master Plan, some proposed interventions of the revised Plan became target of critique by academic experts, research centres, e-NGOs, water users and even public authorities.

A substantial comment was delivered by the NGO Ornithological Society, which pointed out the lack of concrete proposals on the management and administration of the technical restoration works once completed (HOS, 2004).

Comments were also delivered regarding the hierarchy and timetable of the proposed interventions. The Greek Biotopes/Wetlands Centre commented that large-scale, expensive technical works, such as the creation of a wetland, should not proceed before action is taken to halt the main degradation drivers and unsustainable water use practices in the basin (e.g. reduction of water abstractions for irrigation).[104]

Objections were also expressed to the proposed two-way connection between lakes Koronia and Volvi. Many scientists objected this because of the potential degradation of Lake Volvi from the polluted waters of Koronia (Interview Former expert at Information Centre, 23/3/2005).

---

[104] Greek Biotopes/Wetlands Centre, Document Nr. 7950, Addressed to the Prefecture of Thessaloniki, Subject: Comments on the revised plan for the restoration of Koronia, 13/7/2004.

The partial transfer of two torrents from Lake Volvi to Lake Koronia was also criticised for potential impacts on the quantitative and qualitative stability of Lake Volvi (e.g. by the Ornithological Society (HOS, 2004) and the Greek Biotopes/Wetland Centre[105]). Even the EIA procedure for this measure was criticised for not being institutionally legitimate, since public consultation only involved the three municipalities around Lake Koronia and not all affected municipalities of the Mygdonian basin (Statement of Mr. Koukoulekidis, Council member of the Local Union of Municipalities and Communities of Thessaloniki, Newspaper "Thessaloniki", 31/10/2005).

The proposed action to seal off private wells in the area of Langadas and instead create a collective irrigation network fed by the deep aquifer was also criticised. Abstraction from the deep aquifer for hydrological restoration was already rejected in the original Master Plan due to environmental and hydrogeological problems (Interview Former expert at Information Centre, 23/3/2005). Representatives of farmers were thus sceptical about funding possibilities for a collective irrigation network and the difficulty of approving the necessary EIAs (Interview Young Farmers Union, 14/4/2005). Moreover, given the inability of authorities to carry out inspections and impose regulations on farmers, farmers could easily manipulate such a collective network and abstract much more water than that allocated (Interview Consultant to Master Plan, 20/4/2005).

Regarding the proposed measures affecting industries, mainly the issue of saltwater pollution treatment was widely discussed. By the end of this case study, the proposed central saltwater treatment unit seemed unlikely to be implemented. On the one hand, by 2005, most dye industries in the basin operated only occasionally or had already relocated most parts of their production to cheaper host countries (e.g. Bulgaria, FYROM) influenced by the nation-wide crisis of the textile industry. On the other hand, industry owners were reluctant to bear costs for the operation of a central saltwater treatment unit, including costs for infrastructure to separate saltwater from other effluents, for pumping equipment, pipes to transfer saltwater to the central treatment unit and financial contributions for the central unit operation. Industries were only positive to receiving direct funding for the installation of saltwater treatment units in their own premises to desalinate and re-use water in their production systems. However, considering the bad record of industries in the basin regarding compliance with effluents treatment regulations, this approach could lead to a "business-as-usual" situation whereby industries neglect operating the saltwater treatment units. In any case, this appearing as the most acceptable solution, it was proposed to raise public funding contributions, in order to bring the direct financing of saltwater treatment in individual industries forward (Interview Consultant to Master Plan, 20/4/2005). To this aim, the Prefecture in cooperation with the Ministry of Economy and the Ministry of Development, proposed the promotion of a specific measure for subsidising desalinisation and water re-use facilities in the dye industries of the Mygdonian basin through the new Development Law (L.

---

[105] Ibid.

3299/2004 and its amendment L. 3427/2005) and L. 3325/2005 of the Ministry of Development (Interview Prefecture, 11/7/2006).

Renewed environmental collapse and escalation of EU pressure

Although increased rainfall over the winter of 2003/2004 improved the water volume of Lake Koronia (Newspaper "Aggelioforos", 13/4/2004), the ecosystem of the lake collapsed again in 9/2004 when approximately 7,000 birds were found dead in its wetland (Newspaper "Macedonia", 9/9/04). The results of toxicological tests concluded that the birds were poisoned by a toxin, which was produced in large numbers due to the shallow depth of the lake, the immobility of its water and the lack of oxygen (Newspaper "Thessaloniki", 4/10/04). By 12/2004, over 30,000 dead birds of 39 species were recorded (Newspaper "Thessaloniki", 4/0/04).

In 4/2005, a scientific team of the Aristotle University of Thessaloniki made alarming statements on the dangers for aquatic life in Lake Koronia due to the enormous increase of phytoplankton and high number of bacteria decomposing organic compounds (Newspaper "Macedonia", 20/4/2005). Scientists warned of a possible collapse of the food web of the lake due to low dissolved oxygen and high pH.

The 2004 incident of mass bird kills activated several actors. Directly after the incident, the Mayor of Langadas sued any possible responsible entity to the Attorney of the district court of Thessaloniki in order to accelerate actions to deal with the disaster on the lake (Newspaper "Macedonia", 10/9/04). The Prefecture prohibited fishing in the lake and interventions were made at the European and the Greek Parliament.

In 1/2005, the EC announced that it would refer Greece to the ECJ for the pollution and degradation of Lake Koronia on the grounds of the EU Birds Directive, the Habitats Directive and the Directive on water pollution from hazardous substances (EC, 2005). The EC required specific legal acts from the Greek State to protect Lake Koronia. The first legal act (CMD) issued in 2004 for the designation of the National Wetland Park of Koronia corresponded to the requirements of the EU Birds and Habitats Directives. The EC also requested a legal act for a special programme for the reduction of pollution from hazardous substances. This legal act was issued as a CMD in 10/2005 and was formulated largely on the basis of the revised Master Plan. The qualitative objectives of this CMD were to allow the use of Lake Koronia for fishing and irrigation and improve the food web by increasing oxygen concentrations and reducing eutrophication. The CMD required the definition of the problems of Lake Koronia, the determination of pollutants, the definition of quality objectives, the description of a monitoring programme, administrative measures, special regulations to reduce pollution from non-point agricultural sources and a timetable for a series of actions. The CMD also set a concrete timetable to a) revise the emissions limits of industries in the basin within 1 year from issuing the CMD, b) complete the treatment works for urban and industrial effluents proposed by the revised Master Plan within 3 years and c) revise water permits for wells within 2 years.

In June 2006, the case of Lake Koronia was re-examined and the EC, for the moment being, withdrew the threat of referring the case to the ECJ after taking note of additional progress made on its legal requirements (Interview Prefecture, 11/7/2006). Specifically, in April 2006, another CMD was issued (CMD 16175/824) establishing an Action Programme to reduce nitrate pollution from agricultural origin in a wider area including the Mygdonian basin, which was characterised as a nitrates vulnerable zone according to the EU Nitrates Directive. The EC was also pleased with yet another development, namely the signing of a Programme Contract for the execution of the restoration programme of Lake Koronia (see next section for details).

EU approves the revised Master Plan and an Actors' Programme Contract is signed

In 12/2004, DG Environment of the EC approved the revised Master Plan (as far as environmental aspects were concerned) and sent the funding application to DG Regional Policy for funding approval (Newspaper "Eleftherotypia", 19/1/05). At that stage, the funding application was still incomplete in two aspects.

Firstly, the EIAs for two proposed projects were still not approved (for the creation of deep water habitats and for the partial diversion of two torrents from Lake Volvi). The relevant EIA approval was given later by the Environment Ministry in 10/2005 (Newspaper "Makedonia", 18/10/2005).

Secondly, DG Regional Policy requested a Programme Contract among all involved actors with a role in the future management of the proposed measures and works of the Master Plan. The requirement for this Programme Contract was judged as significant also by several authorities and environmental NGOs. The Programme Contract was envisioned as a contract committing a priori all involved institutions on a legal basis. It should define how and who should take decisions on the operation and management of the constructed hydraulic works. Additionally, it should allocate responsibilities and commit actors to secure resources for their tasks. The Programme Contract should also be formulated in such a way to ensure compliance with the approved environmental terms of the different technical projects (Interview Head Expert of the revised Master Plan, 29/3/2005).

The Master Plan was finally approved for funding by the EC in 12/2005 on the condition that some pending requirements of the EC would be fulfilled, especially the required Programme Contract. Funding for the various proposed projects to restore Lake Koronia would be given by the Cohesion Fund, national public investment funds as well as the Region Central Macedonia (Newspaper "Makedonia", 13/1/2006). In the meantime, due to a cut-down of the financial contribution from the Cohesion Fund, it was decided to initiate the realisation of the revised Plan with national funds (Interview Head Expert of the revised Master Plan, 29/3/2005). Already in 1/2005, the Prefecture announced the allocation of 10 million Euro from national public investment funds for the construction of the sewerage network of Langadas (after agreement with the Ministry of Economy) (Newspaper "Thessaloniki", 5-6/2/05). The construction of the sewage network of Langadas was thus the first measure of the revised Master Plan which progressed in terms of practical realisation. In April 2006, the start of relevant works for the sewage network was inaugurated.

The requested Programme Contract was finally signed by all parties concerned, i.e. the Ministries of Environment, of Development, of Agriculture and of Economy, the Thessaloniki Prefecture, the Region Central Macedonia, the Municipality of Langadas and the Management Body of the National Wetland Park of Lakes Koronia and Volvi, in May/June 2006. Thus, the last requirement of the EC for allocating Cohesion funding to the restoration of Lake Koronia was fulfilled. The General Secretary of the Region, which is the last branch of the central government on the regional level, was defined as head of the Joint Committee of the contracting parties for the management and realisation of the Programme Contract.

The Programme Contract on the execution of the restoration programme of Lake Koronia, whose duration is limited until 2010, focused on the allocation of responsibilities and the commitment of resources for the realisation of the technical works and measures to restore Lake Koronia. More specifically, the Contract binds its contracting parties to cooperate on the realisation of the main following activities:

- Realisation of technical works funded by the Cohesion Fund: works for the establishment of wetland and creation of deep water habitats, works for the improvement of hydraulic characteristics and reversible operation of the ditch connecting lakes Koronia and Volvi, works of maturation ponds, management of reeds of the wetland system of Koronia, monitoring of management interventions and development of a decision support system, actions for public information, awareness and volunteer work.

- Construction of the sewerage network of Langadas, units to receive effluents from domestic cesspits of small settlements and effluents from cesspits of small industries and the operation of the UWWTP of Langadas.

- Implementation of the agrienvironmental policy measure.

- Implementation of the CMD on the reduction of water pollution from hazardous substances.

- Implementation of the CMD on the reduction of nitrate pollution of agricultural origin.

- Promotion of subsidies for private investments in the treatment of salt effluents of the dye industries.

- Re-examination of the environmental terms for the operation of the polluting industries in the area.

- Drafting a management plan for the operation of the restoration works.

- Drafting a management plan and a monitoring plan for the National Wetland Park of Lakes Koronia-Volvi and the Macedonian Tembi.

The Contract at present does not define who will be responsible for the management of the technical works once they are completed. This is bound to be defined in the near future, since this is still a requirement of the EC (pers.comm. Ornithological Society, 11/7/2006).[106]

Chronology summary

The process of this case study on the Mygdonian basin, with specific focus on the Koronia subbasin, started off with increasing water degradation of Lake Koronia due to rising economic development, especially in the agricultural and industrial sector. Initially, there was lack of any action to deal with the gradual but obvious water degradation as well as lack of effective regulations and controls on behalf of public authorities. First reactions of the public administration were mainly reactive in the form of certain regulations issued by the Prefecture and studies assigned by the Environment Ministry to comply with national obligations for the international Ramsar wetlands treaty.

After an incident of mass fish kills in 1995, which indicated the environmental collapse of the lake, the Sub-prefecture of Langadas initiated the drafting of a first Rescuing Plan for the lake which was the first effort to describe its environmental problems. Shortly afterwards, the Prefecture convinced the EU Cohesion Fund to fund a Master Plan for the Restoration of Lake Koronia. This Master Plan was drafted in 1998 involving only actors of the public administration mainly on a local level and its dominant focus was on large-scale technical works. Several of the proposed technical works, and especially a proposed water transfer scheme from the River Aliakmon in western Greece, faced opposition from different parties including the central administration, international institutions, e-NGOs, academics and technical experts. Under pressure from the official rejection of several proposed projects and following a change of the political leadership of the Prefecture of Thessaloniki, the Master Plan had to enter a process of revision which was also more open to environmental groups and the broader scientific community. In fact, the restoration paradigm proposed to restore the Lake was modified favouring "softer" restoration principles instead of hard engineering solutions. By the end of this case study, not much progress was yet made in terms of implementing management actions proposed in the revised Master Plan since funding for the Plan was approved by the Cohesion Fund only in 12/2005.

In the meantime and under the threat of an EU infringement procedure for the degradation of Lake Koronia, the Mygdonian basin (including the Koronia subbasin) was designated as a National Wetland Park and a Management Body was set up to coordinate actors involved in its management. These developments marked a new era with clearer policy requirements for a more sustainable management model of the Mygdonian basin.

---

[106] In fact, before signing the contract, it was discussed if and how the Management Body of the Park could have a role and relevant responsibilities in managing the water works of the Master Plan, according to the Programme Contract. At a first stage, it was not judged feasible for the Management Body to take on this role, due to its very limited initial capacity. If the Management Body becomes more resourceful in the future, it may gain more responsibility on the management of the Master Plan works via amendments to the Programme Contract (Interview HOS, 28/3/2005, Interview Scientific head expert of the revised Master Plan, 29/3/2005).

8.7 Actor analysis

Table 8.10 gives an overview of the roles played by the main actors involved in the case study process, their perceptions over the resource problem, their objectives and resources. Table 8.11 summarises the key relationships between the main interacting actors, focusing thereby on the development of more or less cooperation over the case study process.

The actor relationships illustrated in Table 8.11 are related to a variety of issues described in the case study chronology. Many actor interactions concern the development of the Master Plan, but other interactions are also relevant to other issues, such as the development of a legal act to designate the National Wetland Park (e.g. see interaction of MEPPW with farmers), the implementation of the agrienvironmental policy measure (see interactions of the Ministry of Agriculture with farmers) or the implementation of Prefectural regulations. Furthermore, most actor interactions were related to the overall status of water resources concerning both quality and quantity issues. Actors, who interacted with other actors only over water quality or only over water quantity issues due to their interests' profile, are clearly indicated in Table 8.11. Actors who were interested mainly in water quality were industries and the local non-governmental associations which became legally active over water pollution. On the other hand, the Prefectural Division of Water Resources and Land Reclamation and the Regional Water Management Department (RWMD) were exclusively interested in water quantity issues.

In some actor interrelationships, some improvement of cooperation could be observed over the case study process. Especially, the broader consultation carried out over the revised Master Plan and the set-up of a participatory Management Body for the National Wetland Park were important developments for the rise of cooperation between certain actors. In a few cases, the rise in cooperation has also involved additional actors who are not explicitly included in Table 8.11. For instance, the indicated improvement of cooperation between farmers and the Ministry of Agriculture towards the end of the case study took place only indirectly. In reality, farmers cooperated more closely with the Prefectural Division of Agricultural Development to propose in common amendments to the Ministry regarding the agrienvironmental policy measure.

In many actor interrelationships, however, no rise in cooperation was observed. Important non-cooperative actor interrelationships were between industry and several actors (local associations, farmers and fishermen), between local associations and the Prefecture, between farmers and authorities implementing abstraction regulations (see also emphasis in framed cells in Table 8.11). It is also noted that, because Table 8.11 focuses on few selected key actors involved in the process, not all actor interactions suffering from lack of cooperation could be illustrated in detail. For instance, the table indicates that the (National Wetland) Park had a rather cooperative relationship with many actors. However, as described in section 8.9.2.3, the Park faced also some obstacles in its mission due to inadequate cooperation with some sectoral law-enforcing authorities in its territory.

Important conflictual actor relationships are discussed in more detail in later sections, which deal with the outcome of actor interaction processes (section 8.8.2.3 linking actor interaction with the regime change process and section 8.9.2 linking actor interaction with implementation processes of measures and policies). Suggestions on how the cooperation potential between key actors could improve are made in sections 8.8.2.3 and 8.9.2 but also indirectly in the synthesis chapter 9.

Table 8.10 Problem perceptions, roles, objectives and resources of actors in the Mygdonian case study process

Actors	Problem perception & roles played in the process (detailed institutional competence in Table 8.2)	Objectives & resources
EU	**Problem:** Greece did not comply with EU directives concerning the ecological value of the Koronia wetland and the pollution of the lake with hazardous substances Additionally, Greek authorities did not translate into practice EU-funded studies for the restoration of Lake Koronia. **Role:** It had an important but ambivalent role as funding source in the process: it funded the Master Plan of Lake Koronia but also provided agricultural subsidies leading to unsustainable water use It acted as driver for increased actor coordination by requiring a Programme Contract of all relevant actors to manage completed works of the Master Plan	**Objectives:** To ensure the implementation of EU directives **Resources:** Economic, legal
Ramsar Conference of the Parties (Ramsar COP)	**Problem:** The non-compliance of Greece with principles of the adopted Ramsar Convention concerning the preservation of the wetland of international importance of lakes Koronia and Volvi **Role:** It served as the first incentive to start preparing a legal protection framework for the wetland since the 1980s. The COP included the wetland in the Ramsar Montreux Record (black list) pressurising thus the Greek State to take action to reverse the wetland degradation. It demanded a change in the restoration philosophy of Lake Koronia, especially with respect to a proposed water transfer from another basin in the original Master Plan.	**Objectives:** To influence the Greek State in amending the restoration philosophy for Lake Koronia. To apply the Ramsar principles for wetland restoration **Resources:** Pressure based on international prestige
Ministry of Environment, Physical Planning and Public Works (MEPPW)	**Problem:** The MEPPW criticised the weak implementation of environmental policies in the Koronia basin by Prefectural and local authorities. Even after the revision of the Master Plan, MEPPW representatives doubted the Prefecture's capability to take over the coordination and execution of the Master Plan, considering its poor record of compliance control of its own Prefectural environmental regulations (especially for industries) in the Koronia subbasin (Newspaper « Macedonia », 6/5/2005). The delayed involvement of central level authorities, including the MEPPW, resulted in the formal rejection of several proposed projects of the original Master Plan. The MEPPW however considered the drafting procedure of the revised Master Plan improved compared to the closed procedure of the original Plan (Environment Ministry phone comm., 18/11/2005) **Role:** Being national competent authority for environmental policy, it was directly involved in the EU infringement procedure for Lake Koronia It also blocked the realisation of the original Master Plan through its power in rejecting EIA studies of proposed	**Objectives:** To promote the implementation of environmental policies in the Mygdonian basin and avoid EU infringement procedures before the ECJ **Resources:** Political/Policy-making, technical (in the approval of technical projects)

Actors	Problem perception & roles played in the process *(detailed institutional competence in* Table 8.2*)*	Objectives & resources
	large-scale technical works. Its Department for the Management of Natural Environment, which is in general supportive of sustainable water resource management in the Koronia basin, was active in the process in as far as the appropriate instruments were provided by the MEPPW political leadership, especially the adoption of the CMD for the National Wetland Park in 2004 (Interview Ornithological Society, 28/3/2005). After the creation of the Park Management Body, the MEPPW handed over most competence to promote management actions in the Mygdonian basin to the Management Body (Environment Ministry phone comm., 18/11/2005).	
Ministry of Agriculture (MINAGR)	**Role:** It was active in policy formulation for an agrienvironmental measure for the basin	**Resources:** Economic, political/policy-making
	It also used to have a dominant role in water management prior to the 1987 Water Law. Until the late 1980s, it was head of the regional Service for Land Reclamation which was competent for water management in agriculture and for water abstractions for different uses	
	Until 1995, it was also head of the Fisheries Department in the region, which was directly responsible for the proper management of fish resources in Lake Koronia (Lymperopoulou, 1994). Since 1995, the Ministry of Agriculture did not play this role actively	
Regional Water Management Department (RWMD) of Central Macedonia	**Problem:** The excessive water abstractions in the Koronia subbasin are related to the lax policy of issuing permits for wells by the Prefecture in the past (even after the introduction of the 1987 Water Law). The RWMD could not coordinate abstraction policies effectively with the Prefectural Division for Water Resources. At start, it was difficult to agree on the water restrictive measures proposed by the RWMD. Over time, the Prefectural Division for Water Resources became more cooperative and coordination with RWMD improved to some extent (pers. comm. RCM, 31/3/2005)	**Resources:** Political/regulatory, technical but reduced resource basis in terms of human resources. Its human resources increased since it became part of the Regional administration in 1997
	Role: It had a role in approving all water permits issued by other authorities based on the 1987 Water Law and in formulating policy proposals for water restrictive measures to be issued by the Prefecture	
Regional Directorate for Environment and Physical Planning (RDEPP) of Central Macedonia	**Role:** It was the main Regional authority actively participating in the case study process due to the environmental importance of Lake Koronia as a Ramsar site. It also participates in the Management Body of the National Wetland Park	**Resources:** Political/regulatory, technical
	As participant to the Supervisory Committee of the Prefecture on the original Master Plan, it objected to the proposed large-scale water transfer from River Aliakmon	
Prefecture of Thessaloniki (PoT) – especially its political leadership	**Problem:** The environmental problems of Koronia were not on the agenda of the Prefecture until the mass fish kills incident in 1995. After 1995, the Prefecture was made aware of the problems and initiated the Master Plan. However, being an elected self-administration (since 1995), the PoT was also sensitive to protect sectors of production such as industry and agriculture linked to employment opportunities in the area of Langadas	**Objectives:** To secure funds (from the EU and the central government) to implement the proposals of the Master Plan
	A greater change of attitude of the PoT took place after the rejection of several proposed projects of the Master Plan on environmental grounds and the rise of a new coalition in its political leadership in 2003. The new	To provide solutions to the problems of the Koronia basin without undermining the interests of the productive sector (industry, agriculture)
		Resources: Political/policy-making

Actors	Problem perception & roles played in the process *(detailed institutional competence in* Table 8.2*)*	Objectives & resources
	leadership promoted the revision of the Master Plan towards a more environmentally-friendly direction (Interview Former expert at Information Centre, 23/3/2005). The new PoT leadership seemed to be more active and conscious of the increasingly severe environmental problems of the basin (Interview Ornithological Society, 28/3/2005) **Role:** It was leader of the initiative to carry out a Master Plan for the restoration of Lake Koronia It was also target of critique for delays in the promised restoration. Due to the one-sided approach followed, the Prefecture's original Master Plan was rejected by several other actors	
Division for Water Resources and Land Reclamation of the Prefecture (DWRLR)	**Problem:** The deficient water balance of the Koronia basin is related to the deficient implementation of regulations on abstractions. Deficient policy implementation is partly due to the lack of scientific trained personnel in the DWRLR Control over water abstractions became a complex task after the 1987 Water Law introduced a new water permit procedure and dispersed permit issuing competence to a number of authorities The interpretation of the 1987 Water Law was subject of conflict between the DWRLR and the RWMD. For instance, the DWRLR did not require the submission of an EIA to issue individual water permits, contrary to the instructions of the RWMD. The DWRLR considered this a superfluous (and costly for farmers) requirement, while a holistic water basin management is still missing (Interview Prefecture, 1/4/2005) **Role:** It had a dominant role in authorising the drilling of wells (mainly for irrigation). In the future, its role will be limited due to the prohibition of new wells in the Mygdonian basin after 2004 and the transfer of most competence on water permits to the Regions by the 2003 Water Law (Interview Prefecture, 1/4/2005)	**Resources:** Political/Regulatory, technical
Sub-prefecture of Langadas (Sub-pref.)	**Problem:** The sub-Prefecture of Langadas was one of the first actors which realised the urgency of the environmental problems of Lake Koronia **Role:** It initiated the first effort to record the problems of Lake Koronia in a systematic way (1996 Koronia Rescuing Plan) It received many local complaints on the problems in the Koronia basin and thereby supported the positions of local society to a certain extent (Interview Langadas Sub-Pref., 5/4/2005) It objected the proposal of the original Master Plan to transfer water from River Aliakmon It enjoys an important role in the Park Management Body. The Body is chaired by the sub-Prefect, who is keen to find sustainable solutions to the problems of the Mygdonian basin (Interview Ornithological Society, 28/3/2005)	**Objectives:** To incorporate local society in the implementation of the Master Plan (Interview Sub-Prefect, in Oiko Kathimerini, 12/11/2005) To achieve the appropriate operation of the Park Management Body **Resources:** Political, technical
Local municipalities & communities (Mun/comm)	**Problem:** Many local political leaders do not recognise yet the water problems in the area (Interview Former expert at Information Centre, 23/3/2005). Even those who participate in the Management Body of the National Wetland Park are concerned about unemployment in the	**Objectives:** To actively participate in forum with decision-making potential, and thereby become involved in measures and activities considered for their

Actors	Problem perception & roles played in the process (detailed institutional competence in Table 8.2)	Objectives & resources
	area (Interview Ornithological Society, 28/3/2005)	territory
	Towards the end of the case study, there was a change of attitude of few local political leaders in a more positive direction (Interview Former expert at Information Centre, 23/3/2005). More and more leaders started participating in activities for the protection of the area. Involvement increased with the gradual worsening of problems, especially after the ecological disasters were widely covered by the press (Interview Ornithological Society, 28/3/2005)	**Resources:** Social and political pressure, use of legal resources against industrial polluters
	Local political leaders at times argued that not enough was being done to save the lake by higher political levels	
	Role: Local municipalities were involved in some legal procedures (court cases) over pollution in Lake Koronia	
	Their role in the management of the Mygdonian basin became more significant once they became members to the Park Management Body	
Farmers	**Problem:** Farmers are now more aware of the water scarcity problem and the lack of water to irrigate their crops (Newspaper „Aggelioforos", 13/8/2002). Nonetheless, they also blame industries for abstracting large amounts of water all year round contrary to farmers who abstract water only over the summer (Dimakis, 1999). Farmers favour solutions which can secure their production and the future of their area (Blionis, 1999). Over time, even large-scale farmers became aware of the growing water scarcity (Interview Young Farmers Union, 14/4/2005) and farmers' representatives became more open to discussion. In 4/1999, the Union of Agricultural Cooperatives of Langadas even coorganised seminars on agriculture and the environment to inform farmers on the environmental impacts of agriculture and new agrienvironmental policies	**Objectives:** To impose measures and restrictions not only on farmers but also other users, especially on industry Many farmers favour increased production and income rather than the long term welfare of the basin, which involves restrictions upon their activities (Koutrakis & Blionis, 1995)
	Concerning the problem of water pollution, farmers do not accept sole responsibility and blame mainly the industry	In the past, farmers aimed for technical solutions such as the transfer of water from another basin (Dimakis, 1999). Now, however, farmers increasingly argue that the solution is to subsidise the replacement of spraying with drip irrigation for all linear cultivations (Interview Young Farmers Union, 14/4/2005)
	Farmers are nonetheless still considered as the most difficult and most conservative interest group in the basin (Interview Langadas Sub-Pref., 5/4/2005). Although most are aware of the water resource problems, one cannot speak of a uniform change in their attitude considering the large size of farmers as a group. Some farmers accept the problem and their own responsibilities, but still request from the State to take main action (Interview Young Farmers Union, 14/4/2005). Other farmers simply remain reactive and want to continue with business-as-usual (Interview Former expert at Information Centre, 23/3/2005)	**Resources:** Political pressure
	Role: Farmers' representatives participated in meetings on the Master Plan and in consultations on the CMD of the National Wetland Park	
	In the implementation process of the Master Plan, farmers resisted the installation of water meters	
	As concerns the agrienvironmental measure of the Ministry of Agriculture, farmers were active in proposing improvements for its revision. To this aim, farmers cooperated with the Prefectural Division for Agricultural Development	
Industry	**Problem:** Industries argue that they should not be given the main blame for the degradation of Koronia (Interview	**Objectives:** To continue with business as usual and avoid loss

Actors	Problem perception & roles played in the process *(detailed institutional competence in* Table 8.2*)*	Objectives & resources
	Former expert at Information Centre, 23/3/2005). Especially in terms of the lake drop, industries argue that they consume much less water than farmers (Interview Langadas Sub-Pref., 5/4/2005)	of income from increased operational costs due to environmental standards
	Given the unstable financial situation of the fabric-dye sector (due to the rise of international competitiveness), industries threaten to close down if more limitations are imposed on their production. So, there are also exogenous market factors which hinder a substantial change of industries' attitude (Interview Former expert at Information Centre, 23/3/2005)	Some industries, which were approached with financial incentives, would be willing to install tertiary treatment, if they are convinced that this is not too costly for them (Interview Former expert at Information Centre, 23/3/2005)
	Role: Industry was reactive to most proposed measures of the Master Plan affecting its activities (Interview Former expert at Information Centre, 23/3/2005). Industries objected the installation of water meters to measure water consumption (Interview Young Farmers Union, 14/4/2005). They were also negative to proposals for central salt effluents treatment to avoid their financial contribution	**Resources:** Economic, political pressure
Fishermen	**Problem:** Fishermen acknowledge both industry and agriculture as causes of the problems of the basin	**Objectives:** To restore ecosystem health and fish populations
	They became more trustful and collaborative towards scientists in the context of an information and awareness campaign in 1995 (Koutrakis & Blionis, 1995)	**Resources:** Social pressure
	Role: Fishermen were the first water users of a productive sector to be affected by the Koronia collapse. After the 1995 mass fish kills, they made some protests but had little effect due to their lack of lobbying power	
	By the end of the case study, they remained few in number and rarely protested against environmental degradation. Nevertheless, they acted as watchdogs of the lake, being the first to notify the authorities in case of pollution incidents (Interview Langadas Sub-Pref., 5/4/2005)	
Local associations (Union for the Environmental Protection of the wider Region of Langadas)	**Problem:** Corruption of the competent authorities is a major issue in the process of Koronia	**Objectives:** To make industries comply with regulations
	Pressure on water resources is exercised more by the industries and less by farmers. Even though the Union recognizes the share of farmers' responsibility for the excessive use of fertilizers and unsustainable irrigation practices, it is less supportive of industries which can relocate their production once water resources are exhausted (Interview Local Union for the Environment of Langadas, 6/4/2005)	To make authorities implement more adequately water protection regulations **Resources:** Legal, social pressure
	Role: The Union took legal action against polluters, with preference for legal procedures on the level of the Council of State (superior national court) or the European Court of Justice. It trusted less the effectiveness of local courts (Interview Local Union for the Environment of Langadas, 6/4/2005)	
	Local associations were not active continuously, because they depend on the motivation of only few individuals (Interview Former expert at Information Centre, 23/3/2005)	
Hellenic Ornithological Society (HOS)	**Problem:** Problems in the basin are largely linked to the lack of political decisions to modify the development model of the area	**Objectives:** To protect birds and wetland habitats in the basin of Koronia (Interview Ornithological Society, 28/3/2005)
	The fact that the Management Body of the National Wetland Park has no decision-making competence is an	To lobby for an alternative

Actors	Problem perception & roles played in the process *(detailed institutional competence in* Table 8.2*)*	Objectives & resources
	institutional disadvantage (Interview Ornithological Society, 28/3/2005) **Role:** It collects data on birds, monitors pressures, raises public awareness and lobbies for the interests of the wetlands. It actively participated in a Programme Agreement enforced for the protected area of lakes Koronia and Volvi in 1997-2000 (Tsougrakis, 2000) Its access to decision-making (especially on a local level) clearly improved with its participation (as NGO representative) in the Park Management Body. Before participation in the Management Body, it could affect decision-making mainly through lobbying directed at the MEPPW level (informal and indirect participation) (Interview Ornithological Society, 28/3/2005)	development model of the region of Langadas (transition from industry and intensive farming to more sustainable agriculture and small-scale processing of food products of high quality) (HOS, 2003) **Resources:** Technical/scientific, political pressure
External experts	**Role:** External experts were not involved initially in the making of the Master Plan. Academic experts and the Technical Chamber of Greece exercised critique on the original Master Plan, especially on the proposed water transfer from River Aliakmon. Later on, external experts became involved in the supervision of the revision of the Master Plan. The academic community and the Geotechnical Chamber of Greece are also represented in the Management Body of the National Wetland Park	**Resources:** Technical
National Wetland Park – mainly referring to its Management Body	**Problem:** The lack of political support to strengthen the role of the Management Body for the protection of the Koronia basin is a main problem Several public authorities perceive the competence given to the Management Body as intrusion in their fields of competence (Interview Former expert at Information Centre, 23/3/2005) **Role:** The Park Management Body was the first structure to coordinate different authorities with competence within the National Wetland Park of the Mygdonian basin (pers. comm. RCM, 21/4/2005). The Management Body functions as a round table for the exchange of relevant stakeholders (Interview Ornithological Society, 28/3/2005). The Management Body is also consulted prior to licensing any kind of activity or intervention in the National Wetland Park. The expert opinion of the Management Body is necessary for any large interventions such as water transfers from other basins or setting up a new industry (pers. comm. RCM, 21/4/2005)	**Objectives:** To promote exchange between different stakeholders and to supervise the drafting of a Management Plan for the National Wetland Park **Resources:** Political/regulatory, technical, but still lacking financial and personnel resources

Table 8.11   Relationships among key actors in the Mygdonian case study process

Actors	Administration									Users				Others			
	EU	Ramsar COP	MEPPW	MINAGR	RWMD quantity interest	RDEPP	PoT	DWRLR quantity interest	Sub-Pref.	Mun/comm	Fishermen	Farmers	Industry quality interest	Local ass. quality interest	HOS	Experts	Park
EU	▨																
Ramsar COP		▨															
MEPPW	↘/(↘)	→/(↘)	▨			+/+	–/+							→--↘--	⌂+/+		+/+
MINAGR				▨	–/+	+/+						0/(+)					+/+
RWMD quantity interest					▨												
RDEPP			+/+	+/+	+/+	▨	–/+						→↘		0/+	0/+	++/++
PoT	€/€	(↘)/(↘)	–/+	–/+	–/+	–/+	▨	+/+	–/+	↘/+		(+)/(+)	→↘	→--↘--	–/+	–/++	++/++
DWRLR quantity interest					–/+	+/+	+/+	▨				(+)/(–)					
Sub-Pref.						+/+	(–)/+		▨	++/++	++/++	++/++			0/+	0/+	++/++
Mun/comm.						0/+	+/+		++/++	▨					0/+	0/+	++/++
Fishermen									++/++		▨	–/–	–/–				
Farmers			–/0	0/€(+)			(+)/(+)	(+)/(–)			–/–	▨	–/–	–/–			
Industry quality interest						→↘	→↘				–/–	–/–	▨	–/–			
Local ass. quality interest							–/–						–/–	▨			
HOS			⌂+/+			0/+	–/+		0/+	0/+					▨	0/+	++/++
Experts						0/+	–/++		0/+	0/+					0/+	▨	++/++
Park			+/+	+/+		++/++	++/++		++/++	++/++					++/++	++/++	▨

Note (a): The table should be read as the relationship of the actor in the vertical column to each of the actors in the horizontal rows. Blank cells indicate no particular relation observed. Actor abbreviations are explained in Table 8.10.

Note (b): Symbols in the table are explained in the legend below. The presence of symbols left and right of the slash (/) aim at indicating noteworthy changes in actor relationships over the observed process. If left and right symbols are identical, no significant change in the actor relationship was observed. If symbols are different, a certain change took place. For instance, symbols (++/++) indicate close collaboration at first, which was reduced over time. Symbols (--/+) indicate an initial conflict which was followed by certain improvement (formal interaction or sporadic collaboration).

LEGEND OF SYMBOLS:

++	close collaboration (problem-solving interaction)	--	conflict
+	only formal relationship or sporadic collaboration	⌂	information supply or exchange
0	no noteworthy interaction observed	€	financial contribution
		→	One-sided pressure

8.8 Process of water basin regime change

This section analyses the process of change in the institutional water regime of the Mygdonian basin, based on the analytical framework presented in chapter 4/4.1.2. The relevant hypothesis (Hypothesis 2) was that:

"The less favourable the contextual conditions and the characteristics of the interacting actors for integration attempts, the more likely that Greek water basin regimes fail to reach a more integrated phase."

8.8.1 Dimensions of regime change

8.8.1.1 Extent

Since the start of the case study process in the 1980s, gradually most uses with a claim on the water resources of the Mygdonian basin were considered by the water basin regime. An exception remained the use of Lake Koronia for recreation which was not considered at all. Later, claims to use the lake for recreation faded away. An important aspect not considered by the regime was the use of water as medium for the disposal of salt effluents. After salt pollution from dye industries was recognised as the most important source of water pollution of the lake in 2002, the issue was put onto the agenda by the measures proposed in the revised Master Plan. However, by the end of the case study, this pollution type was still not regulated on the policy level.

Table 8.12   Extent of uses in the Mygdonian basin regime

Water uses	Presence in the water quantity domain	Presence in the water quality domain
Agriculture/irrigation	Regulated through Prefectural restrictive measures since 1989. Also the agrienvironmental policy measure adopted in 2003 for the area around the lakes of the basin offered voluntary instruments (incentives) to reduce water consumption for irrigation	Agricultural non-point pollution was considered by the basin designation as Nitrates vulnerable zone and by the 2006 CMD to combat nitrate pollution of agricultural origin. Also the agrienvironmental policy measure of 2003 offered voluntary instruments to reduce water pollution from agriculture
	Prior to 1989, agricultural use was regulated only through authorisations for the construction of wells and connection to power supply	
Urban use	Regulated through Prefectural restrictive measures since 1989	A series of prefectural decisions regulated the disposal of urban wastewater
	Prior to 1989, the regional office for land reclamation gave authorisations on drilling or extending wells for non-agricultural use	
Industry	Regulated through Prefectural restrictive measures since 1989	A series of prefectural decisions regulated the disposal of industrial effluents
	Prior to 1989, the regional office for land reclamation gave authorisations on drilling or extending wells for non-agricultural use	The 2005 CMD on pollution reduction in Lake Koronia from hazardous substances set new quality standards for industries
		Salt pollution put onto the agenda of the Master Plan measures (since 2002) but still not on the policy level
Fisheries	Use of Lake Koronia for fisheries regulated through national fishing codes and

Water uses	Presence in the water quantity domain	Presence in the water quality domain
		legislation.
		Water quality targets set by Prefectural decisions also aimed at safeguarding fisheries as a use. This was set as an aim also by the 2005 CMD on pollution from hazardous substances
		Prohibition of fishing in Lake Koronia since 1995
		Some actors consider that the regime extent remains incomplete for the use of fisheries due to lack of regulations for the fish water quality of torrents and streams mouthing into Lake Koronia
Environment		The ecological value of the basin was explicitly recognized in 2004 with the official designation of the National Wetland Park
Recreation		No specific regulations ever existed for recreation as a water use in the basin

8.8.1.2 Coherence of property rights

Lack of coordination among property rights was made obvious in the case study from the persistent water use rivalries between agriculture, industry, fisheries and the environment.

Rights conceded to water users (mainly industry and farmers) for abstracting water often contradicted each other, since water use permits were issued without a holistic water management plan. In practice, too many abstractions have been permitted in the basin, without considering the water basin balance. Thus, the present use of water exceeds the total water available. Irrigation rights were at times also affected by water pollution rights given to industries. Namely, the quality of water abstracted from the aquifer for irrigation was degraded in some locations of the basin by the effluents of nearby industrial activities. Such cases of contradicting abstraction and pollution rights often led to conflicts of local character.

Additionally, it was criticised that rights to pollute water were conceded to industries (via effluent disposal permits) without calculating the total buffer capacity of the water ecosystem, especially of Lake Koronia and its drainage network. The overall coherence of rights to use water in the Mygdonian basin was also reduced by the fact that, despite some progress in the official concession of rights to abstract or pollute water, much water use took place illegally both in terms of water abstractions and of disposal of effluents (see also section 8.8.3 on the implementation of measures and policies).

8.8.1.3 Coherence of public governance

Levels and scales

As far as the "fit" between administrative levels and the water basin natural scale is concerned, it could be considered an advantage that most of the Mygdonian basin is in the administrative competence of the Prefecture of Thessaloniki and of the Region Central Macedonia. However, despite the involvement of only one Prefecture and only one Region in the management of the basin (cf. 3 Prefectures and 2 Regions in the management of the Vegoritida basin in chapter 7), there was still difficulty in administrative coordination. A standing example was the lack of coordination between the RWMD of the Region and the Prefectural Division for Water Resources and Land Reclamation in the process of implementing the 1987 Water law (lack of intrapolicy coordination of actors).

Initially, there was also little coordination of local with higher administrative levels, despite the national importance of the Koronia wetland. The early focus of the process on the local level (Prefecture) resulted in a conflict with the central level (Environment Ministry) over the proposed works of the original Master Plan. Gradually, except for the Prefecture, also the central government and the Region became involved. The preparation of the revised Plan was considered an improved process, since it developed into a more open and more multi-level actor interaction scheme including also the central administration.

The creation of the Management Body for the National Wetland Park in 2002 also set the basis for improved coordination of different administrative levels involved in the management (especially environmental management) of the broader Mygdonian basin. Also the revision of the Master Plan extended the consideration of water management and restoration issues from the narrower scale of Lake Koronia to the complete scale of the Mygdonian basin.

Another encouraging development was the signing of a Programme Contract for the execution of the restoration programme of Lake Koronia by all actors involved in the restoration works and measures. The Programme Contract was a positive step expected to facilitate the coordination of actors from different levels of the administration (central, regional, local).

Actors in the policy network

Before the establishment of the Management Body for the National Wetland Park in 2002, there were no policy-based participatory arrangements to involve both public and non-public actors interested in the water resource problems of the Mygdonian basin. A Regional Water Committee set up in 2000 (on the basis of the 1987 Water Law) was occasionally active on a formal basis but had no significant role in the management of the Mygdonian basin.

The established Park Management Body aimed at promoting participation and serving as an instrument for exchange between stakeholders of the Park. However, this participatory approach also seems incomplete, if one considers that water users (industry, farmers, fisheries) are not directly represented in the Management Body. Participants of the Management Body Administrative Board include the administration (all levels), e-NGOs and academic experts. The industry is only indirectly represented by the Ministry of Development. Agriculture is not represented at all (except for a substitute member of the Board from a local forest inspectorate pertaining to the Ministry of Agriculture) and neither are fisheries as a water user of the lakes in the basin. Nevertheless, the Management Body brought about a significant improvement in the participation of actors representing the environment as a user of the Mygdonian basin. Before the establishment of the Management Body, e-NGOs had no formal access to decision-making concerning the Mygdonian basin. The admission of one e-NGO representative (Ornithological Society) in the Park Management Body helped e-NGOs to exercise influence on decision-making also on a more local level. Prior to the Management Body establishment, the Ornithological Society could exercise pressure solely on the level of the Environment Ministry in a more informal and indirect way (Interview Ornithological Society, 28/3/2005). As far as local environmental organisations are concerned, these still have no access to decision-making and mainly try to be involved in the process by using the legal judicial system. By the end of the case study, there were still no substantial communication links established between the Ornithological Society (supra-regional NGO in the Management Body of the Park) and local environmental groups.

As concerns the making of the Koronia Master Plan, the process started with limited interaction and consensus of different actors. The drafting process of the original Master Plan lacked a participatory approach and its supervisory Working Committee involved only parts of the administration. Participatory arrangements slightly improved later when a short consultation process also with actors outside the administration took place for the revision of the Master Plan. The drafting of the revised Plan was also characterised by a higher degree of interaction and consensus among several actors, especially the Prefecture, the central administration, e-NGOs and the external scientific expert community. Even the working committee set up by the Prefecture for the revision of the Master Plan (Committee of Guidance and Coordination) did not involve only the administration, but also professional technical chambers, representatives of local authorities and the local agricultural sector. However, by the end of the observed process, there were still no significant indications of increased consensus between the administration and key water users such as farmers and industries.

Perspectives and objectives

The Master Plan for the restoration of Lake Koronia is not equivalent to a policy-based water basin management plan but it can be considered as a water vision for the Mygdonian basin. However, the vision presented in the original Master Plan did not include multiple actor perspectives due to the closed drafting process followed. The original Master Plan was indeed viewed by several actors as one-sided perspective of the Prefecture. Specifically, the Environment Ministry, e-NGOs and even farmers and industry had objections to the Plan. It is also worth emphasising that the Environment Ministry was in parallel promoting its own vision for the management of the basin from a nature protection perspective, by drafting the legal framework for the to-be National Wetland Park. Unfortunately, the Master Plan of the Prefecture and the legal framework for the Park were being drafted in parallel but not consistent with one another.

In the process leading to the revision of the Master Plan, there was, firstly, a change of perspective of the Prefecture itself after a change of the political leadership in power. In the revised Plan, the Prefecture favoured softer measures compared to the engineering large-scale approach of the original Plan. Secondly, the process of the revised Plan was more open to the perspectives of actors other than the Prefecture. There were however also actors, whose perspectives were not reflected to an adequate extent in the revised Plan either. For instance, several objections were expressed by the Greek Biotopes/Wetlands Centre as well as by industries and farmers.

The 2004 CMD designating the basin as National Wetland Park was an official policy also supporting a vision for the sustainable development of the basin with a focus on its environmental values. However, the CMD altogether was also criticised for not being based on a consensus of perspectives, since for instance farmers had different expectations from its content. A more complete vision for the protection of the National Wetland Park is expected to be produced in the future in the form of a Park Management Plan.

In 2003, yet another management plan relevant to the Mygdonian basin entered the picture, when the Ministry of Development assigned consultants to prepare a water management plan for the water district of Central Macedonia (to be completed by the end of 2006). This water district management plan should be carried out according to the principles set out in the national 1987 Water Law and the recent 2003 Water Law. The RWMD of the Region Central Macedonia is the main regional actor formally involved in the drafting process of this management plan. Its role is however limited to the provision of data on abstractions, water uses and hydrogeology. By the end of the case study, there was still no coordination between this water district management plan and the Koronia Master Plan of the Prefecture. Such coordination could prove necessary on several issues such as the definition of a level for Lake Koronia. While the definition of minimum level for lakes of Central Macedonia was assigned as a task for the water district management plan, the revised Koronia Master Plan already made a concrete proposal to restore the Koronia level at 4m (pers. comm. RCM, 31/3/2005).

Finally, the recognition of important rivalries in policy visions concerning the Mygdonian basin is also worth briefly discussing. So far, there seems to be a certain degree of policy recognition of the rivalry between agriculture and water resources, bearing in mind that a specific agrienvironmental measure was issued to protect water resources in the basin. However, the rivalry between industrial development and water resources is only dealt so far in the context of water quality regulations. In this respect, there is further need for interpolicy coordination between water protection policy and policies determining the set up of industries in the basin (including development policies and spatial planning policies).

Strategies and instruments

Policy instruments available to authorities to restrict rights for abstracting or polluting water in the basin have had mainly the character of permits. These systems of permits did not give the possibility to water authorities to re-distribute existing use rights in order to minder existing use conflicts. For instance, decisions to restrict wells were applicable only to new wells and were issued only reactively after the excessive drop of Lake Koronia. The only regulation discontinuing existing use rights was the prohibition since 1995 to use water directly from Lake Koronia, which affected the operation of existing collective irrigation networks.

Most instruments used to manage water resources in the basin were of command-and-control type. Even non-policy measures proposed by the Koronia Master Plan had a top-down character, such as the proposed installation of water meters for the control of water use and collection of data on real water consumption patterns by water users in the basin.

Towards the end of the case study, there was some tendency to enrich the pool of available instruments with less command-and-control instruments such as:

- A voluntary agrienvironmental measure providing financial incentives to farmers to reduce water use in irrigation (introduced in 2004).

- Active promotion by the Prefecture and the Region of a mechanism for direct subsidies to farmers in order to replace water-demanding irrigation techniques with drip irrigation (initially within a pilot area), linked to the fulfillment of agrienvironmental principles during production.

- Proposal for increased direct subsidies to dye-industries of the basin to install saltwater treatment units in their individual industrial premises. To this aim, the Region and the competent ministries were pursuing the inclusion of a suitable mechanism in the national Development Law.

Resources and responsibilities for implementation

The implementation of policies aiming at more sustainable water management in the basin was deficient in resources such as adequate personnel in numbers and qualifications (e.g. in the Prefectural Division for Water Resources and Land Reclamation) and financial support (e.g. for the operation of the Park Management Body).

Additionally, the fragmented distribution of competence on water resources and the lack of suitable coordination mechanisms contributed to the lack of integrated management of the basin. A characteristic example of missing coordination was the case of the implementation of the 1987 Water law by the Prefecture and the Region. Responsibilities on authorisation of water abstractions were held until 1990 exclusively by the Prefecture (Division of Land Reclamation). After the adoption of the 1987 Water Law, part of these responsibilities was passed onto the Region (RWMD) and other authorities (according to the type of water use). The Division of Land Reclamation disliked the dispersal of competence for water abstractions among a number of authorities. Only towards the end of the case study, there was improvement in the coordination between the Region and the Prefecture. Coordination gaps may be completely eliminated soon by making the RWMD competent for issuing all types of water permits according to the 2003 Water Law.

Similarly to the Vegoritida basin examined in chapter 7, responsibility on water quality and water quantity issues was separated between different authorities on all levels of the administration (the Prefecture, the Region and the central government). As a result, regulations adopted by different authorities to control water abstractions and water pollution in the Mygdonian basin were not integrated. Nonetheless, some improvement in terms of integrating quality and quantity issues was observed in the combined consideration of these issues in the Koronia Restoration Master Plan, in the planned extension of competence of the RWMD to include quality issues by the 2003 Water Law and in the treatment of both pollution and scarcity issues in recent policies adopted for the basin (agrienvironmental policy measure and 2004 CMD on the National Wetland Park).

The CMD on the National Wetland Park also made an attempt to coordinate responsibilities of authorities, especially for environmental issues, by setting up the Park Management Body as coordinator of different authorities in the Mygdonian basin. However, there has been resistance to this "coordination concept" by some sectoral authorities who feel that their competence is threatened.

Finally, a positive step at least for the purpose of clarity concerning the resources and responsibilities for executing the restoration works and measures for Lake Koronia was made by signing a Programme Contract in 2006 between the main central, regional and local actors involved.

8.8.1.4 *External coherence between property rights and public governance*

A certain degree of external coherence in the water basin regime resulted from the fact that several changes in public policy were also reflected as changes in the property rights system, such as the introduction of permits for water abstractions, policy restrictions on wells and the introduction of permits for the disposal of effluents. However, satisfactory external regime coherence could not be achieved.

In terms of the water scarcity problems in the basin, although the prefectural restrictive measures and even the CMD for the designation of the National Wetland Park restricted the rights of farmers to irrigate (restrictions on new wells and prohibition of direct abstraction from Lake Koronia), illegal use of water in the basin was frequent. This implies that illegal water users have not been effectively targeted by the public governance system, resulting into lack of external regime coherence due to inadequate policy implementation (see also section 8.8.3). Additionally, holders of pre-1989 abstraction rights, especially for irrigation wells constructed prior to 1989, were not targeted by policy at all and were often a point of disagreement between the Region and the Prefecture, thus further reducing the external coherence between policy and the property rights system. Furthermore, the quantitative right of the environment to water has not been established yet, despite the relevant requirements by public policy (requirement of defining minimum lake levels of the 1987 Water Law).

As concerns the external coherence between policies and rights for water pollution, policy seems to cover all important users with rights to pollute water. Nevertheless, external coherence between policy and rights was reduced in practice by deficient policy implementation (see also section 8.8.3). Namely, although most industries have acquired by now effluent disposal permits, most urban users and animal farms have not applied for formal rights (permits) to pollute, despite the relevant requirement of regulations. Additionally, the holders of rights to pollute (especially industries) are still not the target of appropriate spatial planning policy, to avoid the disperse settlement of newly-established industries in the basin. In this case, the lack of coordination between spatial planning and water quality protection resulted in administrative weakness to manage sources of industrial pollution in a more consistent and spatially-focused way.

8.8.1.5 General assessment of regime change

Table 8.13 summarises the development of the regime elements over the case study process, referring thereby to indicators for a more coherent (and more integrated) water basin regime.

Table 8.13   Change variables of the Mygdonian basin regime

Extent	Indicators for complete extent	Assessment
	All water uses and users are considered by the regime elements	Most uses and users considered by the regime, except for recreation and the use of water as medium for salt pollution
Public governance	**Internal coherence indicators**	
Levels of governance (Where? – Multilevel)	Administrative levels of management which fit with the natural scale of water basins	Satisfactory fit between administrative boundaries and water basin boundaries
	Coordination at a river basin scale, even when various administrative levels are involved	Lack of administrative coordination but potential for improvement by the Park Management Body
		Encouraging development was the signing of the Programme Contract for the execution of the restoration programme of Lake Koronia
	Processes including actors from different levels	Initially, mainly the local level was involved; a multi-level process evolved later
	Coordination and cooperation within each administrative level and between different administrative levels (within the river basin)	Satisfactory coordination within each level but lacking coordination between levels; Improvement expected with the Park Management Body and the Programme Contract on the restoration of Lake Koronia
Actors in the policy network (Who? – Multiactor)	Participatory arrangements for actors, especially non-public actors, with an interest in water resources (formal arrangements such as planning and/or informal consultations)	The Park Management Body ensures participation of several public and non-public actors but without direct involvement of water users
	Increasing cooperation, interaction and consensus among actors in the relevant networks	Improvement in cooperation and interaction within the administration and between the administration, e-NGOs and the scientific community during the revision of the Master Plan
		Consensus remained low between administration and water users
	Open networks instead of closed policy communities	Not fulfilled
	Establishment and involvement of local associations	Local associations established but have no access to decision-making yet

Perspectives and objectives (What? – Multifaceted)	Policy visions and objectives (official visions and visions of stakeholders) recognise new water uses and administer justice to rivalries between different water uses	Official policies recognise some key water use rivalries
	Recognition of relations with other policy fields as coordination topics	Some coordination of water and agricultural policy in the context of the agrienvironmental policy measure
	Development of a (river basin) management plan or a water basin vision (preferably official, but also consider informal initiatives)	The Master Plan of the Prefecture represented a water basin vision
		The CMD on the National Wetland Park is a policy vision for sustainable development and nature protection in the basin
		On-going development of an official water management plan for the water district Central Macedonia since 2003 (steered by the central government – Ministry of Development)
	Incorporation of multiple perspectives in the plan or vision	Original Master Plan did not include multiple perspectives but this partly improved over the process
	Change of water management paradigm, e.g. from hydraulic interventions to "softer" management principles	Change of restoration paradigm from hard engineering to softer restoration in the revised Master Plan
Strategy and instruments (How? – Multi-instrumental)	Availability of water management tools to redistribute use rights and reduce water use conflicts	Systems of permits did not redistribute rights but only restricted the allocation of new rights
	Incorporation of flexible instruments in policy; increasing preference for indirect procedural instruments (e.g. planning, participatory EIA procedures) instead of command-and-control tools and technical works	Dominance of command-and-control instruments
		More indirect flexible instruments were the agrienvironmental measure and the active promotion of subsidies for the installation of drip irrigation and individual saltwater-treatment units
Responsibilities and resources for implementation (With what? – Multiresource-based)	Policy implementation process is sufficiently concerted and equipped (authority, legitimacy, finance, time, human capacity, information etc)	Insufficient resources for policy implementation at all administrative levels
	Clear institutional arrangement for implementation (instead of fragmented arrangement of competence)	Fragmented institutional arrangement which could be improved in the future; In this respect, especially encouraging is the coordination role of the Management Body and the Programme Contract signed for the execution of the restoration programme of Lake Koronia

Property rights	Internal coherence indicators	
	Coordination, not contradiction, of use and ownership rights when competing uses evolve	Contradiction of rights of industry and farmers to abstract and to pollute water with rights of the environment and fisheries
Public governance/ property rights	**External coherence indicators**	
	Changes in policy and public governance are reflected in changes of the property rights (e.g. via restrictive regulations such as restrictions on wells, restriction of private rights, establishment of permits for abstractions or pollution)	Several changes in policy were reflected in changes in property rights, but external coherence was often kept low due to inadequate policy implementation
	Match between actors targeted by policies and actors with use and ownership rights	Satisfactory match between many but not all actors in policy and actors with rights

Altogether, changes in the institutional water basin regime as a whole did not qualify for a regime shift to a more integrated phase. The coherence of the Mygdonian basin regime tended to increase only towards the end of the observed case study process, especially with respect to the public governance system. Aspects related to raised coherence in public governance included the potential created for better coordination between administrative levels through the Park Management Body and through the Programme Contract for the restoration of Lake Koronia, the increased interaction between public and non-public actors in the Park Management Body and the opening of the revised Master Plan to new perspectives and actors. Less progress or no progress was made in the coherence of other regime elements, such as the distribution of resources and responsibilities for policy implementation as well as in the internal coherence of the property rights system. The lack of coherence in certain regime elements can be partly explained by the weak implementation of certain policies and measures adopted during regime change (see section 8.9).

8.8.2 Factors affecting regime change

The following discusses the external impetus for change in the water basin regime as well as the contextual conditions and actor characteristics and interactions which influenced the process of regime change.

8.8.2.1 External change impetus

EU and international impetus

Pressure exercised on the Greek State by the EU contributed to the designation of the basin as National Wetland Park and to the adoption of water protection principles. In particular, pressure from an infringement procedure initiated by the EC against Greece (with the threat of bringing Greece before the ECJ) induced the issuing of the 2004 CMD setting up the National Wetland Park, the 2005 CMD on a special pollution reduction programme of Lake Koronia from hazardous substances and the 2006 CMD on an Action Programme to reduce agricultural nitrate pollution in a broader area.

Additionally, the EU influenced the process of the Master Plan for restoring Lake Koronia. Although the EU Cohesion Fund initially did not object in principle the proposed water transfer scheme from River Aliakmon and other technical measures of the original Master Plan, it refused to fund them once they were rejected on the national level of Greece. Additionally, several requirements linked to the EU funding procedure of the Master Plan induced the execution of a consultation process over the revised Plan and the signature of an actors' Programme Contract for the restoration of Lake Koronia.

As concerns international impetus, the Conference of the Contracting Parties of the Ramsar Convention (COP 7 in 1999 and COP 8 in 2002) exercised pressure on the Greek State to revise the restoration principles of the Master Plan (Koronia being a Ramsar wetland of international importance) and to reconsider the proposed water transfer from River Aliakmon.

Resource problem pressure

Pressure from the alarming water resource degradation of Lake Koronia, especially after the mass fish kills of 1995, was a significant external impetus for issuing several Prefectural regulations and for initiating the Master Plan to restore Lake Koronia.

New policy actor constellations

The change of political leadership of the Prefecture in 2003, combined with other impulses from the international and EU level, influenced the revision of the Master Plan towards a new restoration paradigm away from the previously proposed large-scale engineering measures.

Impetus from national policy

In general, national policy did not act as a significant impetus for change in the regime of the Mygdonian basin. In some cases, however, national policy and especially environmental and nature protection policy had a supportive role in certain developments such as the designation of the National Wetland Park and the creation of a Management Body for the Mygdonian basin (see section 8.8.2.2 on supportive institutional interfaces).

8.8.2.2 Contextual conditions

Characteristics of the actor network

Tradition of cooperation

There was no tradition of previous cooperation among the actors involved in the case study process to support a water basin regime change to more integration. There was also no history of integration of different interests and no former structures to support actor coordination in the basin. The unfavourable cooperation tradition was also linked to the lack of trust among certain actors, especially between the main water users, i.e. industries and farmers, as well as between the Prefectural and central level of administration.

Joint problem

In the early 1990s, almost no actors acknowledged the water resources problem of the Mygdonian basin, except for some local e-NGOs and members of the academic/scientific community. Environmental degradation information started to rise only after the 1995 mass fish kills in Lake Koronia. Since then, several studies were carried out to investigate the environmental and water resource problems of the Mygdonian basin (Koronia Rescuing Plan, Koronia Master Plan and the Specific Environmental Study to designate the National Wetland Park).

Over the process of the case study, the majority of actors (including industries and farmers) admitted the clearly visible problem of both the drying out and the pollution of Lake Koronia. More and more farmers, the local population and even some industries realised that the unsustainable use of water would have negative consequences on their income and life quality (partially shared perception of risk). Especially farmers, who earlier on used to favour the reclamation of the lake to gain more farmland, increasingly acknowledged that water degradation also threatened their future ability to irrigate their fields. However, it cannot be claimed that the common problem perception of water users was translated into will for common action. A sense of shared responsibility of water users over the causes of water degradation could not be developed (Interview Ornithological Society, 28/3/2005). Even in the context of the Master Plan process, despite increasing convergence between authorities, e-NGOs and scientists on the proposed restoration measures, farmers and industries continued resisting parts of the proposed action.

Joint chances

Except for the realisation of joint gains between actors supporting the environment and the revival of fisheries, there was no notion of joint gains from more integrated water management among most actors in the process. The Management Body of the National Wetland Park, which was the most likely forum to convince actors of joint opportunities from integration, was not fully operational within the period of this case study. Towards the end of the case study, at least some joint chances were realised by supporters of the environment and by farmers, thanks to the introduced agrienvironmental policy measure and the environmental principles of the revised EU Common Agricultural Policy.

Alternative threat by a dominant actor

In the management process of the basin, no actor proved powerful enough to impose an alternative strategy to his benefit and thus pose a credible threat to other actors. Even though the Prefecture attempted to impose its own strategy for the management of the basin (in the originally proposed Master Plan), it was not powerful enough to dominate the governance system and was eventually forced to change its strategy by other actors (e.g. Ministry of Environment).

Institutional interfaces

Institutional position of brokers

Only the Management Body set up for the National Wetland Park could realistically claim the role of a broker in the process. Considering that the Management Body was operational for only a short time before the end of this case study, the impact of its potential broker function could not yet be evaluated.

Institutional representation of rival interests

In the observed process, there were effective "umbrella" organisations mainly for the large-numbered group of farmers. The same was not valid for industries, which were involved in the process on an individual basis, and for fishermen who had practically no collective institutional representation in the process. Additionally, the large number of authorities with diverse responsibilities in the management of the basin was translated into a high number of actors involved from the administration.

Alert mass media

Mass media (both local and national) were quite alert in the process and contributed to public awareness on water resource degradation in the Mygdonian basin. Mass media were especially active following the two environmental collapse incidents of Lake Koronia in 1995 and 2004.

Alert public

The majority of the local public in the basin was not organised on issues of environmental and water protection (Interview Langadas Sub-Pref., 5/4/2005). Other social problems such as unemployment were considered more urgent than water problems. In locations such as the city of Langadas, people avoided conflict with industries out of fear of unemployment. Moreover, local perception to a great extent was that development could be achieved on the cost of the environment (Interview Former expert at Information Centre, 23/3/2005). Only towards the end of the case study, a part of the local population became more conscious of environmental problems fearing the effects of water resource degradation on human health and local life quality (Interview Young Farmers Union, 14/4/2005).

Legal leeway for more integrative approaches

The legal and policy framework was relatively favourable to attempts for more integration, especially due to laws coming from the nature and environmental protection sector rather than the water management policy sector. The 1986 Environment Law facilitated some important changes in the water basin regime through the designation of the basin as National Wetland Park and the establishment of a participatory Management Body for the Park.

The 1987 Water Law could have also facilitated more integration in the regime. It gave some impulse for integration by setting up the RWMD of the Region Central Macedonia, by introducing the water permit procedure and by promoting water restrictive measures in the Prefecture since the early 1990s. However, several of its important principles for more integrated water management were inadequately implemented on the regional level all over Greece (see chapter 5 for a discussion on the 1987 Water Law).

8.8.2.3 Actor characteristics and interactions

This section discusses the influence of the characteristics (motivation, information and resources) and the interactions of actors on the main attempts to change the water basin regime (see chapter 4/4.1.2.3 for relevant theory-based assumptions on actor interaction). In the regime change process, the actor characteristics and interactions are themselves influenced by the set of explanatory contextual conditions discussed above. The main case study developments which were relevant to attempts to change the regime included: processes of policy adoption to protect water resources by the Prefecture and later on by the central government, the drafting process of the Master Plan for the restoration of Lake Koronia and the process of designating the basin as a National Wetland Park. Most of these developments however encountered obstacles which delayed their completion.

All in all, many actors in the process were not positively motivated to initiate or to support other actors' initiatives for a water basin regime change towards more integration. Low overall motivation was partly conditioned by the low tradition of cooperation among actors in the basin, the lack of a joint problem and joint chances perception during most of the process. Even the lack of pressure from the local public to protect water resources contributed to low actor motivation for action in the direction of resource protection and integration in the water regime.

The regulatory attempts of the Prefecture since the 1980s to control abstractions and water pollution in the Mygdonian basin were largely the result of policy requirements by national legislation. Moreover, specific restrictive measures to reduce water consumption in the basin were introduced only in a reactive manner, after extreme water problem pressure was made apparent in 1995. The lack of initial motivation of the Prefecture to push for more sustainable water management was partly linked to the low motivation of farmers and industries in the basin to comply with any regulations imposed. As a result, there was either no actor interaction towards a more integrated regime or (even worse) negatively motivated actors acted together to prevent any regime change towards more integration (interaction assumptions I-2 and I-3).

Shortly after the 1995 environmental collapse of Lake Koronia, the Prefecture initiated the drafting of the Master Plan to restore the lake. Inaction was no longer a sensible alternative for the Prefecture, given the wide publicity devoted to the environmental collapse and the mass fish kills observed in the lake. However, considering the restoration approach of the original Master Plan based on large-scale water transfers, it cannot be claimed that, even then, the Prefecture aimed at a sustainable water management model for the basin. The drafting of the Master Plan is viewed by some as an attempt to secure more water and in the same time serve political objectives. As described, the Prefecture came up against several actors (Environment Ministry, e-NGOs, scientific and technical experts) who were not in favour of the proposed large-scale engineering measures and were thus not motivated to support the Master Plan process. Through its competence to approve the relevant EIA studies, the Environment Ministry (and other Ministries) was powerful enough to block the Master Plan of the Prefecture on the national level and thereafter to force the Prefecture towards a more commonly acceptable restoration alternative (interaction assumption I-6). Indeed, given the power of the central authorities in approving large technical works and under pressure also from the international level, the Prefecture was left with no other alternative but to revise its approach.

The realisation of the original Master Plan was also slowed down due to the lack of cooperation of important water users on certain proposed measures. Industries and farmers did not favour cooperation because they could suffer extra costs or loss of production through the measures proposed. Instead, these users continued following the alternative of non-cooperation, trusting in their role as sources of labour and economic development to gain the support of politicians (especially, in the case of industries). Additionally, as an excuse, they often shifted the blame for water degradation to other users (farmers to industry and vice versa).

After the rejection of the original Master Plan by the central government and by the EU Cohesion Fund and following a change in the Prefecture political leadership, the Prefecture became more motivated towards aspects of a more integrated water regime. For instance, it enhanced the participation of external experts and it organised a consultation process for the revision of the Master Plan. Lately, it has also actively worked towards the adoption of a measure to encourage private investment in drip irrigation in the basin. Considering that the more positively motivated Prefecture was interacting also with more positive or at least neutral actors, the revision of the Master Plan developed into a more cooperative and transparent process (interaction assumption I-5).

In parallel to the Master Plan, the designation process of the Mygdonian basin as a National Wetland Park was taking place as another effort to increase the integration of the water basin regime. Although the competent Department for the Natural Environment of the Environment Ministry as well as e-NGOs were positively motivated to bring this policy process forward earlier, the designation of the Wetland Park was delayed for over a decade. Under pressure by an EU infringement procedure, the alternative of further delay became potentially too costly for the political leadership of the Environment Ministry, which then finally designated the basin as National Wetland Park in 2004. Considering that the lack of information was probably not the reason for this delay (enough information was made available through a series of studies carried out for the Mygdonian basin by the Environment Ministry), the delay could be due to the lack of political willingness to adopt a nature protection regime that could reduce local development rates (interaction assumption I-7) (in fact, development interests were a common reason for delays in most Greek regions nominated for designation as nature protected areas of the 1986 Environment Law).

8.8.2.4 Summary

In summary, some contextual conditions were favourable but most remained unfavourable to regime change attempts towards more integration in the Mygdonian basin. The influence of these conditions was exercised also by forming part of the context for the motivation, information and resources of the actors interacting in the process (see Table 7.15).

Table 8.14 Influence of contextual conditions on actor characteristics and the Mygdonian regime

Contextual conditions	Facts	Influence on actor characteristics	Implications for the water basin regime
Selected characteristics of actor network			
Tradition of cooperation	Absence of cooperation tradition	M (↓)	Low degree of cooperation between actors
	Absence of mutual trust between actors		
Joint problem	Joint problem perception of water resource degradation only towards the end of the case study process	M, I (↑)	Farmers and industry recognise water resources problem
	Partial shared risk perception		
	Disagreement over problem causes	M, I (↓)	Unwillingness of farmers and industry to cooperate, unless strict measures are applied to all users (shift of blame)
	Lack of common sense of responsibility		
Joint chances	Joint gains realised only by small part of the actors	M, I (↓)	Unwillingness of farmers and industry to cooperate
			Lack of support of the Management Body by some authorities in the basin
Alternative threat	No actor proved powerful enough to impose an alternative strategy to his benefit and thus pose a credible threat to other actors	M (-)	
Institutional interfaces			
Actors as brokers	Management Body towards the end of the process	M, I, R (-)	No obvious impact yet
Few actors or strong representatives	Strong representatives present for farmers	R (↑)	Effective representation of farmers when requested
	No strong representatives present for industries and fishermen Large number of actors involved from the administration	R (↓)	Despite lack of representatives, industries could effectively defend their position on an individual basis. However, the unorganised fishermen were not able to defend their stakes in the process
			Uncoordinated interference of too many public authorities in the water management issues of the basin
Alert mass media	Satisfactory presence of alert mass media	I, M (↑)	Contributed to raising awareness especially after severe environmental collapse

303

Contextual conditions	Facts	Influence on actor characteristics	Implications for the water basin regime
			incidents
Alert public	Lack of alert public	M, R (↓)	Lack of sufficient local social pressure on authorities to take early action
Legal leeway for more integrative approaches	1987 Water Law implementation	R (-)	RWMD set up, whose management functions however were not activated
		R (↑)	Establishment of water permits and Prefectural restrictive measures on water use
	1986 Environment Law	R (↑)	Creation of the National Wetland Park and its Management Body

M= motivation, I= information, R= resources

(↑) = positive influence, (↓) = negative influence, (-) = no significant influence

8.8.3 Conclusions and outlook on the regime change process

Significant external impetus, especially EU and international pressure as well as pressure from the ecological collapse of Lake Koronia, pushed the institutional water regime of the Mygdonian basin to change. However, some tendency of the water basin regime towards more coherence could only be observed towards the end of the observed case study process. Altogether, it was concluded that changes in the Mygdonian water basin regime so far did not qualify for a shift to a more integrated regime phase. The case study analysis explained that the regime coherence could not increase adequately, under the influence of contextual conditions which were partly unfavourable to integration and under the influence of unfavourable motivation-information-resources constellations of the actors involved during most of the regime change process.

Given that the institutional water regime of the Mygdonian basin has not reached an integrated phase so far, it is worth discussing whether chances for this to happen are better in the near future, also in view of the WFD policy requirements for more integrated water management in river basins.

First, it is observed that external impetus pushing the Mygdonian water basin regime to improve in terms of integration will be further present. In specific, the EC is expected to further act as an external source of change impetus through its compliance-check role on the application of EU legislation for water and nature protection in the Mygdonian basin. International pressure from the commitment of Greece to the Ramsar international wetland treaty will also continue to act as change stimulant given the inclusion of the Mygdonian basin wetlands still in the Ramsar "black list". Problem pressure impetus from water resource degradation will also be further present and even new environmental collapse events may take place in the future. The intensity of change impetus from national policy will depend on the degree of adequate implementation of the recent 2003 Water Law on the regional level. The above shows that there will be sufficient external impetus to induce future attempts towards a more integrated regime phase. In this context, it will be crucial whether also the contextual conditions and actor characteristics are bound to become more integration-favourable than in the past.

Indeed, certain aspects of the contextual conditions became somehow more integration-favourable towards the end of the case study process. For instance, unsustainable water management practices became increasingly acknowledged as a joint problem by most actors in the basin, even though the lack of shared responsibility for action to secure future water resources in the basin could not be overcome. The Park Management Body became better operational and, in the future, it is expected to act as a more efficient process "broker". As concerns a legal leeway for more integrative approaches, some new chances could be provided by the 2003 Water Law and the EU WFD.

Regarding the actors involved in regime change attempts, more actors were positively motivated for more integration in the end than in the beginning of this case study. Increased motivation was especially observed for the Prefecture of Thessaloniki (at least its political leadership) and for most actors participating in the Management Body of the National Wetland Park. In specific, the Management Body acts as an umbrella platform for the exchange of several motivated actors such as the Environment Ministry, e-NGOs, the scientific community, the Sub-Prefecture of Langadas and the local municipalities.

All in all, certain aspects of the contextual conditions appear to be somehow more integration-favourable now than in the past and could exercise their future influence on regime change via a constellation of several more motivated actors. Nonetheless, it cannot be claimed that, in total, contextual conditions in their present state would be very favourable to the establishment of a more integrated regime in the Mygdonian basin. Similarly to the case study on the Vegoritida basin, it is concluded that the institutional water regime of the Mygdonian basin would have a greater adaptive potential towards a more integrated phase, if its contextual conditions became more integration-favourable according to the relevant recommendations of the synthesis chapter 9.

8.9 Process of implementation of measures and policies adopted

This section discusses, first, the water use sustainability implications of new incentives (measures and policies) provided by the regime change process and, secondly, their implementation effectiveness which in essence also determined their sustainability implications on the ground. The relevant hypothesis (Hypothesis 3) was that:

> "The less new incentives provided from regime change and the less adequate the implementation of any new incentives, the more likely that there are less positive sustainability implications of regime change in water use."

8.9.1 Sustainability implications of measures and policies

Implications for water resources and the environment

By the end of this case study, there were no positive sustainability implications observed of measures and policies adopted on the water resources of the basin. The water resource and environmental situation of the water basin showed no signs of improvement. On the contrary, a severe environmental collapse event took place in autumn 2004. It is worth noting that, despite the first environmental collapse incident of 1995 which alarmed the public and authorities, water use practices in the basin did not become more sustainable thereafter.

Nonetheless, despite the absence of any visible improvement in water resources, environmental pressure in the basin started to decrease slightly towards the end of the case study. On the one hand, the level of the lake rose slightly due to heavier rainfall. On the other hand, the disposal of effluents into the lake was reduced due to the reduced operation of the declining textile industry sector. These developments however were rather the result of factors external to the water regime. Overall, the danger (for the ecosystem) was still judged as continuous since most drivers creating the problem had not been eliminated (Interview Fisheries Department of Sub-Prefecture of Langadas, in Oiko Kathimerini, 15/01/2005).

Some positive impact on the status of water resources could result in the near future from policy developments in the agricultural sector. Except for the likely impact of the adopted agrienvironmental policy measure (if implemented adequately), a reduction of irrigated crops is also expected from the implementation of the revised EU CAP requiring crop rotations and overall crop reduction due to 100% decoupled subsidies from crop production (Interview Young Farmers Union, 14/4/2005).

Nature protection aspects of the water ecosystems are also expected to be considered more explicitly in the future, if the legal framework for the National Wetland Park is effectively implemented.

Implications for economic development

Due to the designation of the Mygdonian basin as National Wetland Park, part of the local population feared obstacles to the local economic development. Some financial benefits are however expected from subsidies to farmers through the agrienvironmental policy measure, especially at a time when direct crop subsidies will be discontinued as a result of the revised EU CAP.

The revision of the Master Plan excluding costly large-scale engineering projects (especially the water transfer scheme from River Aliakmon) is also translated to financial gains for the public budget which is allocated to the restoration of the Mygdonian basin.

Implications for social development

The modernisation of agriculture in the basin is likely in the future because of the implementation of the agrienvironmental policy measure and other measures to support more sustainable farming practices.

8.9.2 Influence of actor characteristics and interactions on implementation

In the previous section, it was concluded that within the period of this case study there were no positive sustainability implications in water use from new incentives provided from regime change. In fact, even further degradation of water resources could be observed by 2004. In the following, it is shown that, although several new incentives (policies and measures) to promote more sustainable water use were adopted towards the end of the process, these were insufficiently implemented. According to Arabatzi-Karra et al. (1996), the main problem in protecting Lake Koronia and its basin was difficulty in implementing the relevant institutional framework.

The inadequate implementation of adopted regulations on abstraction and effluent disposal permits also partially explains the low regime coherence observed (in section 8.8.1) within the system of property rights and between the systems of property rights and public governance.

The process of implementation is here assessed as an actor interaction process steered by the motivation, information and resources of the key actors involved (see chapter 4/4.1.2.4 for theory-based assumptions on actor interaction in implementation). Some causal links are also made to the public governance elements of the regime which can influence the actor characteristics and ultimately the outcome of the implementation process.

The measures and policies discussed in the following concern Prefectural restrictive measures to protect water quantity and water quality, the agrienvironmental policy measure adopted by the Ministry of Agriculture as well as implementation issues related to the National Wetland Park and its Management Body.

8.9.2.1 Restrictive measures to protect water quantity

After the introduction of the water abstraction permits system in 1989 and of Prefectural restrictive measures to reduce water consumption, fewer new wells were authorised in the Koronia subbasin than before. However, in the period since 1989, many new wells were drilled illegally (pers. comm. RCM, 31/3/2005). Additionally, there has been sparse compliance-check of conditions set in authorised abstractions.[107] Farmers were at times fined for illegal abstraction but fines were often not paid for or ignored. Most industrial plants also abstracted groundwater without any quantity control (Kazantzidis et al., 1995).

The motivation of the target group (mainly farmers but also industries) to implement the policy measures in question for water abstractions was, in general, low. As a result of the development model followed in the basin throughout the 1980s, well-established farmers and industries considered it as their right to continue their production activities in the same manner as in the past (Knight Piesold et al., 1998).

Farmers disliked being limited in their ability to drill new wells for irrigation and they disliked the command-and-control character of abstraction permits (→influence of the public governance element of "strategies and instruments"). Even non-policy measures proposed by the Master Plan, such as the proposed installation of water meters for the control of water use in irrigation and collection of data on real water consumption patterns, were rejected. Due to its top-down character, this proposed measure faced resistance by farmers and was

---

[107] According to the Presidential Decree 256/89, responsibility for compliance-check lies with the local government and rural police assisted in this task by the authorities which issued the permits (e.g. the Prefectural Division for Land Reclamation for irrigation permits).

not implemented. Industries also resisted the installation of water meters to monitor their water consumption.

Given the negative motivation of the target group for the implementation of policies and measures to protect water quantity, it is important to examine the motivation of the implementing authorities in order to interpret the outcome of the implementation process.

Implementation competence was shared between the Prefecture (Division for Land Reclamation) and the Region (Regional Water Management Department, RWMD). Although the RWMD was motivated for a correct implementation of the policies in question, the Prefectural Division for Land Reclamation followed for quite some time different principles than the RWMD with respect to the issuing of irrigation permits. As a result, at start, it was difficult to agree on the restrictive measures proposed by the RWMD according to the 1987 Water Law (pers. comm. RCM, 31/3/2005). There was lack of agreement on several issues such as the EIA requirement for issuing water permits, the legality or not of pre-1989 wells running on power supply and the criteria that should be met to replace pre-1989 wells (Interview Prefecture, 1/4/2005). Finally, the Prefecture conformed to the procedure defined by the RWMD and coordination improved (pers. comm. RCM, 31/3/2005).

At times, it was proposed to seal-off illegal wells in the Koronia subbasin but the political leadership of the Prefecture refused to bear the political cost of such an action. The Prefecture also hesitated to enforce penalties (Knight Piesold et al., 1998) under pressure by the interests of local water users.

To sum up, in the implementation process of restrictive measures to protect water quantity, the unmotivated target group (farmers in their majority) interacted with implementing authorities which can be characterised rather as neutral for the adequate implementation of the relevant measures (only becoming more positively motivated towards the end of the process). As a result, the adequate implementation of the policy measures in question was often obstructed (*interaction assumption II-11*). The non-cooperative behaviour of farmers was made possible throughout the process, because of the weak enforcement capacity and motivation of the implementers. In order to protect the basin's water resources, the enforcement motivation and capacity of the implementing authorities should be enhanced (see also relevant recommendations for the governance element of resources and responsibilities for policy implementation in chapter 9). In the same time, instruments other than command-and-control should be further promoted for the reduction of abstractions, e.g. support for the better implementation of the voluntary agrienvironmental measure and for the rise of private investments in drip irrigation systems.

8.9.2.2 Regulations to combat water pollution

Given the top-down nature of the Prefectural effluent disposal regulations and the raised costs associated with effluent treatment units, the motivation of the target group (mainly industries but also human settlements in the Koronia subbasin) was not positive to comply for an effective policy implementation (i.e. users disliked the command-and-control character of this type of instrument; →influence of the public governance element of "strategies and instruments"). Additionally, industries did not cooperate on proposals of the Master Plan for investing in the improvement of their effluents treatment systems on an individual basis, for installing devices to measure their effluents volume discharge and, later on, for installing a central saltwater treatment unit. In fact, due to the failure of command-and-control measures and top-down proposals, it is increasingly argued that, to solve the problem of industrial pollution, industries should be given more financial incentives to install and maximise treatment in their own premises (Interview University of Thessaloniki, 29/3/2005).

Given the lack of motivation on behalf of the target group, the following paragraphs concentrate on the characteristics of the implementing authorities to explain the implementation process of the effluent disposal regulations.

On the one hand, the process of the initial application of the effluent disposal permits regulations turned out quite successful by the end of this case study process, at least as far as industries are concerned. Namely, as a result of international pressure to protect the nature and environmental values of the Mygdonian basin, much progress was achieved from 1995 to 2005 in the formal procedure of issuing final effluent disposal permits for industrial polluters in the basin. In this respect, the Prefecture (motivated actor in this formal permit issuing procedure) held the necessary information (information on who should get a permit) and the necessary administrative power to bring the permit issuing procedure forward (*interaction assumption II-6* of forced cooperation with great likelihood for the initial implementation of the policy).

On the other hand, progress was much slower as concerns the degree of adequate implementation of permit conditions on the ground. In general, inadequate control was exercised on the industries of the Koronia subbasin (Arabatzi-Karra et al., 1996). Even industries with an effluent disposal permit did not operate their treatment plants on a regular basis and some only put their plants into operation in case of an upcoming control pre-notification by the implementing authorities (Interview Consultant to Master Plan, 20/4/2005).

The low degree of adequate implementation is also linked to the occasional lack of motivation of the implementers to strictly follow their own regulations. Due to the proximity of Lake Koronia to the large urban centre of Thessaloniki, financial interests of some industries are high. Thus, industries around the lake often exercised pressure on the political level of the Prefecture, complaining about inspections and warning that higher operational costs due to effluents treatment could be translated into loss of local jobs in the sector (Interview Langadas Sub-Pref., 5/4/2005). The non-compliance issue became more complex by the fact that many industries in the basin were indeed not self-sustained economically, while several dye-industries had to shut down due to financial problems (Interview University of Thessaloniki, 29/3/2005). In this context, except for few cases of very obvious breaches of regulations, the Prefecture hesitated to enforce penalties on industries (Knight Piesold et al., 1998). Inspections have also been restricted to industries still at the stage of a temporary disposal permit (Tsagarlis, 1998), while industries with final permits are rarely controlled.

As industries failed to comply with effluents standards claiming that they were an anti-incentive for new investments and an obstacle to market competitiveness, the authorities developed tolerance of non-compliance with regulations. This in its turn enhanced further the non-compliant attitude of industries, ultimately contributing to the pollution and collapse of the system of Lake Koronia (Kotleas, 1988).

All in all, also in this case, the adequate implementation of the regulations in question was obstructed in practice (*interaction assumption I-11*) given a constellation of a negative target group and an occasionally non-motivated (neutral) implementer.

Most industries in the basin chose not to cooperate on the implementation of water quality regulations, because chances were good that they could get away with pollution or at least pay only a small fine. The Prefecture itself chose to remain in conflict with other actors such as local non-governmental associations over the issue of industrial pollution, rather than exercise strong pressure on industries. This alternative was chosen under the powerful influence of industries, which are a source of employment in the area. Industries could only become more cooperative, if the alternative of non-cooperation becomes too costly (e.g. higher fines and stricter enforcement are imposed) or if they are involved themselves in the selection of (not too costly for them) anti-pollution measures (e.g. by receiving subsidies to install treatment).

Even in the ideal case of a positively motivated Prefecture, the implementation process of the pollution-control regulations would still encounter difficulties due to the lack of the necessary information and resources of the implementing authorities.

Although the lack of necessary information did not appear as the main issue for the lack of implementation in this case (Interview Former expert at Information Centre, 23/3/2005), a few deficiencies in terms of data can be noted. Namely, there is lack of information on the actual pollution loads discharged by industries, since field inspections are insufficient and short of appropriate on-site sampling methods (Interview University of Thessaloniki, 29/3/2005). As far as industries themselves are concerned, many lack the necessary devices for monitoring pollution parameters in their effluents (Interview Consultant to Master Plan, 7/4/2005). Furthermore, the Prefectural Division for Environmental Protection lacked a specialized laboratory for correct scientific analysis of the samples collected (Tsagarlis, 1998).[108]

Additionally to the lack of adequate information, the implementing authorities of the Prefecture also lacked complete sanctioning power in case of penalties imposed on industries. When penalised, industries maintained the right to appeal to courts according to Greek law. Indeed, several penalised industries referred their cases to court and achieved a reduction of fines through legal processes of 3-4 years.

---

[108] The Prefectural Division for Environmental Protection uses the laboratory facilities of the Ministry of Macedonia and Thrace, the General State Laboratory for Chemical Analysis or private labs (Interview PoT, 12/4/2005).

8.9.2.3 National Wetland Park and its Management Body

In the case of implementing the policy requirements related to the National Wetland Park and its Management Body, it is here mainly distinguished between contra-actors and pro-actors rather than between a target group and implementers.

On the one hand, there were several actors who were positively motivated to support environmental protection within the National Wetland Park and the appropriate operation of the Management Body. Such actors included the Region (especially its Environment Directorate), some of the local municipalities, the e-NGO Ornithological Society and the Sub-prefecture of Langadas, which were all motivated participants of the Administrative Board of the Management Body. Also the Environment Ministry gave high priority to the National Park of Lakes Koronia-Volvi, probably also due to the pending EU infringement procedure. On the other hand, some authorities in the area of the Park, which are responsible for the enforcement of different sectoral policies, viewed the coordination role of the Park Management Body as a threat to their fields of competence, hindering thus indirectly its optimal operation. Also, parts of the local society and representatives of the local agricultural sector disliked certain Park prohibitions (mainly on the construction of farming and other buildings), which could halt local development.

Information held by the interacting actors did not seem to be a major issue for the implementation problems of the policies related to the Park and its Management Body. Members of the Administrative Board of the Management Body seemed to be well-knowledgeable of the situation of the Park and the issues at stake. Nonetheless, the lack of motivation of certain authorities to accept the role of the Management Body, as mentioned above, led to some lack of openness in the exchange of information.

The lack of resources, however, was an important issue in this implementation process. In general, very scarce financial and personnel resources were made available to the Management Body in the first 2 years of its operation (→influence of the public governance element of "resources for policy implementation"). This was partly due to delays in the issuing of legal decisions necessary for the activation of resource allocation (Interview Ornithological Society, 28/3/2005). As a result, in the initial period after its set-up, the Management Body could operate to a limited extent, based only on the motivation and abilities of its Administrative Board members. Even so, the Management Body managed to make important interventions for the protection of the Park mainly by delivering its expert opinion on the environmental licensing (approval of environmental terms) of works and activities bound to settle or expand within the Park (Zalidis et al., 2006).

Since September 2005, the financial resource basis of the Management Body improved when funding was allocated by the Environment Ministry (3-year funding from the 3rd Community Support Framework)[109] for, among others, a monitoring programme of

---

[109] 1,170,000 Euro were allocated in September 2005 for a period of 3 years, which is quite a substantial amount compared to the 25,000 Euro allocated to the Park for the first 2 years of its operation (Zalidis et al., 2006).

environmental data, the execution of a 5-year management plan of the protected area and a sustainability study on the future operation of the Management Body (Zalidis et al., 2006). Additionally, the Programme Contract signed in 2006 between all parties involved in the restoration programme of Lake Koronia provided the opportunity for the gradual involvement of the Park Management Body in the realisation of the various restoration activities.

In summary, in the implementation process of policies relevant to the National Wetland Park and its Management Body, a constellation of positively motivated actors appears to have been interacting with a limited set of negatively motivated actors. Given that the positively motivated actors seem to hold the necessary information to support the initial application of the relevant policy principles, the process depends on the balance of power of the actors involved. Towards the end of the case study process, more resources and political support were made available to the Management Body, which may soon become better operational and lead to a positive turn of the process towards more cooperation (*interaction assumption II-6* leading to great likelihood for initial policy implementation). If it appears that there are still politically powerful actors opposing the aims of the Management Body in the future, the process of implementing the relevant policy requirements could develop into opposition between the motivated and unmotivated actors (either in the form of negotiation or conflict) (*interaction assumption II-8* leading to an intermediate likelihood for initial policy implementation).

Another factor that may lead to more effective future implementation of the policy concerning the National Wetland Park and the Management Body is the proposed amendment of certain policy requirements based on the initial implementation experience. After issuing the 2004 CMD on the National Wetland Park, many discussions took place locally on the CMD and on the role of the Management Body. Representatives of the local society and of the local agricultural sector argued that certain prohibitions would halt development in the protected area and, thus, pursued the revision of the CMD (Newspaper "Macedonia", 9/12/04). After the first 3-year operation of the Park Management Body, even participants of its Administrative Board admitted that the implementation of the CMD did not create all necessary conditions for the sustainable development of the Park. To this aim, the practical problems of the initial CMD implementation were assessed and several amendments were proposed for its better implementation[110] (Zalidis et al., 2006). Some of the proposed amendments are expected to also reduce fears of local society on the potential negative impact of the Park policy on local development, making thus local actors more motivated to support policy implementation in the future.

---

[110] Proposed amendments concerned certain prohibitive regulations on building constructions and the expansion of human settlements, which are no longer compatible with the present rates of human population increase in the area. Secondly, they concerned regulations prohibiting the development of the agricultural and animal-farming character of the area (related to the construction of agricultural storage houses and the establishment of new animal farming units), as well as strict limitations to the establishment of new or the expansion of existing economic activities with questionable positive results on the protection of lakes Koronia and Volvi (e.g. the prohibition of all new activities without liquid effluents in zone B of the Park).

8.9.2.4 Agrienvironmental policy measure

The agrienvironmental policy measure for the Natura 2000 area around lakes Koronia and Volvi was activated in 1/2004. The initial response of farmers to the first call for applications of 3/2004 was very sparse considering that only five farmers applied for voluntary participation in the measure (ETHIAGE, 2004).

Given the voluntary nature of the measure, it is important to focus on the motivation of the target group, i.e. farmers, for effective implementation. In fact, the measure can only be applied at the request of farmers, who thus enjoy a very strong position in this implementation process.

In the following, it is explained that the target group did not find the measure initially attractive (negative target group) and followed a rather non-cooperative behaviour alternative. The main reason for this was that the measure was not financially attractive to many farmers. The rent paid by most farmers for their farmland[111] exceeded the subsidy which they would receive, if they entered the measure for switching their crops from irrigated (e.g. maize) to non-irrigated ones (e.g. dry wheat) (Interview Young Farmers Union, 14/4/2005). In order to make the measure financially more attractive to farmers, the Union of Agricultural Cooperatives of Langadas proposed several amendments to the measure rules. For instance, the reference period for selecting fields which could be part of the measure was not considered optimal. Fields, which could be candidates for the measure, were irrigated fields cultivated at least once in 2002-2003 with maize, alfalfa or clover (as registered in the records of the integrated administration and control system for crops subsidy payments). Farmers proposed to modify this reference period, so that it would start one year prior to each call for applications. This way, farmers could also include their alfalfa cultivations in the programme, which until 2002/3 were not registered with the crops administration and control system, because they were not eligible for any crop subsidies (Interview Young Farmers Union, 14/4/2005).

In 12/2005, the Ministry of Agriculture indeed issued a new CMD for the application of the agrienvironmental measure including several amendments and simplifications of the measure rules (for instance, it simplified and reduced the cultivated crop categories which are eligible for entering the measure). The amendments, however, did not meet all the requests of farmers. For instance, the reference period for selecting fields for the measure remained the period of 2002-2003.

Nevertheless, it is believed that there is potential for many farmers to participate in the amended agrienvironmental policy measure. Especially elderly farmers may enter the programme, since they would then only need to plough their fields once a year and get paid for it (option of fallow land) (Interview Young Farmers Union, 14/4/2005). Also small-scale farmers could be attracted to the measure for the option of fallow land, since they probably do not earn more by renting their land to large-scale farmers. Indeed, those objecting the

---

[111] 80% of cultivated farmland in Greece is rented.

agrienvironmental measure at most are large-scale farmers. Thus, the measure would be more effective, if the small-scale farmers were targeted with more awareness activities (Interview Langadas Sub-Pref., 5/4/2005).

All in all, although interviews showed that it is politically difficult to reduce irrigation water use in the basin and that farmers are the most reluctant interest group to change their practices[112] even if their income is not reduced (Interview Ornithological Society, 28/3/2005), it is considered necessary to try and make the agrienvironmental measure attractive to farmers. The measure is seen as a unique chance to reduce water consumption and change farming practices in the basin (Interview Former expert at Information Centre, 23/3/2005).

Except for the motivation of the target group, the motivation of implementers also plays a role in the implementation of the agrienvironmental measure. The Prefectural Division for Agricultural Development, which is competent for the administrative procedure of implementation, appeared rather positively motivated to bring the implementation of the measure forward. After the initial sparse response of farmers to the new measure, it even cooperated with farmers' representatives in the basin and submitted joint suggestions to the Ministry of Agriculture for the amendment of the measure. However, the Ministry of Agriculture, which has a decisive role on the policy formulation level, could be characterised as a neutral (rather than positive) implementer. In this respect, it is noted that, in the design of the agrienvironmental measure at the Ministry of Agriculture, the real needs and behaviour of farmers in the basin were not fully considered. For instance, the first call for applications came too late in the cultivation period for farmers to apply. The call was published in the early spring, while cultivation planning (for seed purchase etc) had already started in autumn (Interview Young Farmers Union, 14/4/2005). All in all, however, given the less active role of the Ministry of Agriculture in the practical application of the measure, it should be considered that implementers (Prefecture) competent in the application of the measure on the ground were rather positively motivated.

Based on the above, the implementation process of the agrienvironmental policy measure involved so far a negative target group (farmers) and positive implementers (Prefectural Division for Agricultural Development). The information in the hands of the implementers for the initial application of the instrument can be considered as sufficient, considering the long experience of the competent Prefectural Division with farming practices in the Mygdonian basin. Given, however, the power enjoyed by the unmotivated target group (due to the voluntary nature of the measure), the interaction process so far was one of obstruction leading to small likelihood for any initial implementation of the measure (*interaction assumption II-7*). If farmers react positively to the amended agrienvironmental measure and decide to seize the offered financial benefits (also in view of the discontinuation of subsidies for traditional irrigated crops in the new EU CAP), there is potential for them to become more

---

[112] There could also be practical difficulties in changing the pattern of crops in the basin, considering that the dominant water-demanding crops (90% alfalfa and maize) are largely used as animal food within the Mygdonian basin itself (Interview University of Thessaloniki, 29/3/2005).

positive and cooperative for better implementation (*interaction assumption II-5* with a potentially great likelihood for an initial implementation of the measure).

8.9.3 Conclusions on the implementation process

In summary, it is concluded that the new incentives (measures and policies) provided by the regime change process of the Mygdonian basin, failed to have the expected positive impact on sustainability in water use, due to their inadequate implementation in practice. Some reduction observed in the pressures impacting water resources, towards the end of the case study, was mainly due to influences external to the institutional water regime (changing rainfall patterns and decline of industry due to external market processes).

The inadequate implementation of the regime incentives (measures and policies) examined above was explained by the unfavourable motivation-information-resources constellation of the actors interacting in the respective processes. The inadequate implementation of permit systems to control abstractions and water pollution also explains the observed low coherence within the property rights system of the regime and between property rights and public governance.

All in all, the motivation, information and resources of the interacting actors need to become more supportive to the effective implementation of the adopted policies and measures. Especially, more effort is needed to raise the motivation of authorities, which implement water protection policies, and to protect their resources basis from the external pressure of water users' interests. In the case of policy implementation relevant to the National Wetland Park, efforts are needed to increase the awareness (and motivation to cooperate) of sectoral law-enforcing authorities in the Park, which so far often viewed the Park Management Body as a kind of threat to their own fields of competence. All actors must be convinced of the potentially positive role of the Park to support its mission for the sustainable development of the Mygdonian basin. For the mission of the Park and its Management Body, it is also essential to make resources readily available on a continuous basis. Regarding the implementation of the agrienvironmental policy measure, it remains to be seen whether farmers' motivation to support the measure will increase following the issuing of the amended measure in the end of 2005 and in view of the discontinuation of subsidies for several crops through the new EU CAP.

9 Changing institutional water regimes on a basin level: Synthesis discussion

9.1 Introduction

The present chapter starts by illustrating key issues identified with regard to the process of water regime change in the two examined Greek basins of Vegoritida and Mygdonian (see section 9.2). Key issues identified concern both the development of the main regime institutional elements as well as the regime change causal factors (external change impetus and contextual conditions). Subsequently, this chapter uses existing literature on regime changes in selected water basins of other nations, with a two-fold objective:

- Firstly, to provide external evidence on the validity of key theory-based variables and causal assumptions, which were used for discussing regime change in the Greek cases. To this aim, a brief comparison is made between the two Greek basins and a larger sample of previously investigated European basins in terms of regime change and change causal factors (see section 9.3). In this context, it is checked whether the findings in the two Greek basins reinforce key variable correlations observed in previous comparative research which used a similar set of variables.

- Secondly, to formulate research-based recommendations on how to change the institutional water regimes of the two examined Greek basins towards more integration (see section 9.4). On the one hand, recommendations are supported by relevant observations from the brief comparison of the two Greek cases and the larger sample of European cases mentioned above. On the other hand, recommendations are enriched with qualitative statements from a small sample of case studies on water basins of other nations, whose findings on regime change were considered quite relevant for the two Greek cases. Altogether, it is attempted to learn from other real-life examples of attempted regime change towards integration, whose outcome was explained on the basis of similar theoretical variables.

9.2 Learning from the two Greek institutional basin regimes

In chapters 7 and 8, it was concluded that the institutional water regime both in the Vegoritida and in the Mygdonian basin does not qualify yet for a shift to a more integrated phase. The regime of the Mygdonian basin, however, revealed greater tendency to reach a more integrated phase towards the end of the observed case study processes. Opportunities for higher progress towards integration in the Mygdonian basin arose mainly under strong EU and international impetus to consider nature protection issues in a more coordinated manner.

Regime extent

In both Greek basins, it was observed that the regime extent increased but remained in some aspects incomplete, especially with regard to nature and environmental protection of the water ecosystem in the Vegoritida case, and with regard to the use of water as medium for the disposal of salt industrial effluents in the Mygdonian case.

Coherence of the property rights system

In both basins, the internal coherence of the property rights system remained low. In the case of abstractions, lack of coordination between the use rights permitted was related to the multitude of permit-issuing authorities and the absence of cross-sectoral water basin management plans. Especially, the absence of management plans led to a situation where permits were awarded without considering the limits of the respective basins and abstractions could exceed the total water available. Chances to coordinate the property rights system were also reduced by the presence of several illegal water abstractions (deficient implementation of regulations). Implementation problems also reduced the coherence of rights to pollute water resources. In principle, the existing effluent disposal permit system and the water quality regulations offered the potential for a coordinated allocation of rights to pollute water. However, the fact that several polluters still do not hold formal rights to pollute or they do not respect the terms of their rights (permits), at the expense of the "good quality rights" of the environment and other uses like fisheries, reduced coherence of "pollution rights" in practice.

Coherence of the public governance system

In the processes observed, only minor increase of the coherence of the public governance system could take place, with the Mygdonian basin showing tendency for more progress in this respect than the Vegoritida basin.

Levels and scales

Lack of fit and coordination of administrative levels with the natural basin scale was observed, linked also to the suboptimal operation of the Regional Water Management Departments (RWMDs) created by the 1987 Water Law. Some better prospects for basin-wide administrative coordination rose in the Mygdonian basin after the creation of a National Wetland Park covering the entire basin and a Park Management Body involving all relevant administrative levels.

Actors in the policy network

In the basin of Vegoritida, no official participatory arrangements for actors outside the administration could be set up. Even in the case of the few attempts made for partial participation, the appropriate legal framework was missing to ensure their continuity. In the Mygdonian basin, the Management Body of the National Wetland Park ensured the involvement of some actors outside the administration in the Park management, such as environmental groups and the academic community, but not of water users of the basin. It should be kept, however, also in mind that the Park and its participatory Body have their roots in nature protection and not water management policy.

Perspectives and objectives

In both basins, no policy-based water basin management plans or visions could be produced, also due to the inability of the national policy framework to activate a cross-sectoral water planning process nation-wide. The only attempts for water basin visions made were in the form of scientific studies with management-relevant conclusions and proposals. However, since these studies were not supported by a policy framework, the implementation of their conclusions was not obligatory. They were, moreover, formulated in a rather closed administrative process with poor incorporation of multiple perspectives. Only in the case of the Mygdonian basin, the formulation of the Master Plan to restore Lake Koronia opened up, shortly before its finalisation, to non-public actors and to more environmental perspectives. An additional positive development in this case was that the science-based Master Plan became more obligatory to implement by signing a relevant Programme Contract in 2006 among its main implementing actors.

Strategies and instruments

The two Greek cases examined were characterised by the dominance of command-and-control instruments (water abstraction and effluent disposal permits). The main more flexible and incentive-based instrument adopted was in the form of agrienvironmental policy measures providing subsidies for more water-friendly farming.

Responsibilities and resources for policy implementation

Firstly, there was insufficient support of policy implementation with financial and human resources in both basins. Secondly, it was observed that the sanctioning power of certain implementing authorities was in some aspects incomplete. Thirdly, responsibilities for water policy and measures implementation were not well coordinated neither between different levels of the administration nor between different services within the same level (e.g. for water quantity and quality issues). An encouraging development took place in the Mygdonian basin, where responsibilities and resources for carrying out the planned restoration measures for Lake Koronia were clearly allocated in a Programme Contract signed by the key implementing actors in 2006.

External coherence between public governance and property rights

In both basins, the institutional water regimes could have become quite coherent between the policy and property rights systems, as concerns rivalries over water pollution, due to the introduction of effluent disposal permits for all important water polluting uses. However, this policy-driven external regime coherence remained low in reality due to deficient policy implementation. As concerns rivalries over the quantitative status of water resources, external coherence between relevant policies and water property rights was limited due to several reasons. Firstly, holders of pre-1989 established abstraction rights, especially for irrigation wells, were not targeted by policy changes at all. Secondly, the environment and nature were not allocated quantitative water use rights in practice, although they were targeted by public policy (e.g. environmental river flows and lake levels required by the 1987 Water Law). Thirdly, illegal water users (especially of groundwater) were not effectively targeted with behavioural incentives by the public governance system to comply with abstractions regulations (issue of deficient implementation of regulations).

External change impetus and contextual conditions

In both examined Greek basins, strong external impetus for regime change was mainly related to problem pressure from the degraded water resources of the basins and to EU-originated pressure (infringement procedures on water pollution in the Vegoritida basin and on nature protection and water pollution in the Mygdonian basin). National policy-originated pressure was very little present. These regional observations actually confirm the important role of the EU policy framework as a driver for change in Greek water management and indicate the weak impact of the Greek national water regime on Greek basins (linked to the ineffective implementation of national water policy). An additional strong change impetus in the Mygdonian basin was the Greek State's commitment to international agreements (Ramsar treaty) for the protection of wetlands. Indeed, the Vegoritida basin regime showed lower tendency towards a more integrated phase than the Mygdonian regime, largely due to lower external impetus for coordinated nature protection of water ecosystems. As a result, the protection of economic interests in the Vegoritida basin could not be prevented from dominating the observed process. These economic interests also largely contributed to low political motivation for the promotion of a long-awaited nature protection legal framework and of a management body for the water ecosystems in the basin. Delays in this process were only partly related (and not above all, as argued by some) to the administrative split-up of the Vegoritida basin.

All in all, despite the presence of external change impetus, both basin regimes failed to develop substantially towards a more integrated phase. This was explained by the presence of integration-unfavourable contextual conditions, exercising their influence via a constellation of actors with largely inadequate motivation, information and resources to support more integration in the regime change processes.

More specifically, a pre-existing *tradition of cooperation* among water basin actors was missing. The development of a *joint perception of water resource problems* was also only partial, only concerned part of the actors, or only occurred too late in the observed processes to act as a favourable condition. Furthermore, in both basins, there was no substantial perception of win-win situations from a more integrated regime by the actors involved (at least not by the key water users). Moreover, in neither of the two basins, were there *dominant actors,* who were powerful enough to impose an alternative governance strategy to their benefit and thus pose a credible threat to other actors. In the Mygdonian basin, even though the Prefecture attempted to impose its own strategy for the management of the basin (original Master Plan), it was not powerful enough to dominate the governance system and was eventually forced to change its strategy by other actors (e.g. Ministry of Environment). As far as *institutional interfaces* are concerned, an actor with a kind of institutional broker role was only present in the Vegoritida basin, where the Regional Water Management Department coordinated the administrative negotiations over the level of Lake Vegoritida. The institutional representation of key stakeholders in participatory mechanisms was, however, weak. For instance, in the Vegoritida basin, there were no strong representatives for the large-numbered group of farmers and the same was observed in the Mygdonian basin for fishermen. As for the presence of alert mass media, this was satisfactory in both Greek basins, although their presence was largely linked to incidents of extreme environmental degradation. A supportive alert local public was present only in the Vegoritida basin. This can be partly attributed to the more focused nature of rivalries in this basin against the water-degrading activities of few sources (especially, the Public Power Corporation). On the contrary, in the Mygdonian basin, use rivalries had a more diffuse character with several abstractors and polluters involved. Thus, due to diverse and disperse vested interests (e.g. regarding employment) in local industry and agriculture, most of local society remained rather inactive. As concerns legal leeway for more integrative approaches, there were some instruments present in both basins but not all were functional in a satisfactory way, such as the weakly implemented 1987 Water Law. In the Mygdonian basin, some legal opportunities for more integration in the water regime rose from the activation of the nature protection policy framework, which facilitated the creation of a National Wetland Park in the basin and a participatory Park Management Body.

All in all, partial improvement was observed in only few of the contextual conditions influencing regime change in both Greek basins. In total, by the end of the observed processes, conditions still remained quite unfavourable for the establishment of more integrated basin regimes.

9.3 Comparing the Greek and other European basin regimes: Reinforcing key causal assumptions in the regime change process

This section makes key comparative observations on trends in the regime development of the two examined Greek basins and of 12 basins in six other European nations. The summary data on the 12 basins of other countries stems from recent comparative work of other scholars (Bressers & Kuks, 2004c), whereby an analytical framework was used similar to the one used in this dissertation.[113] In fact, the two Greek basins are here compared with a sample of, not only 12 but, 24 in total European case studies, since in some of the 12 basins previously studied, more than one sub-cases were distinguished to evaluate regime development for independent water use rivalries. A key aim of comparing the two Greek cases with this larger sample of previously studied European cases is to increase confidence on the validity of the theory-based variables and core causal assumptions, which were used for evaluating the two Greek cases.

The boundaries of the 24-European case sample previously compared by Bressers & Kuks (2004c) were defined by choosing cases with clear water use rivalries present and with the

---

[113] The comparison of Bressers & Kuks (2004c) was part of the European research project EUWARENESS. In each of the six European countries participating in the project (The Netherlands, Belgium, Switzerland, Spain, Italy and France), two water basins were studied to analyse specific regime transitions at water basin level during the last decades and up to 2002. The 12 water basins were discussed within their national context and were compared concerning conditions that are important for regime change towards sustainability. Any developments which occurred in the institutional water regimes of these 12 European basins after 2002 are not reflected in the discussions of the present chapter.

(ex-ante) impression that a serious attempt to attain more integration in the regime had taken place. This selection was made irrespective of the fact, whether integration was actually achieved or not, since it was an explicit aim of the study to examine the conditions under which integration can be achieved. In this respect, a random variation strategy was followed since the sample could potentially include a wide variety of cases from complete failure to great success of integration attempts. On the other hand, the approach also had the character of extreme case sampling, considering the argument that "if real integration could not be found in cases where explicit relevant attempts were made, the chances would be minor to find it on any larger scale outside of this sample". Finally, the 24-European case sample deliberately encompassed different situations ("wet" and "dry" cases and different countries) seeking maximum variation in terms of resource problems and national contexts, to identify common patterns but also unique variations.[114]

The addition of the two Greek cases to this previously compared 24-case sample is not considered to disturb the sampling strategy just described. On the contrary, the Greek cases further enrich the variation of the sample, by adding cases with their own resource problems and national context. The selection criterion of existing serious integration attempts is, however, only weakly met in the two Greek cases (and probably also nation-wide in Greece). At least, two Greek cases were chosen, where it was relatively likely to observe some activity towards a more integrated water regime. Namely, both cases chosen suffer from significant resource deterioration and opportunities to resolve the resulting water use rivalries in a more sustainable manner could be missed without more integration. Additionally, both cases involve rivalries largely focusing on important lakes, which potentially could have better prospects for actor cooperation than upstream-downstream river situations.

Comparison within the relatively large sample of 26 cases (2 Greek cases plus the 24 European cases from previous research) is made possible through the allocation of ordinal values to the key variables in each case (regime change variables, external change impetus, contextual conditions and sustainability implications). Thus, instead of carrying out only a qualitative comparative analysis of trends, a more quantitative approach can be followed to reach comparative aggregated conclusions. The ordinal values used were of a scale from (0) to (4). In general, a value of (0) indicates that regime changes and sustainability implications are absent or even negative, and that conditions for regime change are unfavourable for more coherence. On the other end of the scale, a value of (4) indicates the opposite, i.e. changes to a much more integrated regime, positive implications for more sustainable water use and integration-favourable conditions. Annex IX provides general qualitative descriptions of the ordinal values (0) to (4) for each variable. Annex IX also lists the ordinal values actually allocated to the two Greek cases as well as to the 24 European cases as these were previously scored by their own researchers.

---

[114] Source of information on the sampling strategy of the EUWARENESS project is an internal explanatory project note.

In their comparison of their 24-European case sample, Bressers & Kuks (2004c) investigated the influence of regime change causal factors (external change impetus and conditions) on the regime change process. They concluded that the force of external change impetus was directly correlated mainly to changes in the regime extent. Changes in regime coherence (whose presence is decisive for a more integrated regime phase) were closer correlated to the contextual conditions. Also in the two Greek cases, the regime extent increased considerably under pressure by rather strong external change impetus, but the regime coherence remained low under the influence of rather integration-unfavourable conditions. Figure 9.1 illustrates the findings of the 24-European case sample on total regime change and contextual conditions (using the ordinal values) and shows that lower assessments of conditions correlated with smaller total regime changes. This correlation is reinforced by the respective observations in the two Greek cases, which (when positioned on the plot of Figure 9.1) are grouped closer together with other less integration-successful European cases.

Figure 9.2 deals with the causal link between general regime change and its implications for more or less sustainability in water use. The previously studied 24-European case sample showed that there was a rather strong correlation of general regime change, and especially of the variable of public governance coherence, with sustainability implications. This probabilistic correlation was also further confirmed by the findings of the two Greek cases, whose minor regime changes towards more integration were related to only minor or even negative sustainability implications in the period of study (see position of Greek cases in the lower left part of the plot in Figure 9.2).

Figure 9.1   **Relation between contextual conditions and regime changes**

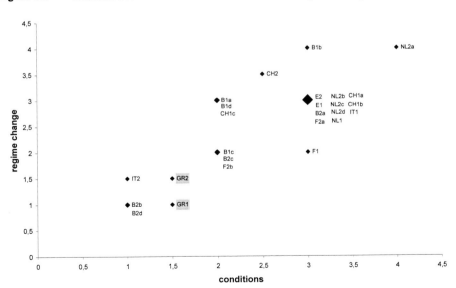

Figure 9.2   Relation between regime change and sustainability implications

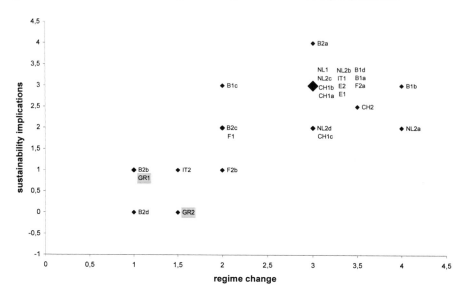

Note on figures 9.1 and 9.2: The abbreviations of the case study names are explained in Annex IX. Several cases share their ordinal values in the plot and, thus, some points are larger than others.

9.4 Shifting the two Greek basin regimes towards more integration: Learning from other regimes

This section aims at discussing how the two examined Greek basin regimes of Vegoritida and Mygdonian could shift towards more integration. As a priority, it is attempted to make recommendations for change in regime institutional elements and contextual conditions, which are considered more feasible to change. In general, recommendations in sections 9.4.2 and 9.4.3 are structured according to the key variables used to assess regime change and the contextual conditions influencing regime change. On the one hand, recommendations are supported by the comparative observations of the previous section 9.3 on trends and the relative importance of variables in the regime change process. On the other hand, they are also supported in a more qualitative way by real-life concrete examples from few cases, which have been selected from the previously studied 24-European case sample (see previous section) for being most relevant to the two Greek cases.

Firstly, the selection of example-cases was oriented to such cases, which have been more successful in achieving a more integrated basin regime than the Greek cases. More successful cases can serve as a pool of "better" examples to explicitly support recommendations towards more integration, instead of failed cases which can only implicitly support recommendations (in the sense of "examples to avoid"). Secondly, from the pool of cases which were more successful towards an integrated basin regime, a further selection was made in favour of those cases which were considered most relevant for the two Greek basin regimes. "Relevance" was related mainly to a most similar water resource problem structure (i.e. scarcity problems in surface and groundwater and only local pollution problems) and similar rivalries (i.e. agricultural and other abstractions vs. environment; industrial and urban pollution vs. environment /fisheries/ recreation). This selection was based on the assumption that, in cases of similar resource problems and rivalries, it is more possible for similar water management strategies to develop (reflected also in the relevant systems of property rights and governance) given all other things equal.[115] Cases with somehow similar problem structure but more success in the development of a more integrated regime than the Greek cases (e.g. due to more favourable conditions) can serve as source of more relevant "better" examples.

Except for the importance of the water resource problem structure to account for differences in the water management strategies in different basins, many other background aspects can also potentially influence developments in any selected example-case, e.g. the centralised or decentralised structure of a State affecting decision-making processes in a basin or the national and regional enforcement capacity. Although such background aspects were not used as a selection criterion for the pool of example-cases, an effort is made to place examples given in the following into context, where this is considered essential.

In total, five cases were selected out of the 24-European case sample previously studied. Two cases involved rivalries relevant to scarcity (one in Spain and one in Italy), two cases involved rivalries relevant to local pollution from industry and/or human settlements (one in Belgium and one in Switzerland) and one case involved both pollution and scarcity rivalries (one basin in France). Furthermore, given the leading importance of scarcity problems in the two Greek cases of interest, it was considered useful to include an additional non-European case with groundwater and surface water scarcity problems in the pool of example-cases. This case concerns the US California Central Valley, whose water management issues were already introduced in chapter 2/2.1. It should be noted that the case of the California Central Valley was not evaluated by its researchers with the analytical framework used for the two Greek cases of this dissertation and for the 24-European case sample. Thus, any statements made in the following on the regime development of this US case study should be considered as provisional and only based on own assessments of the limited literature examined. A concise summary of relevant regime developments in the selected six example-cases is given in the next section (9.4.1).

All in all, it is emphasised that the use of example-cases to support recommendations for the two Greek cases has a purely qualitative and pragmatic character. The intention has not been to prescribe best practice examples from elsewhere, which should be considered as models for adoption in Greece. This is avoided, first, because of the limited number of cases

---

[115]	Differences in the main water resource problems faced are often argued to be related to the type of management strategy adopted. For instance, Hartje (2002) concluded that arid/semi-arid and temperate regions differ in their general water resource problems and, hence, the water management strategies followed.

325

considered and, secondly, because the development of specific institutional aspects of resource regimes towards a more integrated phase is (and should be) adapted to the realities of the specific basin and nation in question. As Shah et al. (2001) argued, a broad understanding of what has worked elsewhere, including success stories in river basin management, may offer "a good backdrop to the design of institutional interventions", but it is unrealistic to expect much more. Also, van der Brugge & Rotmans (2007) argued that particular instruments can be very effective in certain river basin management regimes but be suboptimal in other basins due to local institutional arrangements, problems or societal functions.

9.4.1 Example-cases of other basin regimes

This section provides brief summaries on key regime changes in the six selected cases, which serve as source of qualitative examples to support recommendations for the two Greek basin regimes of this dissertation. A more detailed discussion of the relevant water management processes can be found in the literature referred to in each case.

The Matarraña basin (Spain)

The following information is based on a case study by Costejá et al. (2004). The River Matarraña is a tributary to the River Ebro in north-eastern Spain. Its basin is 1,727 km² large and the main water management problem is one of quantity. Irrigation uses 90% of the available water flow and competes with water supply, tourism and the environment. Water scarcity in the basin has been a source of upstream-downstream conflicts.

Progress towards more integration in the water regime of the Matarraña basin was observed mainly in terms of the coherence of the public governance system. Over the case study process, the actors in the basin established alliances with actors at the regional, State and EU level. The process was, in fact, quite focused on the search for a suitable infrastructure solution to face water scarcity. In 2000, a Water Agreement was signed by all important basin actors (administration, irrigation unions, and environmental organisations) reaching an agreement on a set of hydraulic projects to resolve scarcity. The increase in the number and type of actors involved over the process broke the legacy of power in the hands of traditionally dominant actors.

The external coherence of the regime also improved to some extent after the signature of the Agreement and the rejection of a hydraulic project which would only redistribute uses in a traditional fashion. Less progress was observed in terms of coherence in the property rights system. The 1985 Spanish Water Act attempted to eliminate private property rights on water, to advance the expiration date of such rights and to register all concessions. However, these aspects of the Act were ineffectively implemented in the Matarraña basin (as well as nation-wide); therefore, traditional and *de facto* property rights to water still remained in the basin. Nevertheless, a positive development was that successive regulation works (construction of a dam and diversion tunnel) and the creation of a Central Union of Irrigation Communities of the Matarraña River altered the traditional property rights system, so that use rights can adapt to changes. Furthermore, the Ebro river basin plan established a minimum ecological water flow.

The Chiese-Idro basin (Italy)

The following information is based on a case study by Dente & Goria (2004). The River Chiese flows in northern Italy (crossing 2 Regions and 3 Provinces), generates the Idro Lake and finally flows into the River Oglio. The basin covers 934 km². The Idro Lake is a natural lake (11 km²), which is artificially regulated. The rivalry in the specific case was one of quantity and focused on the maximum variation of the water level of the lake and the minimum downstream flow of the River Chiese. The main rival uses were irrigation, hydropower, tourism, environmental protection, flood protection, soil erosion and land sliding.

After the enactment of the national Law 183/1989 on water and soil conservation, the Po River Basin Authority was created, including the sub-basin of the River Chiese in its territory of competence. In 1992, a local environmentalist group was also created and its first request was the establishment of an adequate minimum downstream flow for the River Chiese. A round of negotiations between most of the actors interested in and affected by the variation of the Idro Lake level concluded in a decision to enforce a new rule for the regulation of the lake. During a 3-year experimental period, the new rule should reduce the lake level variation and ensure a minimum downstream vital flow. The experimental programme was proposed by the Po Basin Authority and acknowledged by the central government. A Government Commissioner was appointed (by the Ministry of Public Works) to enforce the new rule. Additionally, a Users Committee was created in which all water end-users were represented (the irrigation consortia, the hydropower company, the lakefront municipalities, the local environmentalist group and the Province of Brescia). The task of this Committee was to assist the Government Commissioner in defining the daily admissible flows during the experimental period.

Changes in the specific basin regime towards a more integrated phase include, at first, an increased interaction between the levels of government over the process. Moreover, there was an increase of the uses and users entering the governance system. Even the local environmentalist group could take part in the negotiations over the lake level variation and the definition of a minimum downstream flow and was granted the right to participate in the Users Committee during experimentation. In fact, this ensured the early inclusion of the environmental dimension in the public governance system of the regime. It was also important that there was intense debate of the different aspects of the problem at stake in order to set common goals. The experimental programme also proved to be a flexible instrument for developing a new rule for the lake level regulation. There was also some increase in the coherence of property rights. Extra water uses were incorporated into the regime, not by extra consolidated user rights, but mostly by suspending or limiting existing user rights (e.g. by not renewing older concessions for the regulation of the lake waters). In general, the water basin management process gradually became more policy-driven, since the issue of property rights was largely taken over by negotiated agreements in committees.

The experimental programme and the Users' Committee can also be viewed as integrative attempts to link the public governance and property rights systems.

The California Central Valley (US)

A description of efforts towards integrated water basin management in the California Central Valley was made in chapter 2/2.1 (based on Svendsen, 2001). Here, some observations are made on regime development, which should be considered as provisional and only based on own assessments of the limited literature examined.

The California Central Valley is certainly not comparable in terms of size to the other example-cases in this section (two rivers drain ca. 158,000 km² of watershed). The Valley is, nonetheless, an interesting case of innovative solutions to problems of water reallocation and management. California is characterised by the fact that there is very little new water left to develop and by a sophisticated economic environment where water is treated as a commodity rather than a common pool resource.

All in all, it appears that the water regime of the California Central Valley has made considerable progress towards a more integrated phase. The regime extent seems to be quite complete, especially since environmental values were supported by the introduction of the Federal framework Endangered Species Act and the Central Valley Improvement Act (1992). The latter re-allocated a portion of water from irrigation to the Sacramento-San Joaquin Delta, which is very important environmentally.

The specific water regime also appears rather progressed in its coherence of the property rights systems, at least with respect to surface water (see also surface water reallocation from irrigation to nature above). The system of surface water rights is complex but relatively well-specified in law and through cumulative court decisions. Coherence seems to be less developed in the case of groundwater, which is lightly regulated by public policy (e.g. there is no permit process) and is seriously overdrafted.

As far as the public governance system is concerned, the regime appears to be quite coherent. All major interested parties from different administrative levels and user sectors are well-represented in decision-making. The water planning process facilitates the inclusion of multiple perspectives. The 5-year California Water Plans go through an extensive consultation process among a variety of stakeholders. An even more interactive process is the CALFED process managed by a consortium of federal and State agencies with management and regulatory responsibilities in the Bay-Delta system of the State. Its mission is to develop a long-term plan to restore ecological health and improve water management for uses of the Bay-Delta system. It is an open process with participation from the entire range of interests, including environmental NGOs. Last but not least, the policy implementation process is quite concerted and endowed with adequate financing and human resources. Legal authority over water resources is well-established and to a great extent enforceable. The coordination of water responsibilities also greatly improved when, in 1967, responsibilities for water quantity and quality issues were consolidated in the State Water Board.

The Audomarois basin (France)

The following information is based on a case study by Dziedzicki & Larrue (2004). The Audomarois basin in northern France includes the basin of the River Aa and its floodplain, the Audomarois Marsh. The basin (ca. 620 km²) is partly included within the boundaries of the Audomarois Regional Natural Park, which was set up in 1986. The rivalries of interest here are one of quality between industrial pollution and environmental protection and one of quantity between abstractions for water supply and industry and environmental needs of the Park.

In the context of the quality rivalry, the problem of industrial pollution was managed mainly by increasing the coherence of the public governance system of the regime. After a serious industrial pollution accident, a Consultation Committee was set up by the Prefect to support anti-pollution measures, after suggestion of the Park and the Water Agency (one of the six regional Water Agencies set up nation-wide by the 1964 Water Act - an early indicator of river basin management in France). In this Committee (1987-1992), participants included industries, nature protection associations, administration departments and the local authorities. For the first time, industrialists and other water users came together. The work of the Committee supported the implementation of anti-pollution measures, since both the users and many administrative bodies were involved in defining these measures. Significant financial resources, which were offered by the Water Agency, were also very important in bringing the anti-pollution measures forward. All in all, action to deal with industrial pollution was based on financial incentives and compliance with discharge norms in a framework of a collective-based procedure. The Audomarois Regional Park also had an important coordination role. However, it could only produce partial coherence in the public governance system since the Park boundaries did not coincide with the basin boundaries. Additionally, the Park did not have competence to impose itself as a formal coordinator of water policy amidst a multitude of actors involved from too many different administrative levels.

As far as the quantity rivalry is concerned, progress was less significant. The lack of coordination between user rights for abstracting water and land user rights hindered a coherent policy for managing groundwater. Local authorities with power in allocating land user rights (through drawing up town-planning documents) were unwilling to implement perimeter zones for the protection of groundwater, since this would restrict housing developments and economic activities. Additionally, the use of groundwater remained in the hands of actors pre-existing the Park and mostly located outside the Audomarois basin (such as the Water Agency). The Park was, moreover, not involved in decision-making concerning water abstractions. The management of abstractions remained confidential, involving no coordination among water managers and between water managers and users. There was also little progress in discussing this issue in the so-called SAGE planning process for the basin (a SAGE is a Local Plan for Water Development and Management, based on the requirements of the 1992 national French Water Act). Thus, this rivalry noted less progress towards more integration both in terms of governance and property rights coherence.

All in all, the regime in the Audomarois basin is characterised as an ongoing integrated regime in a gradual but slow process. Environmental issues were taken into consideration quite early since the 1980s, thanks to the creation of the Audomarois Regional Park. Indeed, integration was initiated by the creation of the Park and continued with the creation of the Consultation Committee on anti-pollution measures and the drafting of the SAGE which extended the process from the marsh to the whole basin. In the context of the SAGE, also a Local Water Commission (local authorities, local public establishments, State public administration and user representatives) was set up with the task to draw up the SAGE. However, even though the SAGE eventually aimed at the coordination of the various water users, it proved to be a rather slow process. As concerns external regime coherence between public governance and property rights, this could not be achieved yet (e.g. by the SAGE process). In fact, the issue of user rights was not really dealt with so far in the SAGE plan.

The Vesdre basin (Belgium)

The following information is based on a case study by Aubin & Varone (2004b). The Vesdre basin (710 km²) is located in the north-eastern part of the Region of Wallonia. This summary concentrates on two rivalries of main interest: one on surface water pollution and emerging tourism development and one between industrial pollution in the tributary Wayai and anglers.

The pollution of the river Vesdre with domestic and industrial discharges is the most significant problem in the basin. Pollution is historical, with the long-standing industrial activities of the basin, now in decline. The Vesdre was a dead river for years, but fish reappeared since the end of the 1990s. In the rivalry of pollution and emerging tourism, resolution came with a convergence of interests between tourism and wastewater treatment, supported by EU subsidies. The most polluted part of the Vesdre basin was declared an area of economic reconversion and became eligible for the structural funds of the EU. Before that, the intercommunal sewer operator planned the construction of treatment plants but did not have the financial capacity to initiate construction. The partial subsidies of the EU for treatment plants were later supplemented by regional subsidies. Finally, two large treatment plants came into operation and a third one was expected in 2004, under pressure by the EU Wastewater Treatment Directive. The improvement of water quality in the basin and the availability of European Regional Funds made also tourism development possible. The resolution process was reinforced through consultation and cooperation between local users. In specific, a river contract signed for the River Vesdre tries to enhance natural areas and to emphasise tourist activity (the concept of river contracts in the Region of Wallonia was illustrated in chapter 5/5.7.2, as voluntary agreements between stakeholders of a river basin). In this case, the two uses of sanitation and tourism, eventually, became mutually supportive, when it came to sharing the financial burden of water quality improvement (joint chances developed).

In the rivalry of industrial pollution and anglers in Wayai, a mineral water production plant was held responsible for generating pollution that kills fish. The anglers' federation proposed a river contract (for the tributary basin) with the mineral water plant, as a forum of local water users supported by the communes and the Region. The water plant agreed to collaborate. The river contract began with an inventory supported by the anglers and industrialists. New information arose, proving that the mineral water plant was not the main river polluter, but leakage from sewers and other users who did not respect their emission permit (extension of conflict to the whole basin). The identified polluters committed themselves to take measures in the river contract and several actions were fulfilled by 2001. Thanks to the river contract, the parties developed mutual understanding and worked together to overcome the conflict.

In this Belgian case, quality rivalries were resolved through changes in the public policy and governance system. Additionally, the resolution of conflicts should be viewed in the context of the ability to reach local arrangements. In the Wallonia Region, the local level is very autonomous. For instance, the construction of wastewater treatment plants is decided at the scale of the Province by intercommunal associations. This relative autonomy allows more informal arrangements between the local actors. Also local mechanisms of collaboration should be considered (the river contracts). Therefore, this example-case should be considered with caution in view of other cases where decision-making procedures are more centralised.

The Seetal Valley (Switzerland)

The following information is based on a case study by Mauch & Knoepfel (2004), which focused on Lake Baldegg and Lake Hallwil, located in the Seetal Valley. The two lakes are connected by a river (Baldegg drains into a river which then joins Halwill) and their basin is relatively small (ca. 143 km²). The main water management issue is one of quality, involving rivalries of domestic, industrial effluents and agricultural diffuse pollution versus the environment.

In the early 1960s, the canton of Aargovia (northern part of Lake Hallwil) constructed a sewage system around its part of the lake and a wastewater treatment plant. Due to insufficient impact of these measures, the Aargovian municipalities launched a petition to their canton requesting that anti-pollution efforts also take place in the (upstream) canton of Lucerne, which includes in its territory Lake Baldegg and the southern part of Lake Hallwil. Thus, a decision was made that the two cantons would work together in the future. With the construction of five more treatment plants, the sewage system in the basin was complete in early 1980s. In Aargovia, measures were financed by the canton. In the canton of Lucerne, the implementation of water protection policy is the responsibility of the municipalities. Thus, the Canton of Lucerne founded in 1984 the Lake Baldegg and Lake Halwill Association of Municipalities. The aim of the Association was to coordinate municipalities with the aim to restore the two lakes from pollution degradation. The creation of the Association was, in a way, an effort to take the whole water basin into account when considering management measures. Over the process, also the coordination efforts between the two cantons were intensified.

Additionally, there were efforts to coordinate water protection policies with other policies such as agricultural. In the set of instruments used to manage water pollution, gradually more flexible tools were considered. For instance, to deal with diffuse agricultural pollution, agreements on a voluntary basis were developed with farmers to pollute less through a system of subsidies linked to the promotion of ecological services (Phosphorus Project). Additionally, the Lakes Association of Municipalities developed an environmental consultancy for agriculture.

All in all, a shift of the specific water basin regime towards more integration was observed. In specific, there was an increase of the regime extent, reflected in the inclusion of the environment as water use and the development of regulations to target farmers' polluting activities. The development of regulations to restrict agricultural pollution (with polluting farmers as target group of policy) and the policy obligation to implement lake anti-pollution measures by those holding use rights to pollute (e.g. municipalities) also indicate an increase of the external coherence of the regime. The coherence of the public governance system also increased in certain aspects (e.g. set up of Lakes Association, cooperation of cantons and use of more flexible instruments).

In this case, improvements in qualitative water protection were based both on top-down processes (federal subsidies and federal legislation for wastewater treatment and agricultural pollution) and bottom-up processes (use of a petition by the population of Aargovia to push for the cooperation of both cantons on Lake Baldegg).

9.4.2 Changing institutional regime elements towards a more integrated phase

This section proposes the direction of changes needed in the institutional elements of the two Greek basin regimes of Vegoritida and Mygdonian, in order to qualify for a more integrated regime phase. Recommendations are structured according to the key variables used to characterise regime changes in the analytical framework: regime extent as well as coherence within and between the public governance and property rights systems.

Regime extent

Both in the overview of the 24-European case sample previously studied and in the detailed example-cases of the previous sections, part of or the core of changes in the regime extent was the introduction of environment and nature protection considerations in the regime. Also in the Greek Vegoritida basin, the regime would become more complete and meaningful by explicitly considering environment and nature values of the water bodies. In the Mygdonian basin, the regime additionally needs to include in its scope the use of water as medium for the disposal of industrial salt effluents.

Regime coherence

A more complete regime extent alone will not be enough to qualify for a more integrated regime in the two Greek cases. Additionally and most importantly, the coherence of the basin regimes needs to increase. Considering the three forms of regime coherence in the analytical framework, it is here supported that change efforts in the Greek basins examined should focus, as a priority, on increasing the coherence of the public governance system. Also in the example-cases of section 9.4.1, public governance coherence was the regime aspect, which was more frequently improved during attempts towards more integrated basin regimes. In the example-cases dealing with quality rivalries, solutions were mainly sought and found by mobilising the public governance system. In the example-cases dealing with scarcity rivalries, solutions were sought both in the public governance and in the property rights systems. However, coherence in the property rights system appeared more difficult to improve (especially concerning groundwater); thus, improvement of the regime coherence was usually achieved in the public governance system. Furthermore, Bressers & Kuks (2004c) observed that public governance coherence is the aspect of regime change, which more strongly correlated with more positive sustainability implications, in their 24-European case comparison (see section 9.3).

Considering the above, the following starts with recommendations for increase of coherence in the public governance system of the two Greek cases. Secondly, recommendations are made for more coherence in the property rights system and for more external coherence between property rights and governance. Although the improvement of coherence in these two latter aspects is equally important as the improvement of coherence in public governance, it should be kept in mind that such improvement may be more difficult to bring about.

Coherence within public governance

<u>Levels and scales</u>

The governance levels of water management in the two Greek cases need to fit better with and coordinate more within the natural boundaries of basins, as also required by the EU WFD. This especially holds true in the Vegoritida basin, whose territory is divided between three Prefectures. In the Mygdonian basin, it was a favourable coincidence that the greatest part of the basin is in the administrative competence of one Prefecture. To achieve more coordination of administrative levels and natural scales, coordination management bodies on a basin level are needed. This role could not be fulfilled in neither of the two basins by the Regional Water Management Departments (RWMDs) of the 1987 Water Law. Therefore, significant efforts are needed in the future regional implementation of the 2003 Water Law to strengthen the coordination role of the new Regional Water Directorates (RWDs). Even if the RWDs eventually become well-operational, they will have competence on very large territories (river basin districts). Thus, it still remains uncertain how they will take account of specific management issues on the level of individual basins and subbasins. Since it is beneficial for water planning to also take account of more local conditions, other institutional structures such as the Wetland Park Management Body in the Mygdonian basin could in some ways assist the coordination role of large-scale river basin district structures.

In several of the example-cases presented in section 9.4.1, there were many attempts for better coordination arrangements of administrative levels on the basin scale. This was achieved using a variety of ways, such as the signature of a Water Agreement by all basin actors in the Spanish Matarraña basin or the extension of the process to the entire French Audomarois basin after the initiation of the SAGE planning process supported by a Local Water Commission. However, it should be kept in mind that in both Spain and France, there is a much longer tradition of thinking in terms of river basins than in Greece, when it comes to water administrative management.

Actors in the policy network

In the two Greek cases, regular participatory arrangements for actors with an interest in water resources (including actors outside the administration like water users, local and national e-NGOs) should be established. Especially relevant are the participatory bodies foreseen by the 2003 Water Law (Regional Water Councils). However, caution is needed, since similar provisions of the 1987 Water Law which set up Regional Water Committees, did not become operational. Considering the ineffectiveness of the Greek national water policy system to initiate and support regional participatory processes so far, it was interesting to observe how, in the Mygdonian basin, chances for participation rose from nature protection policy (see participatory Park Management Body) and from EU funding instruments' pressure (see ad-hoc participatory process for the revision of the Master Plan to meet Cohesion Fund requirements).

The development of actor participation processes was proven essential for a shift towards a more integrated regime in several example-cases in section 9.4.1. Possibilities were created to ensure also the participation of non-public actors, including both environmental groups and water users, in the relevant water management processes (e.g. inclusion of local environmental groups and water users in the Users' Committee for the experimental programme in the Italian Chiese-Idro basin; the involvement of all basin actors in the Water Agreement of the Spanish Matarraña basin; local consultation and cooperation between users to resolve a water pollution rivalry in the Belgian Vesdre basin; participation of a broad group of public and private actors to influence decision-making in the California Central Valley).

Perspectives and objectives

Except for strengthening participation processes, also an improvement of more top-down planning is needed for a shift of the regime towards more integration. In this context, official cross-sectoral water basin management plans need to be developed in the two Greek cases, especially by strengthening the implementation of the relevant national policy framework (2003 Water Law). For the future formulation of water basin management plans, the incorporation of multiple perspectives needs to be ensured, including the perspectives of water users and environmental groups. The two Greek cases also showed that the incorporation of multiple perspectives in future water basin plans must be extended to recognise relations of water policy with other policies. In the specific cases, the need was recognised to coordinate water policy with mining and energy policies, agricultural as well as spatial and town-planning policy.

In most of the example-cases of section 9.4.1, basin management visions could be formulated (usually supported by national policy) and incorporated the perspectives of most interested parties (e.g. the gradually developing but still incomplete SAGE planning process in the French Audomarois basin or the 5-year Water Plan in California produced after extensive interaction with a variety of stakeholders). Nonetheless, it should be kept in mind that many of the example-cases took place in nations with a relatively long tradition in water planning mechanisms (such as France and the US). Considering that it will take considerable time for Greece as a nation to develop a water management planning tradition, some "temporary" chances are seen in producing basin visions in the form of scientific studies with management-relevant conclusions. This was in fact, already, the case in the two Greek cases examined. In the Vegoritida basin, however, it remained a disadvantage that the basin vision produced in this manner (the stability and purification programme for Lake Vegoritida) was not supported by any policy framework and, thus, had a weak basis for its practical implementation. This barrier seemed to be overcome in the Mygdonian basin, where, upon pressure by the funder-EU, the implementation of the basin vision (Master Plan) was more secured by signing a Programme Contract between the main implementing actors in 2006. Such models of scientific studies, which can be legitimised through actors' contracts, could be an intermediate solution for individual Greek water basins, until the establishment of a policy-based, participatory water planning process nation-wide.

Instruments and strategies

In the two Greek cases, the need was recognised to complement command-and-control water management instruments (abstraction permits, effluent disposal permits) with other types of more flexible instruments. This proposal is especially supported by the observation of the weak implementation of command-and-control instruments by the relevant Greek public authorities. To compensate for this weak implementation capacity, chances are seen in more indirect and incentive-based water management instruments. An example from the Mygdonian basin is the attempt to limit water abstractions through subsidies to change crop patterns into less water-demanding ones (agrienvironmental policy measure) and to replace irrigation systems with water-saving ones, instead of placing all effort in the direct control of the numerous private abstraction points. Certainly, the implementation of such indirect incentive-based water management measures also needs to be supported by enforcement capacity on behalf of the administration. The implementation of such measures can also be supported by investing more in users' awareness and direct training.

More emphasis should also be placed on procedural steering approaches, such as those promoted by the EU WFD (planning processes in combination with more actor participation, the application of the principle of cost efficiency in combination with water pricing) as well as experimental periods on proposed management alternatives.

In several of the example-cases in section 9.4.1, more flexible instruments and management strategies were employed during attempts at a more integrated water regime. In the Italian Chiese-Idro basin, the experimental programme aiming at evolving a new rule for the Idro

Lake level regulation was a rather flexible strategy. In the California Central Valley, there was great willingness to experiment with an alternative model of decision-making and conflict resolution (the CALFED process), in the form of a participatory forum to develop a delta restoration plan. In the Belgian Vesdre basin, the regime could shift towards more integration by using voluntary agreements in the form of river contracts.

<u>Responsibilities and resources for policy implementation</u>

The fragmented distribution of responsibilities for implementing water policy and instruments, observed in the two Greek cases, needs to be overcome. A recent step in this direction (at least on paper) was made by the initiative of the 2003 Water Law to concentrate responsibilities to the Regional Water Directorates for water quality and water quantity management and for issuing all water permits. However, working on better coordination mechanisms could prove much easier and effective than attempting a further re-allocation of responsibilities among authorities. It should be kept in mind that it is a difficult task to remove competence from authorities and transfer it to others (tendency of institutions to resist change). Mitchell (1990) also suggested that in cases of fragmented and shared responsibilities, instead of reorganising the structure of public administration, it is often more productive to concentrate on the effectiveness of co-ordinating mechanisms. A positive coordination example from the Greek cases was the Programme Contract signed in the Mygdonian basin, which more clearly allocated responsibilities and resources to different actors for carrying out restoration works on Lake Koronia.

In the two Greek cases examined, it also became evident that significant commitment and political willingness is needed to equip the water policy implementation process with adequate resources on all administrative levels relevant to the management of water basins. Adequate resources must target the availability of personnel (in number and proper educational backgrounds) and of funds to collect necessary data, carry out necessary studies and management plans.

In the example-cases of more integrated basin regimes (section 9.4.1), the dedication of resources and the clarification of institutional responsibilities for both the formulation and implementation of measures and policies was important (e.g. in the French Audomarois basin or the US California Central Valley).

An additional type of resource which needs to be enhanced and coordinated in the Greek cases is the level of adequate sanctioning power in the hands of implementing authorities. In both cases, it became evident that the authorities issuing water abstraction permits had no complete sanctioning authority in case of policy violation. Non-compliant water users (usually farmers) had the option to appeal to court, which usually declared them innocent due to lack of adequate expert knowledge. Considering that the option to appeal to court could be difficult to abolish, Greek courts dealing with water litigations should be better informed of water management issues and make more effective use of authorities' expert knowledge. In this respect, it should be noted that, in many countries, courts are regarded as inefficient for settling water use disputes because legal suits are costly, take months or years to deliberate, and decisions are prone to be influenced by powerful stakeholders (Molle, 2004). Furthermore, Prefectural authorities implementing water quality regulations in the Greek cases had no complete sanctioning authority either. Penalty decisions for water polluters can, on the one hand, be questioned in court and, on the other, need the approval of the (elected) Prefects. With regard to the latter, due to the potential interference of political electoral interests, it is proposed to make penalty decisions for water polluters independent of elected entities and to restrict this responsibility to the competent Prefectural services.

Coherence within property rights

In the Greek cases, coherence in the system of rights to pollute water depends much on the enforcement of the relevant policy regulations. Although rights to pollute water could theoretically be well coordinated through the existing system of effluent disposal permits, real coherence can only increase when such permits are issued for all users who actually pollute water (including all municipalities, animal farms and illegally polluting industries) and when the conditions of issued permits are respected.

In terms of the quantitative use of water, the activation of cross-sectoral water basin management planning is necessary for the coordinated allocation of property rights. This way, abstraction rights could be awarded (through permits) in a coordinated manner and by considering the total water available in each basin. In fact, water basin management plans would also be important for coordinating rights to abstract and rights to pollute water. Secondly, a better coordination of abstraction rights could be achieved through better control of illegal abstractions (especially irrigation wells).

Looking at the example-cases of 9.4.1 facing water scarcity problems, success in coordinating property rights was limited. In the French Audomarois basin, coherence remained low between groundwater abstraction rights and groundwater nature claims of the Audomarois Park. In the Spanish Matarraña basin, the coherence of property rights remained low in terms of the little-implemented requirements of the 1985 Spanish Water Act to eliminate long-lasting private property rights (especially on groundwater). In the case of the California Central Valley, internal coherence of rights existed for surface water but not for groundwater, whose rights were defined in a vague way leading to overdrafting. The difficulty to increase coherence among rights, especially those linked to groundwater, is partly related to the fact that groundwater lies in the hands of numerous individuals. In general, the regulation of groundwater is a difficult task in many parts of the world. In the Mediterranean region, the share of public inland waters has tended to increase, and abstractions have been increasingly subject to licensing and mandatory regulation. However, groundwater was only later involved in this trend, because it was hidden, poorly known, considered as a local and minor resource, and above all its use often left to the judgement of the owners of overlying land (Dosi & Tonin, 2001). In a comparative study, Shah et al. (2001) concluded that in nations with advanced economies, sustainable management of groundwater is at best problematic, and at worst, as hopeless as in developing countries (e.g. India and Pakistan).

External coherence between public governance and property rights

In the context of quantitative rivalries in the two examined Greek cases, external regime coherence could increase, first, by defining a national policy to target holders of pre-1989 established abstraction rights based on custom or law. Secondly, external regime coherence could increase by allocating real water rights to nature and the environment. This could be linked, on the one hand, to land allocation mechanisms for wetland needs[116] and, on the other, to policy-based definitions of minimum flows and minimum lake tables to serve the needs of the water environment. Thirdly, external coherence would benefit from better implementation of policy regulations for abstractions (control of illegal users). External regime coherence in the context of water quality problems would also significantly improve through better implementation of the relevant regulations. Indeed, in the case of Greek basins, attention is needed on the weak enforcement of water laws. In the case of Greece, Molle's argument applies that defining water rights in public policy should be linked to political will and legal capacity to act against those who disregard them and to control new users (Molle, 2004).

Further chances to increase the external regime coherence in the two Greek cases are also seen in the introduction of (so far, neglected) spatial-planning policy instruments. Spatial-planning could be used for restricting land use rights (informal or formal) in favour of water protection. In the Vegoritida basin, spatial planning can play a critical role to minder rivalries over unclear land and water use rights on the shores of Lake Vegoritida. As mentioned in chapter 7, it is urgent to proceed with policy-based definitions of the lakeshore zone. The issue of spatial planning also needs more attention in the case of the Mygdonian basin. The closer coordination of rights to use land for industrial purposes (spatial planning policy) and rights to pollute water (water policy) could result in more efficient restrictions on the pollution of the basin industries in a disperse manner. Such coordination would strengthen the administration's potential to manage pro-actively industrial pollution. It seems, however, that progress on this issue could be slow since industries, which are holders of relevant rights to pollute water and to use their land for industrial purposes, can still not be forced by the authorities or the Park Management Body to settle in the industrial park of the area (lack of appropriate spatial planning policy).

9.4.3 Towards more integration-favourable contextual conditions

In chapters 7 and 8, it was concluded that the regimes of both Greek basins examined would have a greater adaptive potential towards a more integrated phase, if their contextual conditions became more supportive of integration. This section makes recommendations on how to make conditions more favourable for integration, focusing on the conditions of "joint problem perception", "joint chances perception" and "well-functioning institutional interfaces". As concerns the conditions of "tradition of cooperation" and "credible alternative threat by a

---

[116] For instance, in the Vegoritida basin, the property rights structure over the exposed land on the shores of Lake Vegoritida needs to be clarified. This would help clarify also water abstraction rights on exposed land and minder rivalries between abstractions and the environment.

dominant actor", it appears difficult to make any recommendations for their improvement. A more favourable "tradition of cooperation" can only develop over a long period of time, if the circumstances allow it. The presence of a "credible alternative threat by a dominant actor" highly depends on the specific actor network in place in a specific process.

Joint problem perception

The provision of more information on the water problems of the Vegoritida and the Mygdonian basins to all interested parties in a more symmetrical manner as well as the better sharing of information between interested parties could improve the condition of "joint problem perception". Special effort in this respect is needed to provide information on water problems also to those actors, who do not share any joint problem perception with other actors so far (especially water users in agriculture and industry).

The condition of "joint problem perception" was quite often partially or completely met in most of the example-cases of section 9.4.1. This was usually linked to the existence of monitoring data and adequately disseminated information on the status of water resources. In the Italian Idro-Chiese basin, the development of knowledge on the water basin dynamics through monitoring increased the information available. New information was shared by all actors and increased confidence and trust due to increased actor interaction. The drying-up of the riverbed downstream of the regulated lake became a clearly visible problem to most actors, even to reluctant users (e.g. farmers). In the Spanish Matarraña basin, a shared perception of risk from water scarcity was gradually shaped and the need to preserve the river was commonly acknowledged. In this process, the provision of scientific knowledge to enhance the position of environmental defenders was crucial. Also in the French Audomarois basin, the provision of more information (through studies) and its gradual sharing through participatory institutional formations (Consultation Committee on anti-pollution measures and Local Water Commission for the SAGE plan) were equally important to recognise interdependencies between water uses. In the US California Central Valley, effective water basin management was supported by the availability of technical information and transparency in the decision-making processes.

Joint chances perception

To increase the perception of "joint chances" from a more integrated and cooperative approach, efforts are needed to provide information on relevant benefits and impacts of alternative management strategies to all interested parties (especially reluctant water user groups like farmers and industry). In the Italian Idro-Chiese basin (section 9.4.1), joint chances from a more integrated approach were gradually recognised by most basin actors, even by farmers. In the Spanish Matarraña basin, joint chances were perceived among basin actors in avoiding the environmental and economic impacts of proposed large infrastructures to gain more water for the basin. In the Belgian Vesdre basin, joint chances were recognised in solving a pollution rivalry by both tourism and by those responsible for water purification.

Well-functioning institutional interfaces

Conditions for the two Greek cases could also become more favourable for integration in terms of several institutional interfaces.

First of all, this could be achieved by enhancing the *institutional position of brokers*. Although this role should be played by the Regional Water Directorates for issues like negotiating minimum river flows and minimum lake tables, actors like the Park Management Body in the Mygdonian basin could also act as process-specific brokers. In fact, brokers coming from more independent and participatory organisations could be more easily accepted by a variety of actors, including water users. Also in several of the example-cases of section 9.4.1, effective institutional mediators were present (Po Basin Authority and a Government Commissioner in the Italian Idro-Chiese basin; the Administration of the Ebro River Basin in the Spanish Matarraña basin).

Secondly, the regimes in the two Greek cases would benefit from better *institutional representation of key stakeholders*. A step in this direction is made by the 2003 Water Law requirement to set up Regional Water Councils, where representatives of unions of farmers' cooperatives, industrial chambers and environmental NGOs are admitted. However, when these Councils are activated, it should be ensured that the admitted representative organisations effectively represent the interests of their water user groups on the level of individual basins. For instance, in both Greek basins examined, the industrial chamber was inactive in the relevant water management processes, whereas industries were usually involved on an individual basis. In general, more effort should be invested in establishing networks between actors which share the same interests (e.g. between industries, between farmers and their cooperatives, between national and local NGOs which appeared weakly linked to each other in both cases). The formation of groups of homogenous actors actually allows the participation of a small number of large stakeholders, which is easier to work with. Also in several of the example-cases of section 9.4.1, there were effective representative organisations for large-numbered groups like farmers and a relatively limited number of actors involved in participatory mechanisms. For instance, in the French Audomarois basin, there was almost perfect representation of all stakeholders through their representatives in the Local Water Commission of the SAGE. In the California Central Valley, institutional representation of rival interests is well-developed and groups with similar interests form associations for effective participation.

Thirdly, substantial more effort can be made for building up a more *alert public* in the two Greek cases (especially in the Mygdonian basin). This can be achieved through more intensive public awareness programmes organised by public authorities, NGOs and local grassroots organisations. In several example-cases in section 9.4.1, the integration process was facilitated by the alert public present as well as active advocates of the interests of the local public and the environment. In the Italian Idro-Chiese basin, a local grassroots organisation influenced the process towards more integration. In the Spanish Matarraña basin, an active environmental coalition mobilised the local population to push the water regime towards more integration.

Fourth, *legal institutional interfaces* could become more favourable for integration in the two Greek cases, if the national water policy framework becomes more adequately implemented. If the implementation of national water policy remains a rather delayed process (as in the case of the 1987 Water Law), it may well be the case that institutional support for attempts to further integrate the Greek basin regimes will be sought in nature protection policy (as in the case of the Mygdonian basin). However, this would only be a very partial solution, since nature protection policy provides limited opportunities for more integrated water management. In several of the more integrated example-regimes in section 9.4.1, legal and policy institutional interfaces were quite supportive of integration and were usually linked to the implementation of national water laws. In the Italian Idro-Chiese basin, favourable conditions were offered by the creation of the Po Basin Authority, which provided an effective institutional interface for the governance dimension of the integrative attempts. In the French Audomarois basin, institutional favourable conditions were offered by the 1992 Water Act and the SAGE planning process but also nature protection policy, which ensured the creation of the Regional Audomarois Park.

Finally, windows of opportunity for the *institutional protection of compromises* reached between basin actors are also a condition which deserves attention for facilitating future integration in the Greek basin regimes. First positive signs of such institutional protection were seen in the signature of the actors' Programme Contract for restoring Lake Koronia in the Mygdonian basin in 2006. Indeed, also in more integrated example-regimes in section 9.4.1, institutional protection of actor compromises facilitated the integration process. In the Italian Chiese-Idro basin, this resulted from the formal governmental acknowledgement and involvement in the experimental programme aiming at a new rule for the regulation of the Idro Lake. In the US California Central Valley, the important role of an impartial court system in resolving water disputes and legitimising use compromises was emphasised. In the Belgian Vesdre basin, the protection of the actors' agreement was ensured by the informal regulation of local river contracts.

9.4.4 Concluding note

This chapter discussed how the two Greek case-study regimes of the Vegoritida and the Mygdonian basins could potentially change towards a more integrated phase. To this aim, recommendations have been made for changes in the institutional elements of the regimes and in the contextual conditions influencing regime change. Although the importance of these recommendations for a change towards a more integrated regime is not here questioned, it is reminded that the framework for the analysis of the two Greek basin regimes also used an intermediary set of variables influencing regime change. These intermediary variables were the characteristics of motivation, information and resources as well as the interactions of actors involved in regime change. The same intermediary actor variables were used to explain the process and outcome of the implementation of new incentives from a changed regime (usually policies and measures adopted), which ultimately affect the sustainability implications of regime change in water use. Implementation was, thereby, considered as a process worthy of detailed separate assessment, next to the process of regime change.

The actions proposed in this chapter to make the regime contextual conditions more favourable for integration are expected to also positively influence the characteristics and interactions of key actors in the two basins, in order to work together towards a more integrated regime. At a second stage, the shift of the institutional elements of the regime towards a more integrated phase is expected to have a positive influence on actor characteristics and interactions, to cooperate more towards better implementation of the measures and policies adopted.

Key conclusions on the value of the actor-centered explanatory approach (based on constellations of actor motivation, information and resources) and of the separate assessment of regime change and implementation in the two Greek basins are drawn in the concluding chapter 10. Although it would have been useful to do so, it has not been possible to have a comparative discussion on these aspects of the analytical framework between the two Greek case studies and the 24-European case sample previously compared by Bressers & Kuks (2004c). This was because the actor-centered explanatory variables and the separate assessment of regime change and implementation were not an explicit part of the analytical framework used in that earlier assessment of the 24-European case sample. Nonetheless, qualitative statements made in the assessment of those earlier European cases reveal that the actor-centered variables as well as issues of implementation also mattered in their processes of regime change. For instance, in the Italian Idro-Chiese basin, it was noted that without deliberate attempts of motivated actors in the direction of integration, real integration and environmental benefits would not have occurred (Dente & Goria, 2004). In other cases, progress towards rivalry resolution and towards a more integrated regime was related to the effective implementation of measures adopted during the regime change process. For instance, in the French Audomarois basin, action to deal with industrial pollution was based on financial incentives and compliance with discharge norms. A consultative procedure involving water users facilitated the implementation of the relevant measures adopted (Dziedzicki & Larrue, 2004).

10 Conclusions and outlook

10.1 Context and need for this research

This dissertation was written largely against the background of European policy requirements for the achievement of more integrated water management to reach good quality in European waters by 2015 (see the Water Framework Directive, WFD). Due to the poor record of the EU Member State of Greece so far in terms of a coordinated water management framework, its obligations to meet requirements for more integrated water management seem to be a great challenge for this nation's existing water management system and institutions. To elaborate on this challenge, this dissertation made use of the concept of institutional water resource regimes. The first aim was to identify whether and to what extent there has been any change or attempts to change the Greek institutional water regime towards a more integrated phase to date. Secondly, the dissertation aimed at explaining why there has been change (if any) or lack of change of the Greek regime towards integration.

In the following, this concluding chapter makes closing observations on the analytical framework adopted for guiding the research. The chapter, then, revisits key findings on the main research questions, making thereafter also recommendations on how to achieve more progress towards integrated institutional water regimes in Greece. Finally, the contribution of this dissertation to research, its relevance to policy and potential implications for future research are addressed.

10.2 Theory-based integration and the framework for analysis

Integration in water resource management was a main theme in this dissertation. Therefore, it was necessary to follow a consistent interpretation of this concept throughout. A systematic and structured interpretation of integration was borrowed from institutional resource regime theory, which has been previously applied in national and regional investigations of institutional resource regimes for the use of water and other natural resources in European countries. In specific, this dissertation was largely based on a baseline institutional-resource-regime analytical framework, which was rooted in the work of Knoepfel et al. (2001; 2003). This framework later merged with work on governance of Bressers & Kuks (2003) and was further elaborated and specified with, among others, contextual conditions for regime change by Bressers et al. (2004).

Roots and development of the baseline analytical framework

In detail, Knoepfel et al. (2001; 2003) introduced the thesis that institutional resource regimes do not consist only of property rights or only of public policies but of a combination of both. They also argued that institutional resource regimes matter for resource use sustainability and that, after external stimulation, regimes can change over time towards more complexity. In this context, they discussed two main dimensions to observe regime change towards a more integrated phase: the extent and the coherence of the regime. The extent indicates the scope of uses considered by the regime and the coherence reflects the degree of consistency and coordination within and between the systems of policy and of property rights. A regime can become more integrated, when its extent and its coherence increase. Knoepfel et al. (2003) also argued that, without more coherence, regime changes are not sufficient for more resource use sustainability.

Later on, the "policy" component of the institutional resource regimes of Knoepfel et al. (2001; 2003) was replaced by a model of public governance of Bressers & Kuks (2003) in the context of a European research project (Euwareness). In this model, governance is based on five elements (governance levels and scales, actors in the policy network, perceptions and objectives, strategies and instruments, responsibilities and resources for policy implementation). Furthermore, the explanatory variables of "contextual conditions", which together with external change stimulants influence regime changes, were added to the baseline framework (see Bressers et al., 2004). These "contextual conditions" were applied in the Euwareness empirical work on regime changes in twelve water basins of six European countries. In his comparative work on changes of the respective national water regimes in the same six countries, Kuks (2004c) applied a differentiated explanatory "institutional context" with focus on mechanisms which are decisive for change in national water policy and governance.

Given this theoretical and conceptual background, this dissertation placed analytical emphasis on the systems of public governance and property rights in Greek water management, since these two systems have been defined as the main components of institutional water resource regimes. The application of the particular analytical framework to Greek water regimes aimed, on the one hand, at reinforcing the validity of theoretical and empirical conclusions from the framework's previous application to other European countries and, on the other hand, at identifying the variables whose absence can explain the expected poor record of Greece in integrated water management so far.

Indeed, the specific theory-based analytical framework proved to be a very helpful analytical tool for the research objectives of this dissertation. It was helpful in better understanding important institutional aspects of the Greek water management system. The development of the Greek water regime, nationally and in two specific basins, could be characterised and interpreted in a systematic manner on the basis of the framework's institutional variables. The framework was also chosen for its close conceptual links with the EU WFD requirements for integral water management, making thereby the research more relevant to current EU water policy. Indeed, the key regime change dimensions (extent and coherence) largely reflect key principles of integrated water resource management of the EU WFD, especially with respect to a complete regime extent and a coherent public governance system.

Moreover, the adopted analytical framework provided an approach, which integrated two narrower analytical approaches from classical property rights theory and from public policy analysis. It is believed that the use of a purely property rights-based or a purely public policy-based approach could have overlooked certain institutional aspects, which are important for understanding the Greek water management regime. For instance, a public-policy based

344

approach could have overlooked the encouragement of unsustainable water use due to the uncoordinated distribution of property rights and due to weaknesses in the regulation of rights. Vice versa, important water institutional issues, such as actor coordination and resources for policy implementation, could be overlooked by a sole property rights-based approach. The adopted institutional-resource-regime approach is, indeed, based on the assumption that the (sustainable) use of water depends on institutional rules, which are anchored both in public policy and governance as well as in property rights regulation.

Modifications to the baseline analytical framework

In this dissertation, the baseline institutional-resource-regime framework was applied to and further modified for the analysis of the Greek water regime on two levels: the national level and the water basin level. Also previous researchers using the baseline framework (Knoepfel et al., 2003; Kissling-Näf & Kuks, 2004b; Bressers et al., 2004) carried out their investigations both on national and on regional levels (including water basins). The distinction of the analysis between the national and the water basin level helps to reflect on the different variables which can be "at play" for achieving integrated institutional water regimes on each of these two levels. This distinction may also allow more detailed insight into the process and outcome of any water management decentralisation attempts.

At a first stage of this dissertation, the baseline analytical framework was applied to the Greek national water regime (national level of analysis). In specific, the baseline framework was used in an already modified form, whereby it had been enriched with an explanatory "institutional context" (power distribution, values and perceptions in the water sector) by Kuks (2004c). In this dissertation, the institutional context (especially the "power distribution" context) was further enriched with some variables of Blomquist et al. (2005a) to better reflect decentralisation-supportive conditions and the importance of resources committed to reform. Although these minor modifications were initially made to better reflect the specific Greek national institutional context only, it is considered that the result is a more complete explanatory institutional context, which is suitable for further use elsewhere. Especially the clearer emphasis on the resources committed to decentralisation and integration efforts makes the framework more appropriate for use in less economically developed countries, where resources often act as a limiting factor to national integration attempts in water management.

At the second stage of this dissertation, the baseline analytical framework was modified and adapted, before its application to the analysis of institutional water regimes in Greek basins. In specific, based on insights gained during the first-stage application of the framework to the Greek national level and during the screening of candidate case studies for the water basin level, it was decided to proceed with two main modifications of the baseline framework. Firstly, the baseline framework was re-structured into a phased process approach, to distinguish between the process of making changes to the regime and the process of implementing new incentives resulting from regime change (mainly policies and measures adopted). Such an approach was judged useful for application to the Greek basins, which appeared to suffer from bottlenecks both in terms of making attempts at regime change towards more integration and in terms of subsequently implementing any changes attempted. Secondly, the baseline framework was enriched with an actor-centered explanatory approach, whereby explicit emphasis is placed on the influence of actor characteristics and actor interactions on the processes and the outcomes of regime change and implementation. The contextual interaction theory of Bressers (2004) served as the point of departure for the actor-centered explanatory links added to the framework. Therein, the main factors influencing the course and results of regime change and implementation are the motivation, information and resources of the key interacting actors. The enrichment of the baseline framework with such an actor-centred approach was considered analytically useful for the water basin level of research, where real-life water management processes are at stake. In the European Euwareness project, which also worked earlier with case studies on water basin regimes, such an actor-centered explanatory approach was not used.

Finally, next to the modifications made to the baseline analytical framework for its application to institutional water regimes in Greek basins, an effort was also made to present more explicitly the framework's dynamics of change. The graphical illustration of the modified analytical framework shows how the regime reacts to external change impetus and under what conditions, the actors act towards a more integrated or a more complex regime (see Figure 4.4 below; cf. Figure 3.1 of the baseline framework in chapter 3). The graphical illustration of the framework also shows the differentiated implication of regime change on sustainability in resource use through more or less adequate implementation of the new regime incentives.

Further proposed refinement of the analytical framework

Based on the empirical findings of research in the two Greek basins of this dissertation, it is here proposed to make a further addition to the analytical framework for the water basin level. During research, it became apparent that the ultimate sustainability implications in water use may not only depend on new incentives from regime change and their implementation, but also on external factors acting outside the process of water regime change. Such external factors may include natural factors (e.g. changing rainfall patterns) as well as socioeconomic factors (e.g. the decline of industries due to changing global market forces). The addition of these external factors, which influence the water status and sustainability in water use, will improve the explanatory power of the analytical framework. Figure 4.4 illustrates graphically the proposed addition.

Figure 10.1   Further proposed refinement of the framework for change in institutional water basin regimes

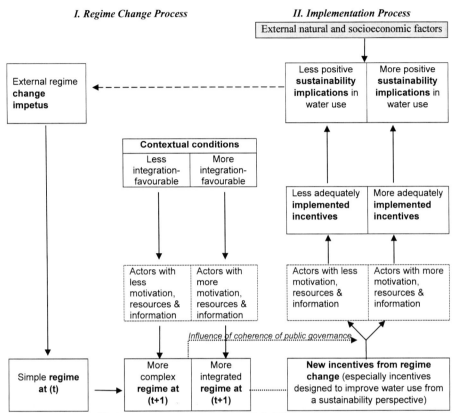

Source: Further modified framework based on Bressers et al. (2004) and Bressers (2004).

10.3 Realising integration in Greek institutional water regimes

A key objective in this dissertation was to conclude whether, to what extent and why institutional water regimes in Greece have changed or have not changed towards a more integrated phase to date. Keeping this in mind, this section briefly revisits the main research questions, linking them to the empirical findings. Secondly, on the basis of the research findings, recommendations are made on how to shift the Greek national water regime and the two examined water basin regimes towards more integration. Recommendations focus on the institutional elements of the regimes in question and on the conditions which influence regime changes. Recommendations have also been partly supported by comparative discussions of the institutional water regimes in Greece and in other nations and basins (see chapters 5 and 9). Finally, this section closes with brief concluding remarks on the role of actors and their interactions in the examined processes of water basin regime change and implementation.

10.3.1 Re-visiting the key research questions

The leading research questions of this dissertation were the following:

"Have there been any changes or attempts to change the institutional water regime in Greece towards a more integrated phase? How can progress or lack of progress towards a more integrated phase be explained?"

National level of analysis

The empirical findings showed that there have been attempts towards a partially more integrated national water regime since the 1980s (especially, after the adoption of the first Greek Water Management Law in 1987). However, these attempts were to a large extent unsuccessful on the ground. This lack of success to change the regime towards more integration could be explained by the unfavourable institutional context of the national water sector, which prevailed at the time of the attempts. This finding also confirmed Hypothesis 1 that *"the less integration-favourable the institutional context of the Greek national water sector, the more likely that attempts at a more integrated national institutional water regime are unsuccessful"*. The unfavourable institutional context was characterised by the unequal power distribution within the administration (especially between central and decentralised levels) and between public and non-public actors, unfavourable water sector values and unfavourable perceptions of water issues (in specific, absence of cooperative water policy style and intersectoral planning, fragmented understanding of water issues and low societal awareness of the environmental dimension of water). The explanatory value of these aspects of the institutional context was further validated through a comparative discussion of the national regime development and its context in Greece and in other European countries.

Especially, the low degree of practical implementation of the Greek 1987 Water Law led to a situation, where few "real" changes occurred in the national water regime, despite the substantial reform required by policy. In the latest phase of the national regime development, this dissertation witnessed the adoption of the 2003 Water Law, which transposed key principles of the EU WFD into national law. The characteristics of this new national water policy framework have been interpreted as some more progress of the Greek national water regime towards integration (at least on paper). However, the 2003 Water Law was also the target of some critique, while Greece also failed meeting several WFD requirements already in its initial implementation phase. All in all, it still remains to be seen whether the national institutional water regime can indeed shift towards more integration in practice. Although some aspects of the institutional context have been evaluated as somewhat more integration-favourable at present than during the attempt of the 1987 Water Law, in total the context remains largely unfavourable for integration. Improvement of the institutional context of the national water sector is, thus, urgently needed to support a "real" shift of the Greek national water regime towards a more integrated phase.

Water basin level of analysis

The leading research questions of this dissertation were examined empirically also in two water basins used as case study areas: the Vegoritida basin and the Mygdonian basin. In both basins, there were intense water use rivalries due to water degradation, indicating the urgent need to develop the water regime towards a more integrated phase. Furthermore, preliminary information gathered during the screening of candidate case studies indicated the presence of some cooperation efforts in these two specific basins, which was considered as a favourable indication for possible attempts at more regime integration.

Despite the presence of external impetus for regime change (including pressure from the intense water problems and rivalries), the institutional water regime in both examined basins has not yet shifted towards a more integrated phase. In both basins, the few cooperation efforts observed could not result in real regime change towards more integration or they took place too late in the process to have an effect yet. The poor performance in reaching a more integrated regime phase could be explained by the presence of integration-unfavourable contextual conditions (low cooperation tradition, weak perception of joint problem and of joint chances from more integration, lack of well-functioning institutional interfaces) as well as by an unfavourable constellation of motivation, information and resources of the actors relevant to integration attempts. The converging conclusions from both examined basins provide evidence for Hypothesis 2 that "*the less favourable the contextual conditions and the characteristics of interacting actors for integration attempts, the more likely that Greek water basin regimes fail to reach a more integrated phase*". Indeed, the similar sets of conclusions arising from these two independent Greek case studies are considered analytically quite powerful, at least compared to the conclusions from a single case.

It should be noted that the regime of the Mygdonian basin showed some higher tendency to develop towards a more integrated phase towards the end of the observed case study processes. Firstly, in the Mygdonian basin, the regime extent could become (almost) complete due to the explicit consideration of nature protection claims of the main water bodies (designation of a National Wetland Park). Secondly, the Mygdonian regime tended to become slightly more coherent in its public governance system, under pressure from strong EU and international drivers which demanded more coordinated action for the nature protection of water ecosystems (wetland protection). Better coordination potential was created by setting up a Management Body for the designated National Wetland Park of the basin and by signing an actors' Programme Contract for restoring the wetlands. All in all, it was observed that most positive changes in the Mygdonian institutional water regime were stimulated and supported by nature protection policy and EU funding procedures rather than the national water policy framework (which remained inadequately implemented). The lower tendency of the Vegoritida basin regime towards a more integrated phase can be attributed partly to lower external impetus for securing the nature claims of its water ecosystems. As a result, economic interests had considerably more free "space" to dominate the observed process in the Vegoritida basin. Economic interests contributed to low political motivation for the promotion of a long-awaited nature protection legal framework for the water ecosystems of the basin. Delays in this process were thus only partly (and not above all, as argued by some) related to the administrative split-up of the Vegoritida basin among three Prefectures and two Regions. The greater administrative simplicity of the Mygdonian basin may have, indeed, partly facilitated coordination efforts in that basin, but it was not the decisive factor responsible for the higher tendency of its institutional water regime towards more integration.

As concerns the WFD influence on the institutional water regimes of the two examined basins, the Directive was little relevant to the actual events observed. Although the WFD was transposed in Greece already in the end of 2003, the relevant 2003 Water Law remained largely inactive until the end of the two case study investigations in 2005.

An additional aspect investigated in both Greek case study basins was the implication of new incentives from regime change (key policies and measures adopted) on the sustainability of water use. In this respect, it was concluded that sustainability implications in water use were only minor or even negative in practice. Firstly, this was attributed to the few new incentives for more sustainable water use provided by the non-integrated regimes. Secondly, it was attributed to the fact that even the few new incentives provided were inadequately implemented under the influence of an unfavourable motivation-information-resource constellation of the interacting actors. The inadequate implementation of key adopted policies and measures was also responsible for the low coherence achieved in practice in some aspects of the institutional water basin regimes. All in all, the above converging case study conclusions provided supporting evidence for Hypothesis 3 that "*the less new incentives provided from regime change and the less adequate the implementation of any new incentives, the more likely that there are less positive sustainability implications of regime change in water use*". Actually, it was concluded that a few minor improvements observed in the environmental status of water resources in the Vegoritida basin and the reduction in the intensity of industrial pressures exercised in the Mygdonian basin were related more to external factors (changing rainfall patterns and decline of industries due to global market forces) and less to new incentives from changes in the institutional water regimes.

In chapter 9, also external evidence was provided for the validity of key variable correlations (causal assumptions) included in the hypotheses for regime change in the two Greek basins and its implications for sustainability. To this aim, the research conclusions from the two Greek basins were compared with conclusions from a larger sample of previously studied European basins. Firstly, findings from the previously studied European basins showed that less integration-favourable contextual conditions correlated with smaller regime changes towards more integration. This correlation was reinforced by the respective observations in the two Greek cases. Secondly, findings from the previously studied European basins showed that there is a rather strong correlation of general regime change, and especially of the variable of public governance coherence, with sustainability implications. This probabilistic correlation was also further confirmed by the findings of the two Greek cases, whose minor regime changes towards more integration were related to only minor or even negative sustainability implications.

Interchange of national and water basin level

This dissertation also posed a set of secondary research questions on the interchange of the Greek national water regime and the regimes on the water basin level. In this respect, the main questions were:

"Did any of the observed changes in the national water regime lead to or influence changes in the water basin regimes, and vice versa? Were there any other regime dynamics at play on the water basin level, which cannot be explained by the development of national water regime determinants?"

The empirical findings revealed no significant influence of changes in the two investigated basin regimes on the national water regime. However, several influences could be observed in the opposite direction, i.e. from the national regime to the water basin regimes. This observation, in fact, well reflects the predominantly one-way top-down character of policy developments in the centralized Greek State, where mainly national processes influence regional processes, instead of vice versa.

In specific, the national policy requirement for the establishment of water abstractions permits exercised important influence on the system of property rights in the institutional water basin regimes. Also, the fact that national policy has not yet intervened in the pre-1989 custom-based or law-based abstraction rights also influences the coherence of the property rights system in the basins. Moreover, some changes or lack of changes in the public governance systems of the two basin regimes were linked to the national water regime, especially to the 1987 Water Law. This concerned, firstly, the creation of Regional Water Management Departments (RWMDs) as water management authorities on the Regional level. Secondly, this was relevant to the lack of a national water plan and the lack of support to activate the process of water district management planning. Thirdly, the low level of institutionalised societal involvement in policy and the lack of participatory mechanisms in water resource management nation-wide were well reflected in the regional findings of the two case studies. Fourthly, the dominant presence of command-and-control water management instruments in the basin regimes was directly related to the respective set of instruments promoted by national policy. Finally, the lack of coordination of water management responsibilities and resources often reflected relevant problems in the national water regime.

Furthermore, it is interesting to note that certain developments in the institutional water regimes on the basin level were not explicitly related to national water policy. Especially, the tendency of the Mygdonian basin regime to become more coherent in its public governance system was not due to impetus from national water laws. Instead, this tendency was stimulated mainly by EU and international pressure for more coordinated wetland protection and by procedural requirements of EU instruments to fund the restoration of Lake Koronia.

All in all, the empirical findings of this dissertation revealed that integration (as this is defined in the theory-based analytical framework adopted but also in the EU WFD) is not achieved so far in the Greek national and in the two examined basin regimes. Moreover, it seems unlikely that significant change towards a more integrated regime phase can be realised in the near future. Integration-relevant external impetus pushing the Greek national and water basin regimes to change will be present in the future. As concerns impetus from the EU level (and in general the supranational level), the EC and e-NGOs remain alert on the compliance of Greece with EU water policy such as the WFD. Change impetus from the degraded status of water resources will also continue to be present or even increase, especially with respect to water scarcity phenomena. Nevertheless, despite the expected presence of external change impetus, the adaptive potential of Greek institutional water regimes towards a more integrated phase will remain seriously compromised, if critical context conditions do not become more favourable for change in the direction of integration.

10.3.2 Improving the institutional elements of regimes

This section synthesises conclusions on key improvements needed in the institutional elements of the Greek national water regime and of the Vegoritida and Mygdonian basin regimes, in order to change towards a more integrated phase. Detailed recommendations were made in separate for the national regime and for the basin regimes in chapter 5/5.7.3 and chapter 9/9.4.2 respectively. The following discussion is structured around the two key regime change dimensions of extent and of coherence. As previously mentioned, a more complete extent alone will not be enough to qualify for more integrated water regimes. Additionally and most importantly, the regime coherence needs to increase.

More complete extent: Environment and nature claims of water bodies

On both the national level and in the two basins examined, most uses with claims on water were somehow considered by the institutional water regimes. An exception were the environmental and nature claims of water bodies, which need to become more explicitly considered by the regimes through the practical application of respective water and environmental legislation. In the Mygdonian basin, nature claims of water bodies could mainly be added onto the agenda so far through nature protection policy.

Coherence of public governance as priority focus for regime change

In chapter 9, it was argued that efforts to increase regime coherence in the two examined Greek basin regimes should focus, as a priority, on increasing the coherence of public governance. Although the improvement of coherence within the system of property rights and of external coherence between property rights and public governance is also important for a more integrated regime, it was noted that these forms of coherence may be more difficult to bring about. Difficulties to increase coherence related to property rights partly relate to difficulties to re-distribute property rights on the quantitative use of water.

More coordination of governance levels and natural scales

The governance levels of Greek water management need to coordinate more within the natural scale of water basins. To this aim, coordination management bodies on a basin level are needed. This role could not be fulfilled so far by the Regional Water Management Departments (RWMDs) of the 1987 Water Law nation-wide, including the two examined basins. Therefore, significant efforts are needed in the future regional implementation of the 2003 Water Law to strengthen the coordination role of the new Regional Water Directorates (RWDs) in water management. Even if the RWDs eventually become well-operational, they will have competence on quite large territories (river basin districts) covering several river basins. It still remains uncertain how the RWDs will take account of specific management issues on the level of individual basins and subbasins. Possibly, smaller-scale institutional structures, such as the Wetland Park Management Body in the Mygdonian basin, could in some ways assist the large-scale RWDs in taking account of more local conditions in regional water management.

Need to include key actors interested in water resources into the policy network

Both on the national and on the regional level, substantial efforts are needed to establish functional participatory arrangements for actors with an interest in water resources, including actors outside the administration (water users, e-NGOs, epistemic community). In this respect, the participatory bodies foreseen by the 2003 Water Law (National and Regional Water Councils) provide a promising platform. However, chances for smaller-scale participation can also arise from nature protection policy processes with relevance to water management issues (see case of participatory Wetland Park Management Body in the Mygdonian basin).

Considering perspectives of different actors and sectors via water planning activation

Except for strengthening participation processes, also an improvement of more top-down planning is needed for a shift of the Greek water regimes towards more integration. In fact, it is urgently necessary to activate national and regional water planning, in order to be able to better coordinate the perspectives of different actors (including water users and e-NGOs) and of different policy sectors relevant to water management.

However, considering that it may take considerable time for Greece as a nation to establish a water management planning tradition, some "temporary" chances are seen in the production of water basin visions in the form of scientific studies with management-relevant conclusions. If such scientific studies can be legitimised through contracts between key implementing actors (as was the case with the Programme Contract for restoring Lake Koronia in the Mygdonian basin), these studies could be an intermediate solution for coordinating water management measures until the establishment of intersectoral water planning nation-wide.

Less command-and-control, more indirect and incentive-based instruments

In this dissertation, the need was recognised to complement Greek command-and-control water management instruments (e.g. abstraction permits, effluent disposal permits) with more flexible instruments. This proposal is especially based on the observation of the weak implementation of Greek command-and-control instruments. To compensate for this weak implementation capacity, chances are seen in more indirect and more incentive-based water management methods. An example is the reduction of water abstractions through subsidies to change crop patterns and/or to replace irrigation systems, instead of placing all effort in the top-down direct control of numerous private abstraction points. Nonetheless, even such less command-and-control instruments need to be supported by enforcement mechanisms but also by more investment in water users' awareness and training (e.g. for the promotion of incentive-based reduction of irrigation). New opportunities for more flexible instruments could also be offered by procedural steering instruments required by the EU WFD, such as full cost-recovery of water services, water pricing and more participative management.

Need to coordinate responsibilities and resources for policy implementation

Firstly, the fragmented distribution of responsibilities and resources for implementing water policy needs to be overcome via a combination of competence concentration to fewer authorities and more functional coordination arrangements between authorities. A recent step towards the concentration of competence was made by the 2003 Water Law, which gathered responsibilities for water quality and quantity management and for issuing all permits for water abstractions and water works to the Regional Water Directorates. Bearing in mind, however, that it may prove relatively difficult to remove and transfer further water competence especially between different sectoral authorities (due to the well-known tendency of institutions to resist change), there is need to invest more in better coordination arrangements between competent authorities.

Secondly, it became evident that significant commitment and political willingness is needed to equip the water policy implementation process with adequate resources on all relevant administrative levels. Adequate resources must target the availability of personnel (in number and proper educational backgrounds) and of funds to collect necessary data and to carry out necessary studies and management plans.

Thirdly, there is a need to improve sanctioning procedures through better coordination or re-organisation of sanctioning power. In the two Greek basins investigated, weak implementation of water management instruments was partly linked to the intervention of electoral and economic interests of powerful industrial and agricultural water users in order to influence sanctioning decisions. It is, thus, considered sensible to separate sanctioning decisions of policy implementing authorities from politically-elected entities (especially that of the Prefects). Additionally, Greek courts which are involved in water litigations should coordinate better with the implementing authorities which have expert knowledge resources.

Coherence of property rights via re-distribution and via better enforcement

In terms of quantity, water has almost reached its limits of availability in both Greek basins investigated. An increase in the coordination of property rights to reduce water rivalries can be achieved mainly through the redistribution of the existing rights. To this aim, the activation of cross-sectoral water management planning is necessary, so that abstraction permits can be awarded (and if necessary, limited and/or redistributed) on the basis of the total allocation capacity of each basin to support different water uses. Secondly, clearer allocation of water rights could be achieved through better control of abstractions and especially through better control of the private use of water outside the property rights system (mostly through illegal wells). However, especially the regulation of (legal and illegal) groundwater users is a difficult task, considering that groundwater use lies in the hands of numerous private individuals.

In the system of rights to pollute water, the increase of coherence depends much on the enforcement of the relevant policy regulations. Although rights to pollute water could theoretically be well coordinated through the existing system of effluent disposal permits, real coherence can only increase when such permits are issued for all users who pollute water (including all municipalities, animal farms and illegally polluting industries) and when the issued permits are respected.

Coherence between rights and governance: Restricting established rights - Improving enforcement of rights - Rights for the environment - Spatial planning tools

Firstly, it is necessary to adopt national policy to target holders of pre-1989 abstraction rights based on custom or law, who are so far not subject to permits. In general, the existing system of water property rights should not be taken for granted. Instead, redistribution of rights should be investigated as an option in the context of future water planning to reduce use rivalries and increase sustainability in water use. However, it is noted that the restriction of established rights is usually not an easy task (as experience in other countries has also shown). Difficulties to restrict established rights in the Greek context are not only related to the pre-1989 established rights but also to post-1989 rights, which have been established through the 1987 Water Law but do not comply with the defined abstraction conditions.

Secondly, the coordination between public governance and property rights (external regime coherence) could improve, if private groundwater ownership rights of land-owners, which are still rooted in the Civil Code, are eliminated by public water policy. Such an action could have a behavioural impact on water consumption, by increasing the respect of this resource by private users.

Thirdly, external regime coherence could increase in favour of water ecosystem protection, by establishing water rights of nature and the environment via the implementation of relevant policies. In the two Greek basins examined, this could be linked to mechanisms for allocating land to wetland needs and to the implementation of policy-based definitions of minimum river flows and minimum lake tables.

Fourth, external regime coherence would benefit from better implementation of policy regulations for abstractions as well as for water pollution. The deficient enforcement of water protection regulations causes so far uncertainty in the degree of protection of the system of rights to water. In fact, the issue of weak enforcement and implementation has emerged as a dominant problem throughout this dissertation, limiting chances to increase all three forms of water regime coherence (external regime coherence, coherence in the system of property rights and coherence in the system of public governance).

Finally, additional chances to increase external regime coherence, especially in the two Greek basins of Vegoritida and Mygdonian, were seen in the introduction of spatial-planning instruments to restrict informal or formal land use rights in favour of water protection. Around Lake Vegoritida, spatial planning could assist in clarifying rivalries over disputed land and water use rights on the lakeshores. Around Lake Koronia in the Mygdonian basin, spatial planning could assist in controlling (and, if possible, re-distributing) rights of land use for industrial purposes, in favour of more coordinated industrial pollution control and effluent treatment within an industrial park.

10.3.3 Improving the conditions which influence regime change

The change of Greek institutional water regimes towards a more integrated phase cannot be a simple "technical" exercise of changing their institutional components, but a result of long-term social processes and changes in dominant values and ideals. This principle was also supported by the cross-country and cross-basin discussions of this dissertation, which showed that the establishment of integrated institutional water regimes is a long-term historical evolutionary process, which is closely linked to the context of the respective nations or water basins. As previously said, also in Greece, institutional water regimes could change gradually towards more integration, if their relevant context becomes more favourable for integration. This section summarises key recommendations for making context conditions more integration-favourable for relevant change, firstly, in the Greek national water regime and, secondly, in the Vegoritida and Mygdonian basin regimes (see more details in chapter 5/5.7.3 and chapter 9/9.4.3 respectively).

Institutional context of the national water regime

The institutional context of the national water regime was examined in terms of the dominant values in the national water sector, the main perceptions of water issues and the power distribution within the administration and between public and non-public actors. This context reflects key institutional aspects which are decisive for change in national water policy and governance and which can influence the outcome of attempts to change the national regime towards more integration.

To make dominant values of the national water sector more integration-favourable, *political effort is, above all, needed to achieve a more cooperative water policy style*. This will assist in gradually harmonising rival sectoral water use fronts. In this respect, it is important to make use of the opportunities given by the new 2003 Water Law, especially the requirement for the creation of an Interministerial National Water Committee.

Secondly, with regard to main perceptions of water issues, effort and willingness are needed by both public and private actors to develop *more integrated understanding of water problems* and to *be open to new paradigms of water management* other than the traditional engineering approach. Further, resources must be devoted to the *establishment of intersectoral water planning* and *a national water information basis* suitable to modern water management needs. Also the ability of existing water institutions to expand their expertise to a growing range of water use functions (including the environmental ones) should be strengthened. Again, to this aim, *commitment of appropriate resources and political readiness is needed to overcome administrative inertia* on the national, regional and local levels.

Thirdly, power distribution in the water sector should become more balanced to support changes of the national water regime towards a more integrated phase. Especially, *coordination and co-governance between central and decentralised authorities in water management needs to be improved.* Following the steps of the 1987 Water Law, the 2003 Water Law gives the opportunity to set up a network of regional authorities (Regional Water Directorates) to take over water management competence on the regional level. Political willingness and resources need to be committed as soon as possible on the national and on the regional level, in order to avoid a network of non-functional and inadequately equipped authorities with little ability to communicate water management priorities to the central level. *More support of water management policy by the environmental policy sector* will only be achieved, if the new Central Water Agency at the Environment Ministry gains support on a political level and by different sectoral actors. Finally, *to strengthen public debate on water and to ensure a balanced participation of non-public actors such as e-NGOs in national water discussions,* careful use of the new participation opportunities provided by the 2003 Water Law should be made (National and Regional Water Councils).

All in all, the - at first sight - partial similarity between the institutional context influencing change in the national regime and the institutional elements of the regime should not be mistaken for a duplication of the issues considered in the analysis. The institutional context explicitly refers to preconditions (e.g. tradition of participation and debate), which - if existing - support the change of the institutional regime towards a more integrated phase (e.g. towards more coordinated participation of actors in the public governance system).

Contextual conditions of the Vegoritida and Mygdonian basin regimes

Contextual conditions influencing change in institutional water regimes of specific basins were examined in terms of selected characteristics of the actor network in place and in terms of specific institutional interfaces. Key recommendations to improve contextual conditions in the examined Vegoritida and Mygdonian basins focused on the conditions of "joint problem perception", "joint chances perception" and "well-functioning institutional interfaces".

The condition of *"joint problem perception"* could improve through the provision of more information on water problems to all interested parties in a more symmetrical manner and through the better sharing of information among these parties. Special effort in this respect is needed to provide information on water problems also to those actors who have not shared any joint problem perception with other actors so far (especially water users in agriculture and industry).

To increase the perception of *"joint chances"* from a more integrated and cooperative water management approach, efforts are needed to provide information on relevant benefits and impacts of alternative management strategies to all interested parties (including cooperation-reluctant water users like farmers and industry).

Considerable room for improvement is seen in terms of conditions related to institutional interfaces. Firstly, the *institutional position of brokers could be enhanced* in both basins. Although this role should be played basin-wide by the Regional Water Directorates for issues like minimum river flows and minimum lake tables, actors such as the Park Management Body in the Mygdonian basin could also act as process-specific brokers in the future. Secondly, the regimes in the two Greek basins would benefit from better *institutional representation of key stakeholders*. On the one hand, more effort should be invested in establishing networks between actors which share the same interests (e.g. between industries, between farmers and their cooperatives, between national and local NGOs). On the other hand, it should be ensured that representative organisations (especially of agricultural and industrial water users), which are admitted in the future to participative structures like the Regional Water Councils, effectively represent the interests of the respective water user groups on different territorial levels. Thirdly, more effort can be made to build up a more *alert public* in the two basins (especially the Mygdonian basin) through more intensive public awareness programmes organised by public authorities, NGOs and local grassroots organisations. Fourth, *legal institutional interfaces* would become more integration-supportive, if the national water policy framework became more adequately implemented. If the implementation of national water policy remains a rather delayed process (as in the case of the 1987 Water Law), it may well be the case that institutional support for attempts to further integrate these two basin regimes will be sought in nature protection policy (as was so far the case in the Mygdonian basin). However, this would only be a partial solution, since nature protection policy provides limited opportunities for more integrated water management. Finally, the achievement of more regime integration would be facilitated by the *institutional protection of compromises* reached between basin actors, as was the case with the signature of an actors' Programme Contract for restoring Lake Koronia in the Mygdonian basin in 2006.

10.3.4 Role of actors and their interactions in regime change and implementation

As mentioned before, on the water basin level of research and analysis of this dissertation, an intermediate set of actor-centered explanatory variables (motivation, information and resources of interacting actors) was added to the analytical framework. These actor-centered variables served to link the contextual conditions with water regime changes (regime change process) as well as the new incentives from a changed regime with their sustainability implications in water use (implementation process). Indeed, the motivation-information-resource constellations of the main interacting actors provided useful complementary explanations for delay or progress in regime change and implementation in the Vegoritida and the Mygdonian basins. In this respect, the empirical observations made in both basins matched well several of the theory-based assumptions of chapter 4/4.1.2 on actor interactions and their influence on regime change and implementation.

In the regime change process, the motivation, information and resources of interacting actors could explain why certain attempts at partial integration of the two basin regimes could take place, while others could not or were significantly delayed in their realisation. Actor motivation proved to be a crucial explanatory factor in this context. Only a few actors were motivated to support attempts in the direction of more coordination and more integration. There was especially lack of politicians' willingness to put forward water management strategies, which could promote more integrated water management but could undermine the economic interests of powerful water users. Key water users could also afford the alternative of non-cooperation or even resistance to regime change attempts, due to their powerful influence on politicians. In the Mygdonian basin, a breakthrough could be achieved, when a powerful more environmental protection-oriented coalition of actors jointly acted to reject the original Master Plan of the Prefecture for restoring Lake Koronia and thus achieved a wetland restoration paradigm change. The coalition achieving this breakthrough was much supported by strong EU and international external impetus to comply with sustainable wetland protection principles in a coordinated manner.

All in all, the actors interacting in the specific regime change processes may become more cooperative and motivated towards a more integrated water regime phase, under the influence of stronger integration-supportive external impetus (e.g. EU directives or EU funding instruments explicitly demanding more actor coordination). For certain actors, such impetus can make future inaction towards integration a less attractive alternative, pushing them thus to become more cooperative. Moreover, actors may have more motivation, information and resources to work in the direction of a more integrated regime, under the influence of more integration-favourable contextual conditions (especially better-functional institutional interfaces, a more common perception of a joint problem from water degradation and a perception of more joint gains from more cooperative water management).

Actor motivation, information and resources also provided explanations for the inadequate implementation of several new incentives from regime change (key adopted policies and measures), which contributed to the observed minor or even negative sustainability implications in water use. Both in the Vegoritida and in the Mygdonian basins, the motivation of target groups of new policies and measures (mainly farmers and industries) was in general low with respect to adequate implementation. The main instruments targeting farmers and industries were command-and-control instruments of permits and use restrictions. Indeed, the water users were not given early enough more attractive and less command-and-control instruments and incentives to change their water use behaviour (such as environmentally-oriented subsidies to reduce water consumption in agriculture or early participation in the selection of anti-pollution measures in the case of industry). Concerning the command-and-control instruments, farmers and industries remained largely non-cooperative for their implementation. In fact, they did not count with severe consequences from such non-cooperative behaviour, due to the weak enforcement capacity of the administration and due to the pressure that they could exercise on politicians, if needed. Indeed, the limited information and resources (and sometimes even motivation) of the implementers undermined their capacity for an effective implementation of regulations. Furthermore, quite often the formal resources of implementers were balanced or outbalanced by the significant informal resources in the hands of the target groups, in the form of pressure exercised on elected politicians. In fact, these informal resources of key water users appeared as an issue both in the implementation but also in the regime change process. Interestingly, in literature, the persistent failure of Greek governments to introduce and to enforce an effective regulatory framework in general environmental policy is also partly attributed to the resistance of powerful industrial actors. On the national level, a large component of this actor group includes state-owned industry in environmentally sensitive sectors (Koutalakis, 2006), like the Public Power Corporation, which was also an important actor in the water management process of the Vegoritida basin.

All in all, it is concluded that the implementation of new incentives adopted during regime change (usually policies and measures) may improve, if it is supported by more favourable motivation-information-resource constellations of the interacting actors. This involves overcoming barriers of political costs for the infliction of penalties on illegal abstractions and water pollution (motivation of implementers). Additionally, the enforcement capacity of implementers should be enhanced, especially by improving their resource basis and their ability to acquire more data on actual water use patterns and to carry out more frequent field inspections. The increase of "immunity" of politicians to pressure from powerful water users and the better enforcement capacity of the implementers would also force water users (farmers and industries) towards a more cooperative behaviour for better implementation of policies and measures. Water users may also become more motivated and cooperative, if they are more involved themselves in the formulation and selection of attractive water management measures (this is related to a change of the institutional water regime towards more interactive and participatory processes).

10.4 Contribution to research and relevance to policy

Contribution to research

The focus of this dissertation on institutions, in order to elaborate on the challenge faced by Greece in view of requirements for more integrated water management, contributes to the "filling of a gap" in existing literature on Greek water resources. So far, Greek water management issues have been addressed in literature in a rather technical way and rarely by seeking their links to institutions and policies. Institutional issues have not been considered so far as critically relevant, despite international and occasionally national recognition of their importance for the main problems in Greek water management.

Given the baseline institutional-resource-regime analytical framework used, this dissertation placed particular emphasis on assessing the Greek system of public water governance and the system of water property rights. As van Hofwegen (2001) argued, an assessment of the existing water management system (including water rights, existing institutions and water policies, the legal framework, stakeholders and interest groups) should be the first step of any assessment to reach a desired situation of integrated water resources management. This dissertation also modified the baseline analytical framework into an actor-centered phased-process approach for its application to Greek institutional water regimes in specific basins. This way, specific research attention could be paid to implementation issues and to the importance of the human dimension and actor interactions in Greek water management on the ground.

The regional research focus of this dissertation also resulted in the first detailed assessment of the water management processes in two important Greek basins (Vegoritida and Mygdonian) from an institutional perspective. The re-construction of the chronology of events in each basin and the subsequent analysis of change in the institutional water regimes can be a useful reference for future research in these basins.

Except for its contribution to water management research in Greece, the results of this dissertation are, furthermore, of relevance to currently running Mediterranean activities which promote integrated water management in this region (especially the Mediterranean component of the EU Water Initiative[117]). On a more global level, the present dissertation can serve as a source of information on Greece for use in future European and international comparative research on integrated water management, water governance and water institutions.

Relevance to the EU WFD policy targets

From a policy perspective, the occurrence and outcome of attempts towards integrated institutional water regimes in Greece so far was worth assessing and explaining, especially in view of the compliance obligations of this nation with the EU WFD. In fact, the WFD policy vision of integrated water management was related, as far as possible, to the adopted theory-based interpretation of the integrated phase of institutional water regimes.

In general, the adoption of the WFD in 2000 gave clear signals on its potentially far-reaching implications for the institutional systems of water management in EU Member States including Greece. In other words, the WFD gave a top-down impulse for a new configuration of existing institutional water management arrangements across Europe. In this context, the present work contributed to enhancing relevant knowledge on the development of key characteristics of the institutional water management system in Greece, in view of the WFD change impulse. The research findings indicate that the case for far-reaching WFD implications in Greece cannot be taken for granted. Clearly, integrated water management is not fulfilled yet in Greece, and also the potential to make good use of the WFD impulse for

---

[117] At the 2002 World Summit for Sustainable Development (WSSD) in Johannesburg, the EU launched a Water Initiative designed to contribute to the achievement of the Millennium Development Goals and WSSD targets for drinking water and sanitation, within the context of an integrated approach to water resources management. The EU Water Initiative highlights the importance of integrated water resource management, the need to balance human water needs and those of the environment and refers to the EU experience in river basin management (information accessed online on 13/09/2005: http://www.euwi.net).

future change appears seriously weakened. All in all, it appears difficult for Greece to meet the WFD policy requirements content-wise and time-wise.

It is proposed that, in view of future attempts to achieve more integrated water resource management (IWRM) in Greece, including the IWRM model promoted by the WFD, the research conclusions and recommendations of this dissertation on Greek institutional water regimes should be taken into account. Especially the following aspects of the Greek institutional context are identified as important selected issues to be considered for the implementation of the WFD.

Firstly, the unitary nature of the Greek State accounts for the still unbalanced distribution of power between central and decentralised governance levels. A process of strengthening regional governance structures, which could prove favourable for the establishment of more IWRM through the WFD, only took off relatively recently in the mid-1990s.

Secondly, the Greek water administration apparatus suffers from the lack of resources (human and financial) and institutional capacity. This picture is partially responsible for the weak enforcement capacity of new (and existing) water policies, including policies promoting IWRM. In general, the existing Greek water sector is not organised enough at present to adequately support IWRM reform attempts.

Thirdly, policies promoting IWRM (and indirectly more environmental protection regulation) find little support in Greek society, whose needs for environmental amenity still have lower priority over development, compared to other European countries. Having said that, there exist local exceptions, as was the case of the Vegoritida basin, where the lakeshore society appeared particularly aware of the need to protect Lake Vegoritida.

Fourthly, the explicit focus of the WFD on water quality is not necessarily of paramount interest in Greece. Instead, the main Greek water issue is one of water quantity, which is mainly considered by the WFD as a supportive aspect to the promotion of good water quality.

Bearing the above in mind, it seems firstly sensible not to place all expectations for the WFD implementation on the regional level of the Greek administration. The implementation process should be supported by the central water administration, until the regional level is capable of taking over. Secondly, given the weak enforcement potential of the administration, it is wise not to rely only on command-and-control water management instruments but to gradually complement these with indirect, less control-intensive and incentive-based instruments. Thirdly, more effort is needed to make Greek society more aware of the fact that the WFD does not aim at halting development but that it could even promote development through water protection. Finally, Greek water decision-makers should make the best use possible of the few instruments provided by the WFD to tackle water quantity issues, especially through the river basin management plans and the programme of measures adopted for each river basin district.

10.5 Possible implications for future research

Working hypotheses and proposed research in other water basins of Greece

The two Greek basins examined as case studies in this dissertation were not randomly but purposefully chosen, according to a set of criteria defined to fit the adopted theory-based analytical framework. A more random selection of cases as well as a larger number of cases could have increased the likelihood that data are more representative for the whole country, so that some generalisation of the findings can be made thereafter. As Yin (1994) argues, case studies are in principle not suitable as a research approach for discussing their representativeness for a more abstract level, also due to their small sample.

Nonetheless, having selected two Greek basins, which mainly concern important lakes and aquifers and share a similar water resource problem structure, one is tempted to formulate a working hypothesis (even if abstract) for other different or similar inland Greek basins[118]. It is here reminded that one of the reasons for selecting two case studies which are relevant to important lake systems was preliminary evidence that more cooperation efforts had taken place in these "lake" basins than in the other case-study candidate "river" basins. This, in a way, was preliminary evidence supporting the abstract theoretical assumption that water problems with reciprocal externalities (usual in shared lakes or aquifers) have better chances for the development of cooperation than water problems with unidirectional externalities (typical of river upstream-downstream situations) (see Dombrowsky (2005) on the theoretical relation between cooperation potential and resource use problems with different distribution of externalities, mentioned also in chapter 4/4.3.2.2).

Ultimately, the research findings of this dissertation showed that, despite the presence of few cooperation activities in the two examined Greek lakes (more precisely, "lake" basins), no significant development of their water regimes could be achieved yet towards a more integrated phase. In the Vegoritida basin, past cooperation efforts were only short-lived, while, in the Mygdonian basin, cooperation efforts took place only late in the observed process. Slow progress towards a more integrated regime in both basins was explained via the adopted analytical framework, pointing to the lack of integration-favourable contextual conditions and of favourable motivation-information-resource constellations of the interacting actors.

In an abstract manner, it could be argued that if no regime integration could be achieved in the examined "lake" basins, where chances for the development of actor cooperation around the shared lakes were better, then logically chances are even worse for regime integration to be observed in Greek "river" basins, which suffer from upstream-downstream rivalries (as in most of the screened-out candidate "river" basins in chapter 6). From the perspective of the context-specific variables used in the analytical framework of this dissertation, differences

---

[118]   In chapter 6, it was decided to leave Greek islands and coastal waters out of the scope of this dissertation. Islands and coastal water systems are, first, related with difficulty to one specific water basin and, secondly, deserve a research focus of their own to deal with their particular local water problems, in need of locally-adapted management strategies.

between "lake" basins and "river" basins may occur, for instance, in the contextual conditions of joint problem perception and joint chances perception. In rivers with upstream-downstream rivalries, joint problem and joint chances perceptions may take even longer to develop than around lakes, whose geographical limitation often forces users to perceive sooner the common threat from water resource degradation.

Logical abstract generalisation from the two examined "lake" basins to other "lake" basins in Greece seems more difficult, because any given lake can be characterised by a reciprocal distribution of externalities (as argued by Dombrowsky, 2005) but also by a more unidirectional distribution of externalities depending on the specific conditions in each case. For instance, in the Vegoritida basin studied in this dissertation, the distribution of externalities from water problems was not only reciprocal but also unidirectional for some aspects. In the case of water pollution, there was a unidirectional advantage in favour of one Prefecture in the "lake" basin, which hosted many activities that polluted a stream mouthing into Lake Vegoritida, but did not have any part of the lake on its territory. Thus, the discharge of effluents from this Prefecture mainly affected the two other Prefectures of the basin which shared the lake downstream. In this case, the division of the "lake" basin into three administrative units contributed to the unidirectional nature of externalities from the generation of water pollution. During the case study research, it even became obvious that there were unidirectional externalities also in the case of abstractions for irrigation. Aquifer users in one of the two Prefectures sharing Lake Vegoritida had a unidirectional advantage over users in the other Prefecture. Although most wells were drilled (illegally and legally) in one Prefecture, the aquifer dropped faster in the other Prefecture of the lake due to the aquifer shape. This - in part - unidirectional distribution of externalities in the Vegoritida basin can even explain the two following issues in the case study. Firstly, there was differentiated behaviour of farmers in terms of aquifer abstraction intensity and compliance with abstraction regulations in the two Prefectures sharing the lake. Secondly, there were differences in the sensitivity of the population over water pollution in the three Prefectures of the "lake" basin. The population in the two downstream lake-sharing Prefectures was more mobilised to protect water resources than in the upstream non-lake-sharing Prefecture.

All in all, it became obvious that, even though the Vegoritida basin is a "lake" basin, it is not characterised by only lake-typical reciprocal distribution of externalities. This could be the case also in other Greek "lake" basins, depending on the specific administrative and natural conditions in each basin. Therefore, one should be careful when arguing for a better cooperation potential (and regime integration potential) around lakes in any real-life lake or "lake" basin. In fact, the potential to encounter unidirectional externalities rises significantly, when considering a lake from a basin perspective. Furthermore, even if there are other Greek "lake" basins with only reciprocal distribution of externalities among the parties involved, the water management process may still suffer from the lack of other integration-favourable contextual conditions (e.g. well-functioning institutional interfaces) and from a weak enforcement capacity to implement any adopted measures and policies.

In the following, some additional possible research in other Greek basins is proposed, on the basis of the more context-specific variables of the analytical framework used in this dissertation. The relative progress of institutional water regimes towards a more integrated phase in other Greek basins is expected to depend on their variation in terms of the independent variables of the analytical framework, e.g. the force of external impetus for change, the relevance of external impetus to integration as well as the favourable or unfavourable status of contextual conditions for integration. The similarities or differences in the integration progress achieved in the two basins examined in this dissertation and in further Greek basins, on the basis of the same independent variables, would provide more validity for the findings of this dissertation and a broader basis for more generalised conclusions on institutional water regimes in Greek basins.

A first suggestion is to carry out similar research in other Greek basins, where rivalries on water resource degradation are not urgently relevant (yet) to compliance with European and international environmental protection standards. Given the lack of European and international pressure for the environmental protection of water bodies, it is likely that progress of these institutional water regimes towards more integration will be even slower and less resource protection-driven than in the two basins examined in this dissertation. In such cases, the environmental dimension of water resources will probably continue to be excluded from the regime development. Such research would also further emphasise the importance and value of EU and international forces for enacting changes in Greek water management processes towards more integrated and resource protection-driven approaches.

Secondly, in the two basins of this dissertation, water users were mainly served by groundwater, once direct abstractions from the lakes were prohibited.[119] In these basins, there was mainly non-collective irrigation with a multitude of users privately extracting groundwater. In future research, it could be interesting to apply the institutional-resource-regime framework to evaluate regime change in Greek basins, where water users mainly depend on surface water rather than groundwater. Especially in areas where agriculture depends on surface water, Greek farmers are served by collective irrigation networks managed by obligatory farmers' cooperatives (Organisations for Land Reclamation). In such basins, it is expected that some contextual conditions (e.g. with regard to the institutional representation of farmers) and some institutional regime elements (especially regarding the distribution of property rights for water) may be different than in the two basins of this dissertation.

Phased-process and actor-centered explanatory approaches

In this dissertation, it was decided to modify and adapt the baseline analytical framework into an actor-centered phased-process framework for its application to institutional water regimes

---

[119] In Greece, 60% of water consumed annually is surface water, while groundwater abstraction covers 40% of annual water demand, which is double than the EU average (Gorny, 2001).

in Greek basins. This decision was based on insights, which were gained during the initial application of the baseline framework to the national water regime and during the collection of information to screen candidate case study basins (see chapter 4/4.1.2). It is here proposed that future research on the Greek national institutional water regime would also benefit from the application of the actor-centered phased-process version of the framework. This could deliver more in-depth analysis of national water policy formulation and implementation processes, under the influence of interactions of key national water actors.

Moreover, acknowledging that the inadequate implementation of policies and measures has been recognised as an important bottleneck in the development of more integrated institutional water regimes in Greece, it is proposed that more research should be devoted to implementation processes in the Greek water sector. Research focused on implementation could deliver more in-depth findings on the determinants of successful or failed implementation. In this context, actor-centered explanatory approaches, like the contextual interaction theory used in this dissertation, are recommended given the decisive role of the human dimension and personalities in implementation processes in Greece.

International comparative research

Finally, it is proposed that future international comparative research on the achievement of integrated water management should place more emphasis on recording and discussing experience from countries and river basins, which are not very developed economically and suffer from weak enforcement capacity for policy implementation. Successful cases of integrated water management in such national and regional contexts would provide interesting lessons learned and potential remedies towards more integration in countries such as Greece with a tight national budget and low policy implementation record.

11 List of references

Abernethy, C.L. (ed) (2001). Intersectoral management of river basins (Working Group Reports). Proceedings of an international workshop on IWM in water-stressed river basins in developing countries: Strategies for poverty alleviation and agricultural growth. Loskop Dam, S.Africa, 16-21 October 2000. Colombo, Sri Lanka: International Water Management Institute (IWMI) and German Foundation for International Development (DSE), 365-376.

Aggelakis, A. (1993). Legislation of Greece in the field of water. Athens: TEE 5206 (in Greek).

Aggelakis, A. N. and Diamantopoulos, E. (1995). Management of water resources in Greece including the use of marginal waters. Proceecings Conference Tourism and the Environment in island regions, Herakleion Crete, 17-19 March 1995, 100-110 (in Greek).

Amanatidou, E. (1997). The significance of water quality in the management of water potential in the region. Proceedings of Conference on Water Resource Management in the basin of Kozani-Ptolemaida, Amyndeo, 15/2/1997. Ptolemaida: Technical Chamber of Greece, no page numbers (in Greek).

AN.FLO. S.A. (Development company of Florina) (1998). Study on the intention of setting up a management body for water resources management of the basin of Vegoritida. Phase A. Florina (in Greek).

Anonymous (2004). Plan for the restoration of Lake Koronia: The Prefecture of Thessaloniki claims significant funds. Ydrooikonomia, 24, p.47 (in Greek).

Arabatzi-Karra, Ch., Tamvakli-Travasarou, M., Katopodis, G., Gofas, A., Christidis, Ch. and Tsakaleris, P. (1996). Management Programme of the Protected Area of Lakes Koronia, Volvi, the Macedonian Tembi and their surrounding area. Phase A. Athens: Ministry of Environment, Physical Planning and Public Works, Directorate for Environmental Planning, Prefectures of Thessaloniki and Chalkidiki (in Greek).

Argirakis, S. (2002). The causes of level drop of Lake Vegoritida. Technical Report Summary. Edessa: Prefectural Self-Administration of Pella (in Greek).

Assimacopoulos, D., Karavitis, C., Manoli, E., Gerasidi, M. and Katsiardi, M. (2002). Report on Greece: Range of circumstances and region analysis. Deliverable D1 "The range of existing circumstances", EU-Project WaterStrategyMan, Ruhr: University Bochum.

Aubin, D. and Varone, F. (2004a). The evolution of European water policy: Towards integrated resource management at EU level. In: Kissling-Näf, I. and Kuks, S. (eds). The evolution of national water regimes in Europe. Dordrecht: Kluwer Academic Publishers, 49-86.

Aubin, D. and Varone, F. (2004b). Diverging regimes within a recently federalised state. In: Bressers, H. and S. Kuks (eds). Integrated governance and water basin management: Conditions for regime change and sustainability. Dordrecht: Kluwer Academic Publishers, 99-130.

AUTH (Aristoteles University of Thessaloniki, Laboratory for agricultural chemicals) (2002). Final report of results "Programme for surface water quality control in Macedonia-Thrace". Thessaloniki: AUTH (in Greek).

AUTH (Aristoteles University of Thessaloniki, School of Agriculture, Laboratory of Applied Soil Sciences) (2004). 1st Technical Progress Report of the Research Programme "Monitoring of Lakes Volvi and Koronia with Remote Sensing". Thessaloniki: Development Company of the Prefecture of Thessaloniki A.E (in Greek).

AUTH & EKBY (Aristoteles University Thessaloniki & Greek Biotope Wetlands Centre) (2001). Study and proposal for works of protection and restoration of functions of wetlands Zazari-Chimaditida. Thessaloniki: Region West Macedonia (in Greek).

Bandaragoda, D. J. (2000). A framework for institutional analysis for water resources management in a river basin context. Working Paper 5. Colombo, Sri Lanka: International Water Management Institute.

Barraqué, B. (1998). Water rights and administration in Europe. In: Correia, F.N. (ed). Selected issues in water resources management in Europe, Vol. 2. Rotterdam, Brookfield: Balkema, 353-385.

Bartzoudis, G. (1993). The improvement of the function of Lake Kerkini, as irrigation reservoir, in order to serve also other functions. Thessaloniki: Greek Center for Wetlands and Biotopes (in Greek).

Birtsas, P. K. (2005). Chronology of a preannounced death in Lake Koronia. Accessed online on 2/2/2005: http://www.atraktos.net/article.asp?cat=14&article=8671 (in Greek).

Blionis, G. (1992). Lakes Koronia and Volvi: Over the hills of Thessaloniki. Oikotopia, 20, 24-28 (in Greek).

Blionis, G. (1999). Conferences for the information and sensibilisation of farmers. Hydrobio (Quarterly publication of the Information Centre for Lakes Koronia and Volvi), 4, 4-5 (in Greek).

Blöch, H. (2001). Die EU Wasser-Rahmenricthlinie: Europas Wasserpolitik auf dem Weg ins neue Jahrtausend. In: Bruha, Th. und Koch, H.J. (Hrsg.). Integrierte Gewässerschutzpolitik in Europa. Baden-Baden: Nomos, 119-127.

Blomquist, W., Dinar, A. and Kemper, K. (2005a). Comparison of institutional arrangements for river management in eight basins. World Bank Policy Research Working Paper 3636. Available online: http://econ.worldbank.org.

Blomquist, W., Giansante, C., Bhat, A. and Kemper, K. (2005b). Institutional and policy analysis of river basin management: The Guadalquivir River Basin, Spain. World Bank Policy Research Working Paper 3526. Available online: http://econ.worldbank.org.

Bodiguel, M., Bruckmeier, K., Buller, H. I., Fuentes Bodelon, F., Glaeser, B., Gonzales Fernandez, M., et al. (1996). La qualité des eaux dans l'Union Européenne, Pratique d'une réglementation commune. Paris: L'Harmattan.

Bousbouras, D. (2004). Actions in lakes Petron and Vegoritida. Oionos, 20, 6-8 (in Greek).

Bressers, H. (1983). Beleidseffectivitet en waterkwaliteitsbeleid. Enschede: University of Twente, 24-27.

Bressers, H. (2004). Implementing Sustainable Development: How to know what works, where, when and how. In: William M. Lafferty (ed). Governance for Sustainable Development: The Challenge of Adapting Form to Function. Cheltenham, Northampton: Edward Elgar.

Bressers, H., Fuchs, D. and Kuks, S. (2004). Institutional resource regimes and sustainability. In: Bressers, H. and S. Kuks (eds). Integrated governance and water basin management: Conditions for regime change and sustainability. Dordrecht: Kluwer Academic Publishers, 23-58.

Bressers, H. and Kuks, S. (2003). What does "governance" mean? From conception to elaboration. In: Bressers, H. and Rosenbaum W. A. (eds). Achieving sustainable development: The challenge of governance across social scales. Westport: Praeger Publishers, 65-88.

Bressers, H. and Kuks, S. (eds) (2004a). Integrated governance and water basin management: Conditions for regime change and sustainability. Dordrecht: Kluwer Academic Publishers.

Bressers, H. and Kuks, S. (2004b). Governance of water resources: Introduction. In: Bressers, H. and Kuks, S. (eds). Integrated governance and water basin management: Conditions for regime change and sustainability. Dordrecht: Kluwer Academic Publishers, 1-22.

Bressers, H. and Kuks, S. (2004c). Integrated governance and water basin management: Comparative analysis and conclusions. In: Bressers, H. and Kuks, S. (eds). Integrated governance and water basin management: Conditions for regime change and sustainability. Dordrecht: Kluwer Academic Publishers, 247-265.

Bromley, D. W. (1991). Environment and Economy. Property rights and Public Policy. Oxford UK/Cambridge USA: Blackwell.

Cordova-Novion, C., Stokie, M. and Jacobs, S. H. (2000). Background report on government capacity to assure high quality regulation in Greece. Peer reviewed in July 2000 in the OECD's Working Party on Regulatory Management and Reform of the Public Management Committee. OECD.

Costejá, M., Font, N. and Subirats, J. (2004). Redistributing water uses and living with scarcity. In: Bressers, H. and Kuks, S. (eds). Integrated governance and water basin management: Conditions for regime change and sustainability. Dordrecht: Kluwer Academic Publishers, 165-188.

CSEH (Committee for Social and Economic Issues of Hellas) (2003a). Opinion of CSEH on 'Protection and management of water resources – Harmonisation with Directive 2000/60 of the EU'. September 2003. Accessed online on 20.06.2004: http://www.oke.gr/greek/gnomi96.htm (in Greek).

CSEH (Committee for Social and Economic issues of Hellas) (2003b). The draft water bill does not include basic proposals which have been submitted, Ydrooikonomia, 15, 30-37 (in Greek).

Dacoronia. E. (2003). The Development of the Greek Civil Law: From its Roman – Byzantine Origins to its Contemporary European Orientation. European Review of Private Law, 5/2003, 661-676.

De Clercq, M. and Suck A. (2002). Theoretical reflections on the proliferation of negotiated agreements. In: De Clercq, M. (ed). Negotiating environmental agreements in Europe: Critical factors for success. Cheltenham: Edward Elgar, 9-66.

Delithanassi, M. (2004). Drill a well, you can! Newspaper "Kathimerini", 31/10/2004, p. 24 (in Greek).

Del Moral, L., Pedregal, B., Calvo, M. and Paneque, P. (2002). The river Ebro interbasin water transfer. Project Report of FP5 research project ADVISOR (Integrated Evaluation for Sustainable River Basin Governance). Seville: University of Seville.

Delimbasis, K., Katsifarakis, K., Mylopoulos, I. and Economidis, G. (1998). Working Group "On the Master Plan for the environmental restoration of Lake Koronia". Thessaloniki: Technical Chamber of Greece, Department of Central Macedonia (in Greek).

Dente, B., Fareri, P. & Ligteringen, J. (1998). A theoretical framework for case study analysis. In: Dente et al (eds). The Waste and the Backyard. Dordrecht: Kluwer, 197-223.

Dente, B. and Goria, A. (2004). Competing integration principles in a decentralisating state. In: Bressers, H. and Kuks, S. (eds). Integrated governance and water basin management: Conditions for regime change and sustainability. Dordrecht: Kluwer Academic Publishers, 189-212.

Dimakis, N. (1999). Concerns of farmers in the Sub-Prefecture of Langadas. Proceedings of Workshop on Agriculture and the Environment in the Sub-Prefecture of Langadas, Apollonia, May 1999. Apollonia: Centre for Information on Lakes Koronia-Volvi, 10-11 (in Greek).

Dinica, V. and Bressers, H. (2004). Partnerships in implementing sustainability policies – theoretical considerations and experiences from Spain. 12th International Conference of Greening of Inndustry Network, 7-10 November 2004, Hong Kong.

Dombrowsky, I. (2005). Institutionalisation in the management of international waters: The role of problem structure. 9th Annual Conference of the International Society for New Institutional Economics (ISNIE), September 22-24, 2005, Barcelona.

Dosi, C. and Tonin, S. (2001). Freshwater availability and groundwater use in southern Europe. In: Dosi, C. (ed). Agricultural use of Groundwater. Dordrecht: Kluwer, 15-34.

Dziedzicki, J-M. and Larrue, C. (2004). An innovative but incompleted integration process. In: Bressers, H. and Kuks, S. (eds). Integrated governance and water basin management: Conditions for regime change and sustainability. Dordrecht: Kluwer Academic Publishers, 131-164.

EC (European Commission) (2002). Letter of formal notice, Infringement nr. 2002/2289, C(2002) 5211, Brussels, 17/12/2002.

EC (European Commission) (2005). Greece: Commission pursues legal action in three cases for breach of EU environmental law. Press Release IP/05/43, Brussels, 14/1/2005.

EC (European Commission) (2005a). Water policy: Commission takes legal action against Italy, Spain and Greece over key directive. Press release IP/05/1302, Brussels, 18/10/2005.

EC (European Commission) (2005b). Environment: Commission to pursue legal action against Greece over infringements. Press release IP/05/1644, Brussels, 20/12/2005.

Ecological Movement of Thessaloniki (2002). Who is to blame for the disappearance of Koronia? Accessed online on 17/5/2005: http://www.ecogreens.gr/0586KorwneiaPoios.html (in Greek).

EEB (European Environmental Bureau) (2004). EU Water Policy: Making the Water Framework Directive work. The quality of national transposition and implementation – A snapshot. Brussels: EEB.

EKBY (Greek Biotope/Wetland Centre), WWF-Greece and HOS (Hellenic Ornithological Society) (2002). The State of the Ramsar Sites of Greece in July 2002: An overview. October 2002.

EKTHE (National Centre for Marine Studies) (2001). Morphological and Geophysical Survey of the bottom of Lake Vegoritida. Florina: Self-Administration of Florina (in Greek).

ETHIAGE (National Institute for Agricultural Research) (2004). Technical report on Possibilities to change the management of soil-plant-water in the subbasin of Lake Koronia with the aim to protect its water potential (Annex to the Revised Restoration Plan of Lake Koronia). Thessaloniki: ETHIAGE (in Greek).

Fotiou, S. (2001). Nature conservation and conservation development: A case study on small islands in the Aegean archipelagos in Greece, Conference on sustainable development and management of ecotourism in small island developing states and other small islands, Mahe, Seychelles, 8-10 December 2001.

Giotakis, K. I. (2003). Sixteen years later, a few days beforehand. Ydrooikonomia, 15, 12-15 (in Greek).

Goria, A. and Lugaresi, N. (2004). The evolution of the water regime in Italy. In: Kissling-Näf, I. and Kuks, S. (eds). The evolution of national water regimes in Europe. Dordrecht: Kluwer Academic Publishers, 265-292.

Gorny, T. (2001). We have water but we spend it. Internet edition of newspaper "Kyriakatiki", 20/5/2001 (in Greek).

Goulios, G. (1990). The slow but certain death of Vegoritida. Oikonomikos Tachidromos, 1/2/1990, 93-94 (in Greek).

Grammatikopoulou, N., Kehayas, D. and G. Economidis (1996). Rescuing Plan for Lake Koronia. Environmental report. Prefecture of Thessaloniki/Sub-Prefecture Langadas. Thessaloniki: Ministry of Environment, Physical Planning and Public Works (in Greek).

Gravanis, Th., Papahatzis, A. A. and Nasiaras, H. K. (1997). Plan for the environmental management of the Vegoritida biotope. Larisa: Bioefarmoges E.P.E. (in Greek).

Greek Committee for Hydrogeology (2003). Proposals on the draft bill for Water Resources. Accessed online on 10/11/2003: http://www.iah-hellas.geol.uoa.gr/ (in Greek).

Haas, P. M. (1992). Introduction: Epistemic communities and International Policy Coordination. International Organization, 46 (1), 1-35.

Haintarlis, M. (2003). Directive 2000/60 and its correct transposition in Greek legal and administrative reality. Proceedings 3rd Congress for the Development of Thessaly, Larissa, 12-13 December 2003 (in Greek).

Hartje, V. (2002). International dimensions of integrated water management. In: Baz, I. Al and Scheumann, W. (eds). Co-operation on transboundary rivers. Baden-Baden: Nomos Verlagsgesellschaft, 7-34.

HOS (Hellenic Ornithological Society) (2003). Nature and farmers claim Lake Koronia. Accessed online: http://www.interpet.gr/nea/recent_2003/cover_060703.htm.

HOS (Hellenic Ornithological Society) (2004). First assessment and comments on the revised plan for the restoration of Lake Koronia of 16/7/2004. Thessaloniki.

ICWE (International Conference on Water and the Environment) (1992). The Dublin statement and report of the conference. International Conference on Water and the Environment, 26-31 January 1992.

IGEKE & D.A.P./AUTH (Institute for agrieconomical and social studies of the National Agricultural Research Institute and Department of Agricultural Production/Aristotle University of Thessaloniki) (1999). Study on the restructuring of water-demanding crops. Assigned by the Prefectural Self-Administration of Florina. Thessaloniki (in Greek).

Iosifidis, V. and Avgitidis, V. (2000). Management plan of wastewater and effluents of the wider basin of Vegoritida of the Prefecture of Florina. Florina: Prefectural Self-Administration of Florina (in Greek).

Ingram, H. M., Mann, D. E., Weatherford, G.D. and Cortner, H.J (1984). Guidelines for improved institutional analysis in water resources planning. Water Resources Research, 20 (3), 323-334.

Iyer, R.R. (2004). A critical view on integrated water resources management: Definition, implementation and linkage to policy reviewed. Stockholm Water Front, 4/2004, 10-11.

Jaspers F. G. W. (2003). Institutional arrangements for integrated river basin management. Water Policy, 5, 77-90.

Jänicke, M. and Weidner, H. (eds) (1997). National environmental policies: A comparative study of capacity-buidling. Berlin-Heidelberg: Springer Verlag.

Kallis, G. (2003). Institutions and Sustainable Urban Water Management: the case of Athens. Ph.D Thesis. Mytilini: University of the Aegean, Department of Environmental Studies (in Greek).

Kapsi, M.C. (2000). Recent administrative reforms in Greece: Attempts towards decentralisation, democratic consolidation and efficiency. Policy Review Paper, John F. Kennedy School of Government.

Karageorgou, V. (2003). The WFD: An important stop for European environmental law. In: CD-Rom "Hellenic-Bulgarian Cooperation on the Protection and Use of Transboundary Watercourses. The EC Water Framework Directive 2000/60". Athens: Nomos + Physis (digital publications).

Karakostas, I. K. (2000). Environment and Law. Athens: Sakkoula (in Greek).

Kazantzidis, S., Anagnostopoulou, M. and Gerakis, P.A. (1995). Problems of 35 Greek biotopes and activities for their solution. Wetlands Monitoring Programme 1992-94, Volume 2, Region of Central Macedonia. Thessaloniki: EKBY (Greek Biotopes/Wetlands Centre) (in Greek).

Kissling-Näf, I. and Kuks, S. (eds) (2004a). The evolution of national water regimes in Europe. Dordrecht: Kluwer Academic Publishers.

Kissling-Näf, I. and Kuks, S. (2004b). Introduction to institutional resource regimes: Comparative framework and theoretical background. In: Kissling-Näf, I. and Kuks, S. (eds). The evolution of national water regimes in Europe. Dordrecht: Kluwer Academic Publishers, 1-24.

Knill, C. and Lenschow, A. (2000). On deficient implementation and deficient theories: the need for an institutionalist perspective in implementation research. In: Knill C. and Lenschow, A. (eds). Implementing EU environmental policy: New directions and old problems. Manchester: Manchester University Press, 9-35.

Knight Piesold, G., Karavokyris & partners, Anelixi and Agrisystems (1998). Environmental Restoration of Lake Koronia. Final report. August 1998. European Commission, DG-XVI, Cohesion Fund (Greek translation of the English original).

Knoepfel, P., Kissling-Näf, I. and Varone, F. (2001). Institutionelle Regime für natürliche Resourcen. Boden, Wasser und Wald im Vergleich. Basel: Helbing & Lichtenhahn.

Knoepfel, P., Kissling-Näf, I. and Varone, F. (2003). Institutionelle Regime natürlicher Resourcen in Aktion. Basel: Helbing & Lichtenhahn.

Koklis, S. (2003). Management of water resources in the basin of Lake Vegoritida. Xanthi: University of Democritos of Thrace (in Greek).

Kooiman, J. (2002). Governance: A social-political perspective. In: Grote, R.J. and Gbikpi, B. (eds). Participatory Governance. Opladen: Leske und Budrich, 71-96.

Kotleas, S. (1988). Pollution of Lake Koronia: Research around the legal and administrative framework of environmental protection. Proceedings of 1st Conference on the Development of the Region of Central Macedonia, Thessaloniki, 26-27/4/1988, 235-244 (in Greek).

Kousis, M. (1994). Environment and the State in the EU Periphery: The case of Greece. In: Baker, S., Yearly, S. and Milton K. (eds). Protecting the Periphery: Environmental policy in the peripheral regions of the European Union. London: Frank Cass, 118-135.

Koutalakis, Ch. (2006). The single decision-making trap: Explaining Greece's non-compliance with EU environmental law. Accessed online on 24/5/2006: http://www.nomosphysis.org.gr.

Koutrakis, E.T. and Blionis, G.I. (1995). Raising public awareness and information with regard to Lakes Koronia and Volvi (Macedonia, Greece). MedWet, Greek Ministry of Environment, Spatial Planning and Public Works, Greek Biotope/Wetland Centre (in Greek and English).

Kouvelis, S., Bekiaris, Y. and Nantsou, Th. (1995). Water management in Greece. Athens: WWF Greece.

Kuks, S. (2004a). Comparative review and analysis of regime changes in Europe. In: Kissling-Näf, I. and Kuks, S. (eds). The evolution of national water regimes in Europe. Heidelberg: Springer, 329-368.

Kuks, S. (2004b). The evolution of the water regime in the Netherlands. In: Kissling-Näf, I. and Kuks, S. (eds). The evolution of national water regimes in Europe. Heidelberg: Springer, 87-142.

Kuks, S. (2004c). Water governance and institutional change. PhD dissertation. University of Twente: CSTM.

Lenton, R. (2004). A critical view on integrated water resources management: Definition, implementation and linkage to policy reviewed. Stockholm Water Front, 4/2004, 10-11.

Libecap, G. D. (1993). Contracting for Property rights. Cambridge: Cambridge University Press.

Liberopoulou, C. (1994). Environmental study of the Lake Agios Vasilios. Thessaloniki: AUTH, Department of Environmental Chemistry (in Greek).

LVPS (Lake Vegoritida Preservation Society) (2002). Letter of LVPS of 29/8/2002 to the European Commission, Directorate General for the Environment. ENV(2002)A/807452/13.9.2002). Subject: The continuing environmental destruction of the hydrological ecosystem hosted by Lake Vegoritida and the Soulou stream due to the Hellenic Republic's negligence and non-compliance to the confounding verdicts related to the legal cases C232/95 & C233/95. Arnissa.

Maestu, J., Tabara, D., and Sauri, D. (2003). Public participation in river basin management in Spain: Reflecting changes in external and self-reflected context. Accessed online on 15.11.2004: www.harmonicop.info.

Manheim, J. B. and Rich, R. C. (1995). Empirical political analysis: Research methods in political science. New York: Longman.

Mantziou, D. (2003). Implementation of the Water Framework Directive: A practical guide for Organisations of Local Administration. WWF and Central Union of Municipalities and Communities of Greece (in Greek).

Margat, J. (1999). Utilisations et Abuses des Eaux Souterraines Dans les Pays Mediterraneens. Paper presented at the Cursos de Verano de la Universidad Complutense, Fundacion Marcelino Botin, Madrid, 2-6 August 1999.

Mauch, C. and Knoepfel, P. (2004). Rivalry based communities in Europe's water tower. In: Bressers, H. and Kuks, S. (eds). Integrated governance and water basin management: Conditions for regime change and sustainability. Dordrecht: Kluwer Academic Publishers, 213-246.

Massaruto, A., de Carli, A., Longhi, C. and Scarpari, M. (2003). Public participation in river basin management in Italy: An unconventional marriage of top-down planning and corporative politics. Accessed online on 15.11.2004: www.harmonicop.info.

McNabb, D. E. (2002). Research methods in public administration and nonprofit management: quantitative and qualitative approaches. Armonk, New York: M. E. Sharp.

MDEV (Ministry of Development) (1996). Management plan for water resources of the country (Greece). Athens: Ministry of Development (in Greek).

MDEV (Ministry of Development) (2003). Plan for a Management Programme of the Water Resources of the Country. Carried out by Ministry of Development, National Technical University, Institute for Geology and Mineral Exploitation, Centre of Planning and Economic Research. Athens (in Greek).

MEPPW (Ministry of Environment, Physical Planning and Public Works) (2006). WFD Article 3 report-Greek maps. Accessed online on 14.12.06: http://www.minenv.gr/nera/.

Ministry of Industry, Energy and Technology (1988). The Law 1739/87 for the Management of Water Resources: Which gaps it fills, which problems it can resolve, how and who implements it. Athens (in Greek).

Mitchell, B. (1990). Integrated Water Management. In: Mitchell B. (ed). Integrated Water Management, International Experiences and Perspectives. London: Belhaven Press, 1-21.

Mitraki, C., T. L. Crisman and Zalidis, G. (2004). Lake Koronia: shift from autotrophy to heterotrophy with cultural eutrophication and progressive water-level reduction. Limnologica, 34, 110-116.

Mody, J. (2004). Achieving accountability through decentralisation: Lessons for integrated river basin management. Policy Research Working Paper 3346. Washington DC: World Bank.

Molle, F. (2004). Defining water rights: by prescription or negotiation? Water Policy, 6, 207-227.

Moss, T. (2003). Induzierter Institutionenwandel 'von oben' und die Anpassungsfähigkeit regionaler Institutionen: Zur Umsetzung der EU-Wasserrahmenrichtlinie in Deutschland. In: Moss, T. (Hg). Das Flussgebiet als Handlungsraum: Institutionenwandel durch die EU-Wasserrahmenrichtlinie aus raumwissenschaftlichen Perspektiven. Münster: LIT, 129-175.

Municipality of Amyndeo (2004). Lakes Vegoritida and Petron. Broschure, pp.36 (in Greek).

NCESD (National Center for the Environment & Sustainable Development) (2002). A plan for the development of a River Basin Management Scheme in Greece. NCESD Research Paper No 17. Athens.

North, D. C. (1990). Institutions, institutional change and economic performance. New York: Cambridge University Press.

OECD (1983). Environmental Policies in Greece. Paris: OECD.

Ostrom, E. (1986). An agenda for the study of institutions. Public Choice, 48, 3-52.

Ostrom, E. (1990). Governing the Commons: The evolution of institutions for collective action. Cambridge: University Press.

Ostrom, E. (1999). Institutional rational choice: An assessment of the institutional analysis and development framework. In: Sabatier, P. A. (ed). Theories of the Policy Process. Boulder: Westview, 35-72.

O'Sullivan, E. and Rassel, G. R. (1999). Research methods for public administrators. New York: Longman.

Papaioannou, M. (1998). Legal Framework for the management of water resources. Environment and Law 1998, 41-54 (in Greek).

Papakonstantinou, A. and Katirtzoglou, K. (1995). Proposals for the restoration of the hydrodynamic in Lake Koronia, Greece. Thessaloniki: National Institute of Geology and Mineral Exploitation (in Greek).

Papalimnaiou, F. (1994). The legal regime for the management and protection of water resources. Nomiko Vima, 1994, 898-911 (in Greek).

Papanidis, M. (2002). Battles for control over water. Ydrooikonomia, 1, 12-13 (in Greek).

Paraschoudis, V., Georgakopoulos, Th. and Stavropoulos, X. (2001). Hydrogeological study of the wider basin of Vegoritida: Hydrogeological report. Florina, Athens: Prefectural Self-Administration of Florina (in Greek).

Planet Northern Greece S. A. – Planet S. A. (1998). Deliverable 1: The basic outline of the programme. Programme Management: Stabilisation of water levels and purification of Lake Vegoritida. Prefectural Self-Administration of Florina (in Greek).

Planet Northern Greece S. A. – Planet S. A. (2001). Summary of completed projects. Programme: Stabilisation of water levels and purification of Lake Vegoritida. Prefectural Self-Administration of Florina (in Greek).

Planet Northern Greece S. A. and AN.FLO. S. A. (2000). Study on the intention of setting up a management body for water resources management of the basin of Vegoritida – Management Plan. Phase B. Thessaloniki, Florina (in Greek).

PPC Directorate for Studies and Constructions for Hydropower Generation (1979). Study on the water balance of Lake Vegoritida and the wider area of Ptolemaida. Public Power Corporation, Directorate for Studies and Constructions for Hydropower Generation, Department of Screening Studies for Hydropower Generation and the Environment, Centre for the Selection of Environmental Data (in Greek).

PPC General Mines Directorate (1994a). Environmental Impact Assessment of Amyndeo Mine, Lignite Centre Ptolemaida-Amyndeo, 1994 (in Greek).

PPC General Mines Directorate (1994b). Environmental Impact Assessment of Ptolemaida Mines, Lignite Centre Ptolemaida-Amyndeo (in Greek).

PPC General Mines Directorate (1998). Complementary Report to the Environmental Impact Assessment of Amyndeo Mine, Lignite Centre Ptolemaida-Amyndeo (in Greek).

PPC General Mines Directorate (2005). Environmental Impact Assessment of Mavropigi Mine, Northwestern Field (in Greek).

PPC S. A. (2004). Environmental Report 2004. Accessed online on 01.09.2005: www.dei.gr (in Greek).

PPC S. A. (2005). Annual Report 2004. Accessed online on 08.07.2005: www.dei.gr.

Prefecture of Thessaloniki (2000). Plan for the treatment and management of wastewater of settlements in the Prefecture of Thessaloniki. Thessaloniki: Prefecture of Thessaloniki (in Greek).

Pridham, G. (1994). National environmental policy-making in the European framework: Spain, Greece and Italy in comparison. Regional Politics and Policy, 4(1), 80-101.

Pridham, G. and Konstandakopoulos, D. (1997). Sustainable development in Mediterranean Europe? In: Baker S. et al (eds). The politics of sustainable development: Theory, policy and practice in the European Union, 127-151.

Pridham, G., Verney, S. and Kostandakopoulos, D. (1995). Environmental Policy in Greece: Evolution, Structures and Process. Environmental Politics, 4(2), 244-270.

Reading, H. F. (1977). A dictionary of the social sciences. London: Routledge & Kegan Paul.

Rubin, H. J., and Rubin, I. S. (1995). Qualitative interviewing: The art of hearing data. Thousand Oaks, CA: Sage.

Sabatier, P. A. and Jenkins-Smith, H. C. (1999). The advocacy coalition framework: An assessment. In: Sabatier, P. A. (ed). Theories of the Policy Process. Boulder: Westview, 117-168.

Scharpf, F. W. (1997). Games real actors play: Actor-Centered Institutionalism in Policy Research. Boulder: Westview Press.

Schlager, E. and Ostrom, E. (1992). Property- rights regimes and natural resources: A conceptual analysis. Land Economics, 68(30), 249-262.

Shah, T., Makin, I. and Sakthivadivel, R. (2001). Limits to Leapfrogging: Issues in transposing successful river basin management institutions in the developing world. In: Abernethy, C. L. (ed). Intersectoral management of river basins. Proceedings of an international workshop on integrated water management in water-stressed river basins in developing countries: Strategies for poverty alleviation and agricultural growth. Loskop Dam, S.Africa, 16-21 October 2000. Colombo, Sri Lanka: International Water Management Institute (IWMI) and German Foundation for International Development (DSE), 89-114.

Sioutis, D. (1989). Hydrology of Lake Vegoritida. The problem of the level drop – Environmental impacts. Xanthi: University of Democritos of Thrace (in Greek).

Skourtos, M., Kontogianni, A., Lanfgord, I. H., Bateman, I. J. and Georgiou, S. (2000). Integrating stakeholder analysis in non-market valuation of environmental assets. CSERGE Working Paper GEC 2000-22.

Sotiropoulou, E., Hatziavgoustidou, E., Naxaki, X,. Kiki, S., Piperopoulou, E., Mina, I. and Savvidou S. (1992). Water pollution and management of effluents in our region (Ptolemaida). Unpublished manuscript, pp. 43 (in Greek).

Spanou, K. (1995). Public administration and the environment: The Greek experience. In: Skourtos, M. S. and Sofoulis, K. M. (eds). Environmental policy in Greece: Analysis of the environmental problem from the point of view of social sciences. Typothito: Athens, 115-175 (in Greek).

Spanou, K. (1998). Greece: Administrative symbols and policy realities. In: Hanf, K. and A. I. Jansen (eds). Governance and Environment in Western Europe. Politics, Policy and Administration. Singapore: Longman, 110-130.

Svendsen, M. (2001). Basin management in a mature closed basin: The case of California's Central Valley. In: Abernethy, C. L. (ed). Intersectoral management of river basins. Proceedings of an international workshop on integrated water management in water-stressed river basins in developing countries: Strategies for poverty alleviation and agricultural growth. Loskop Dam, S.Africa, 16-21 October 2000. Colombo, Sri Lanka: International Water Management Institute (IWMI) and German Foundation for International Development (DSE), 297-316.

Teclaff, L. A. (1987). International control of cross-media pollution – an ecosystem approach. In: Uton, A.E. & Teclaff, L. A. (eds). Transboundary resources law. Boulder/London: Westview Press.

Terzis, C. (1997). Expertise concerning the introduction of integrated water management structures in Greece. Final report financed by the EC DG XVI B.2 through the Technical Assistance Management Company. Contract number 6807/1-3-96. Athens.

Tremopoulos, M. (2004). On the Restoration Plan of Lake Koronia. Published 18/8/2004. Accessed online on 2/11/2005: http://www.e-ecology.gr/Discussions.asp?forum_id=6& (in Greek).

Tsagarlis, G. (1998). Liquid waste in the Prefecture of Thessaloniki. Thessaloniki: Prefecture of Thessaloniki, Directorate for Environmental Protection (in Greek).

Tsougrakis, Y. (2000). Lakes Volvi and Koronia. In: Maragou P. and Mantziou, D. (eds). Assessment of Greek Ramsar Wetlands. Athens: WWF Greece.

UMEWS (Union of Municipal Enterprises for Water Supply and Sewerage) (2003). Protection and management of water- Harmonisation with Directive 2000/60 of the European Union: Position on the draft water law. Ydrooikonomia, 10, 28-32 (in Greek).

Usman, R. (2001). Integrated water resources management: Lessons from the Brantas River Basin in Indonesia. In: Abernethy, C. L. (ed). Intersectoral management of river basins. Proceedings of an international workshop on integrated water management in water-stressed river basins in developing countries: Strategies for poverty alleviation and agricultural growth. Loskop Dam, S.Africa, 16-21 October 2000. Colombo, Sri Lanka: International Water Management Institute (IWMI) and German Foundation for International Development (DSE), 273-296.

Van der Brugge, R. and Rotmans, J. (2007). Towards transition management of European water resources. Water Resource Management, 21, 249-267.

Van Hofwegen, P. (2001). Framework for assessment of institutional frameworks for IWRM. In: Abernethy, C. L. (ed). Intersectoral management of river basins. Proceedings of an international workshop on integrated water management in water-stressed river basins in developing countries: Strategies for poverty alleviation and agricultural growth. Loskop Dam. S.Africa, 16-21 October 2000. Colombo, Sri Lanka: International Water Management Institute (IWMI) and German Foundation for International Development (DSE).

Varvaressou, Y. (2004). Omissions in new state water policy - Greece has some way to go to meet EU directive deadline. English internet edition of newspaper "Kathimerini", 10.12.2004.

Verani N. and Katirtzoglou, K. (2001). Exploration of possibilities to exploit the deep aquifer of the subbasin of Lake Koronia. For the Prefecture of Thessaloniki. Thessaloniki: National Institute of Geology and Mineral Exploitation (in Greek).

Verani N. and Katirtzoglou, K. (2003). Hydrological balance of the flatland of the subbasin of Lake Koronia. For the Prefecture of Thessaloniki. Thessaloniki: National Institute of Geology and Mineral Exploitation (in Greek).

Weale, A., Pridham, G., Cini, M., Konstadakopulos, D., Porter, M. and Flynn, B. (2000). Environmental Governance in Europe. An Ever Closer Ecological Union? Oxford: Oxford University Press.

Wester, P., Burton, M. and Mestre-Rodriguez, E. (2001). Managing the water transition in the Lerma-Chapala Basin, Mexico. In: Abernethy, C. L. (ed). Intersectoral management of river basins. Proceedings of an international workshop on integrated water management in water-stressed river basins in developing countries: Strategies for poverty alleviation and agricultural growth. Loskop Dam, S.Africa, 16-21 October 2000. Colombo, Sri Lanka: International Water Management Institute (IWMI) and German Foundation for International Development (DSE), 161-181.

World Bank (1993). Water resources management: A World Bank policy paper. Washington DC: World Bank.

WWF (2003), WWF Water and Wetland Index – Critical issues in water policy across Europe, November 2003.

Ydromeletitiki E. E., Montgomery Watson Makte E.P.E. and Speed A. E. Karamitopoulou D. (2002). Study on sewerage, treatment and disposal of effluents of industries and animal farms of the area of Lake Koronia and of wastewater of the Municipality of Langadas. Thessaloniki: Prefectural Self-administration of Thessaloniki, Division of Technical Services (in Greek).

Yiannopoulos, A. N. (1988). Property. In: Kerameus, K. D. & Kozyris, P.J. (eds). Introduction to Greek Law. Deventer, Athens: Kluwer Law and Taxation Publishers, Sakkoulas, 101-114.

Yin, R. K. (1994). Case study research – Design and Method. Applied Social Research Methods Series, Vol. 5. Thousand Oaks: Sage Publications.

Young, O. R. (1982). Resource regimes: Natural resoures and institutions. Berkeley: University of California Press.

Zahariadis, N. (1999). Ambiguity, time and multiple streams. In: Sabatier, P. A. (ed). Theories of the Policy Process. Boulder: Westview, 73-96.

Zalidis G. C., Takavakoglou, V. and Alexandridis, T. (2004a). Revised Restoration Plan of Lake Koronia. Thessaloniki: Aristotle University of Thessaloniki, Department of Agronomy, Laboratory of Applied Soil Science (in Greek with English summary).

Zalidis, G. C., Tsiafouli, M. A., Takavakoglou, V., Bilas, G. and Misopolinos, N. (2004b). Selecting agri-environmental indicators to facilitate monitoring and assessment of EU agri-environmental measures effectiveness. Journal of Environmental Management, 70, 315-321.

Zalidis, G., Michalopoulou, H. and Tsagarlis, G. (2006). First Assessment Report of the implementation of the terms of protection of the National Park of lakes Volvi, Koronia and the Macedonian Tembi (September 2003 – February 2006). Thessaloniki: Ministry of Environment and Management Body of lakes Koronia-Volvi (in Greek).

Zisi, R. (2002). Immediate priority to works for the collection and treatment of effluents. Ydrooikonomia, 1, 14-17 & 77 (in Greek).

Annex I: Assumptions on the outcome of actor interaction in contextual interaction theory

Assumptions on the likelihood of application of a policy instrument:

Mi	Mt	I+	Pi	Sit.	Outcome	Process
+	+/0	+		1	++	Cooperation (O++ → active)
	-			2	--	Learning towards 1
	-	+	+	3	++	Cooperation (forced)
			0	4	+/-	Opposition
			-	5	--	Obstruction
		-		6	--	None / Learning → 3
0	+	+		7	++	Cooperation
	-			8	--	Learning towards 7
	0/-			9	--	None
-	+	+	+	10	--	Obstruction
			0	11	+/-	Opposition
			-	12	++	Cooperation (forced)
		-		13	--	None / Learning → 12
	0/-			14	--	None

Legend:
Mi = Motivation implementers viz. application, Mt = Motivation target group viz. application, I+= Information for application of positive partner(s) (highest level), Pi = Balance of power viewed from position implementer

Assumptions on the likelihood of "adequate application" of a policy instrument:

Mi	Mt	I+/0	Pi	Sit.	Outcome	Process
+	+/0	+		1	++	Constructive cooperation (active cooperation)
	-			2	--?	Learning towards 1
	-	+	+	3	++	Constructive cooperation (forced cooperation)
			0	4	+/++	Negotiation / Conflict
			-	5	+/-	Negotiation
		-		6	--?	Symbolic / Learning → 3/4
0	+	+		7	++	Constructive cooperation
	-			8	--?	Symbolic / Learning → 7
	0			9	--	Symbolic
	-			10	--	Obstructive cooperation
-	+	+	+	11	+/-	Negotiation
			0	12	+/++	Negotiation / Conflict
			-	13	++	Constructive cooperation (forced cooperation)
		-		14	--?	Symbolic / Learning → 12/13
	0/-			15	--	Active obstructive cooperation process

Legend:
Mi = Motivation implementers viz. adequate application, Mt = Motivation target group viz. adequate application, I+= Information for adequate application of positive or neutral partner(s), Pi = Balance of power viewed from position implementer

Source of above figures: Bressers (2004).

Annex II: Personal communications with national water actors

1. Ms. Siatou, Head of the Directorate for Water Potential and Natural Resources at the Ministry of Development (phone, October 2003 and March 2004)

2. Ms. Ghini, Directorate for Water Potential and Natural Resources at the Ministry of Development (September 2003)

3. Ms. Ghioni, Directorate for Water Potential and Natural Resources at the Ministry of Development (September 2003)

4. Ms. Mavrodimou, formerly at the Directorate for Water Potential and Natural Resources at the Ministry of Development, currently at the National Technical University in Athens (March 2004)

5. Ms. Lazarou, Ministry of Environment (September 2003)

6. Mr. Tasoglou, Ministry of Environment (September 2003, November 2005)

7. Ms. Fotopoulou, Ministry of the Interior, Public Administration and Decentralisation (September 2003)

8. Mr. Kouvopoulos, Public Power Corporation, email communication (April 2004)

Annex III: Case study interview partners and personal communications

Interview partners and personal communications in the Vegoritida basin:

Time-period in the field: January-March 2005 and November-December 2005

	Organisation / Position	Name(s)	Date, Location
1.	RWM (Region of West Macedonia), Managing Authority of the Regional Operation Programme / Manager of the Auditing Unit	Mr. Papadopoulos	Kozani
2.	RWM (Region of West Macedonia), Division for Planning and Development, Water Resources Management Directorate	Mrs. Tsagaridou	Kozani
3.	RWM (Region of West Macedonia), Division for the Environment and Spatial Planning	Mr. Gasis Mr. Dontsios	Kozani
4.	RWM (Region of West Macedonia) / Former councilor at the Managing Authority	Mr. Mouratidis	Kozani
5.	Florina Prefecture, Division for Land Reclamation / Director	Mr. Athanasiadis	Florina
6.	Florina Prefecture, Division for Land Reclamation / Geologist	Mr. Nikou	Florina (and follow-up written communication)
7.	Florina Prefecture, Division of Health	Mr. Mavropoulos	Florina (Personal communication)
8.	State Chemical Laboratory, Department of Florina	Mr. Koutsoumpidis	Florina
9.	Pella Prefecture / Special councilor	Mr. Argyrakis	Edessa
10	Pella Prefecture, Division for the Management and Protection of Natural Resources / Geologist	Mr. Papadopoulos	Edessa
11	Kozani Prefecture, Division of Health	Mr. Maggos	Kozani (Personal communication)
12	Kozani Prefecture, Division of Agriculture	Mr. Diamantopoulos	Kozani (Personal communication)
13	Municipality Amyndeo / Mayor	Mr. Liasis	Amyndeo
14	Community Agios Panteleimon / Former Secretary General	Mr. Nikoloudakis	Agios Panteleimon
15	Consultant Planet Northern Greece S.A. / Director	Mr. Hatzidiamantis	Thessaloniki
16	PPC (Public Power Corporation)		
	Mining Directorate	Mr. Zarafidis	Phone communication
		Mr. Filippos	Phone communication
		Mr. Dimitrakopoulos	Phone communication
	Amyndeo steam-electricity plant / Director	Mr. Lovatsis	Phone communication
		Mr. Leris	Phone communication
	Generation Directorate, Hydropower / Director		
	Agras hydropower plant / Director	Mr. Tsakiroglou	Edessa (Personal communication)

17	LVPS (Lake Vegoritida Preservation Society) / member of the Administration Board Engineer for wells	Mr. Tortokas	Filotas
18	LVPS (Lake Vegoritida Preservation Society) / member of the Administration Board Former President of the community of Arnissa	Mr. Patetsinis	Edessa
19	Hellenic Ornithological Society	Mr. Bousbouras	Thessaloniki (and follow-up phone communication)
20	Agricultural cooperative of Arnissa	Mr. Karavitis	Arnissa

Interview partners and personal communications in the Mygdonian basin:

Time-period in the field: March-April 2005; July 2006

	Organisation / Position	Name(s)	Location
1.	Environment Ministry, Department for the Management of the Natural Environment	Ms. Marmara	Phone communication
2.	RCM (Region of Central Macedonia), Division for Environment and Spatial Planning Management Body of the National Wetland Park / Member of the Administrative Board	Ms. Michalopoulou	Thessaloniki (Personal communication)
3.	RCM (Region of Central Macedonia), Division for Planning and Development, Water Resources Management Directorate	Ms. Grammatikopoulou	Thessaloniki (Personal communication)
4.	Prefecture of Thessaloniki, Office of the Prefect / Special Advisor of the Prefect	Mr. Karachalios	Thessaloniki
5.	Prefecture of Thessaloniki, Division of Environmental Protection	Mr. Kalaboukas	Thessaloniki
6.	Prefecture of Thessaloniki, Division of Environmental Protection	Mr. Kotleas, Mr. Chrysohoou	Thessaloniki (Personal communications)
7.	Prefecture of Thessaloniki, Division for Water Resources and Land Reclamation / Geologist	Mr. Tsakoumis	Thessaloniki
8.	Langadas Sub-Prefecture, Department of Fisheries / Ichthyologist	Mr. Oikonomidis	Langadas
9.	Young Farmers Union of the Prefecture of Thessaloniki / Secretary Municipality of Langadas / employee	Mr. Kefalas	Langadas (and follow-up phone communication)
10	Hellenic Ornithological Society Management Body of the National Wetland Park / Member of the Administrative Board	Mr. Tsougrakis	Thessaloniki (and follow-up phone and written communication)
11	Former expert at the Information Centre of Lake Koronia and Volvi Biologist with NGO involvement	Mr. Blionis	Thessaloniki
12	University of Thessaloniki / Professor Head scientific expert of the revised Master Plan Management Body of the National	Mr. Zalidis	Thessaloniki

	Wetland Park / Member of the Administrative Board		
13	Local Union for the Environmental Protection of the Wider Region of Langadas / President	Mr. Hatziparadissis	Thessaloniki
14	Consultant to the Master Plan / Forester and environmental scientist	Mr. Vitoris	Thessaloniki
15	Consultant to the Master Plan / Hydraulic scientist	Mr. Katsonis	Thessaloniki

Annex IV: Documentary sources of case study data

Documentary sources of case study data in the Vegoritida basin:

Source	Description of data and information sources
Official documents	
Official Government Journal	Law 1468/1950/Official Government Journal A 169 on the "Establishment of the Public Power Corporation"
	Law 2773/ Official Government Journal A'286/22/12/99 on the "Liberalisation of the Electricity Market"
	Law 210/1973/ Official Government Journal A 277 on the "Mining Code"
	Presidential Decree 412/89/Official Government Journal A 175 on the "Establishment of Water Resource Management Directorates for West Macedonia, East Macedonia and Thrace"
	Common Ministerial Decision 15782/1849 on a "Special pollution reduction programme of Lake Vegoritida and River Soulou from hazardous substances", Official Government Journal B 797, 25 June 2001
	MEPPW Decision 26295, Official Government Journal B 1472/09.10.2003, on "Approval of a Regional Framework for Spatial Planning and Sustainable Development of the Region West Macedonia"
	Common Ministerial Decision 664/219845 on "Protection of the lakes of West Macedonia (Lakes Vegoritida, Petron, Chimaditida and Zazari). Agrienvironmental Measure 3.16 of the Agricultural Development Programme (EPAA) 2000-2006", Official Government Journal B' 1136, 22.08.2006
MDEV	Decision of the Minister of Development D11/F16.3/3691/153 on the "Establishment of a Regional Water Committee for the water district of West Macedonia (09)", 3 April 2000
Prefectures	Interprefectural decisions on the disposal of effluents and wastewater: decisions 2667/76, 1900/79 and 10032/87
	Decisions of the Prefecture of Florina on the disposal of effluents and wastewater: 292/1987, 555/90
	Various restrictive measures on water use in the Prefectures of Florina, Kozani and Pella
Archives of the Greek Parliament	Records of questions in Parliament on the problems of the Vegoritida basin and ensuing correspondence between MPs, Ministries and the PPC
MINAGR	Proposed policy for an agrienvironmental measure for the Vegoritida basin
RWM	RWMD policy proposals to Prefectures for restrictive measures
	Comments of RDESP on the EIAs of the PPC mines
EU	ECJ decision on case of Vegoritida and River Soulou
	Correspondence between EU and basin actors on the ECJ case
Policy-relevant meetings	Minutes of meetings of the Supervisory Committee of the 2001 CMD on a pollution reduction programme for Lake Vegoritida and River Soulou
	Conclusions' reports of the Interprefectural Committee on water pollution (1970s)
Books-publications-conference reports	
Local associations and environmental NGOs	Publications, broschures, declarations on the activities of local associations in the basin
	Journal publication "Oionos" of the HOS
Conferences and seminars	Scientific papers on water resources status and management in the Vegoritida basin, especially from numerous seminars of the Technical Chamber of Greece

Source	Description of data and information sources
	Research studies and reports
PPC	Annual PPC reports for shareholders
	PPC Environmental Report – 2004
	Environmental Impact Assessment reports of PPC steam-electricity plants and lignite mines in the basin
	Water balance study of the Vegoritida basin (1979)
Prefectures	Press releases
	Various technical reports on Lake Vegoritida
	Various technical reports drafted in the context of the Programme on the Stability and Purification of Lake Vegoritida
	Environmental impact assessment reports on industrial activities in the Vegoritida basin
	Maps of the basin
IGME	Research papers on the hydrogeology of the basin
EKBY	Specific Management Plan for lakes Chimaditida and Zazari
Universities	Dissertations on the Vegoritida basin
	Scientific annals of research departments including water use data and water quality data for the Vegoritida basin
	Primary data records
PPC	Primary data series on water levels and abstractions from Lake Vegoritida for the Agras hydropower plant
Prefectures	Primary data on permits for water use and water works
	Primary data on disposal permits for effluents and wastewater
	Primary data series on water levels of Lake Vegoritida
Universities	Primary data on water pollution levels
	Various archives
RWM	Minutes of meetings of 1994 Lake Committee, including proposals on water quantity, water quality and institutional aspects
Archives of local associations (Lake Vegoritida Preservation Society, MESNA)	Minutes of meetings of local union of municipalities and communities of Florina
	Minutes of meeting of community council of Arnissa
	Texts of denunciations to local courts and the EU
	Letters of correspondence between:
	Region ↔Prefectures over restrictive measures
	HOS ↔MEPPW over exposed land
	Local municipalities ↔PPC
	LVPS ↔authorities, MPs, Land Register
	Agricultural association Vegoritida ↔Ministries, PPC, Prefectures
	Action Committee for Salvation of Lake Vegoritida ↔Interprefectural Meeting
	Prefectures ↔Local communities
	Agricultural cooperative Arnissa ↔MINAGR
	Industry AEVAL ↔Action Committee
	PPC ↔PoP

Source	Description of data and information sources
	MESNA association ↔ministries, Prefectures, MPs, universities
	Press
Archives of regional and local newspapers	Articles on the Vegoritida basin in the following newspapers: "Politis Florinas" "Macedonia" "Proini of the Prefecture of Pella" "Edessaiki"
Archives of national newspapers	Articles on the Vegoritida basin in the following newspapers: "Eleftherotypia" "Ethnos tis Kyriakis" "I Proti" "Aggelioforos"

Documentary sources of case study data in the Mygdonian basin:

Source	Description of data and information sources
	Official documents
Official Government Journal	Common Ministerial Decision 35308/1838 on a "Special pollution reduction programme of Lake Koronia from hazardous substances", Official Government Journal B 1416, 12.10.2005
	Law 849 on "Incentives to support the regional and economic development of the country", Official Government Journal A 232, 22.12.1978
	Law 3044 on "Transfer of constructions coefficient and regulation of other issues in the competence of the MEPPW" (Article 13: Set up of 25 national protected areas in Greece), Official Government Journal A´197, 27.8.2002
	Common Ministerial Decision 6919 on "Characterisation of the wetland system of lakes Volvi-Koronia and the Macedonian Tembi as National Wetland Park of lakes Koronia, Volvi and the Macedonian Tembi and definition of zones of protection, uses, terms and building limitations", Official Government Journal D 248/5.3.2004
	Law 2742 on "Physical planning and sustainable development", Official Government Journal A'207, 7.10.1999
	Common Ministerial Decision 125192/365 on „Definition of number of members of the Administrative Council of the Management Body of lakes Koronia and Volvi", Official Government Journal B´ 126, 7.2.2003
	MEPPW Decision 50547 on „Approval of regulations for the operation of the Administrative Council of the Management Body of lakes Koronia and Volvi, Official Government Journal B´1876, 17.12.2004
	MEPPW Decision 50550 on „Approval of regulations for the execution of works and the assignment of tenders by the Management Body of lakes Koronia and Volvi, Official Government Journal B´1869, 16.12.2004
	Common Ministerial Decision 127795 on „Application of the programme of management of lakeshore areas of the Natura Network 2000 site of lakes Koronia-Volvi (A 1220001) of the Agrienvironmental Measure of the Agricultural Development Programme (EPAA) 2000-2006", Official Government Journal B´ 583, 13.05.2003
	Common Ministerial Decision 642/141608 on „Application of the Agrienvironmental Measure 3.10 on the management of the lakeshore areas of the Natura Network 2000 site of lakes Koronia-Volvi (A 1220001) of the Axis III of the Agricultural Development Programme (EPAA) 2000-2006", Official Government Journal B´ 1996, 30.12.2005
	Common Ministerial Decision 16175/824 on Action Programme for the plain of Thessaloniki-Pella-Imathia which is characterised as vulnerable zone from nitrate pollution of agricultural origin, Official Government Journal B´ 530,

Source	Description of data and information sources
	28.04.2006
Ramsar Bureau	Resolution VII.12 on sites in the Ramsar List, 7th meeting of the Conference of the Contracting Parties, 1999 San Jose Conference, Costa Rica
	Resolution VIII.10, 8th meeting of the Conference of the Contracting Parties, 2002 Valencia Conference, Spain
European Commission	Infringement procedure correspondence with the Greek State regarding Lake Koronia
MINAGR	Invitation (of March 2003) for applications to the Agrienvironmental Programme of management of lakeshore areas of the Natura Network 2000 site of lakes Koronia-Volvi
	Proposed policy (of March 2005) for the amendment of the Agrienvironmental Programme of management of lakeshore areas of the Natura Network 2000 site of lakes Koronia-Volvi
Prefectures	Decisions of the Prefecture of Thessaloniki on the disposal of effluents and wastewater: 103349/68, 56860/72, 15549/83, 12167/86, 22374/94, 1585/2002
	Various restrictive measures on water use in the Prefecture of Thessaloniki from 1989 to 2004
Archives of the Greek Parliament	Records of questions in Parliament on the problems of the Koronia basin
Archives of the European Parliament	Records of questions in Parliament on the problems of the Koronia basin
Books-publications-conference reports	
Conferences and seminars	Scientific papers on water resources status and management in the Koronia basin
Scientific journals (national and international)	Scientific papers on water resources status and management in the Koronia basin
Research studies and reports	
Prefecture of Thessaloniki	Technical reports on wastewater treatment in the Prefecture of Thessaloniki and specifically in the basin of Koronia
	Original (1998) and revised Master Plan (2004) for the Restoration of Lake Koronia and related technical reports
NGOs and research institutes	Various reports on the assessment of Greek Ramsar wetland sites, including the site of lakes Koronia-Volvi
MEPPW	Specific Environmental Study for the protected area of lakes Koronia, Volvi and the Macedonian Tembi and related technical studies
Universities	Research studies and reports on water quality and the environmental status of the Koronia basin
Primary data records	
Prefecture of Thessaloniki	Primary data on permits for water use and water works in the basin
	Primary data on disposal permits for effluents and wastewater in the basin
Various archives	
Management Body of National Wetland Park	Minutes of meetings of the Administrative Council of the Management Body
Information Centre of Wetland Koronia-Volvi	Proceedings of seminars organised by the Centre on the Wetland Koronia-Volvi
	Archive of regular publications of the Centre
Prefecture of Thessaloniki	Press releases on Lake Koronia

Source	Description of data and information sources
European Commission	Press releases on Lake Koronia and the implementation of EU Directives
Archives of non-governmental organisations and professional groups	Press releases and announcements on the status of the Koronia basin
Press	
Archives of regional and local newspapers	Articles on the Vegoritida basin in the following newspapers: "Makedonia" "Thessaloniki" "Aggelioforos" "Egnatia" "Avriani Makedonia-Thraki" "Typos Thessalonikis" "Chortiatis"
Archives of national newspapers	Articles on the Koronia basin in the following newspapers: "Kathimerini" "Ta Nea" "Ethnos" "Eleftherotypia" "Eleftheros typos" "To Vima"

Annex V: Prefectural water restrictive measures in the Vegoritida basin

Year of measures	2004/5			2003			2000/1			1998/9			1996/7			1994/5		
Prefecture	F	P	K	F	P	K	F	P	K	F	P	K	F	P	K	F	P	K
Specific restrictive measures for the Vegoritida basin	'05	'05	'04/5 invalid		'03	'03 invalid d	'00	'01	'00 zA/zB [120]	'98	'98/9	'98 zA/zB		'96/7	'96 + '97		'94/5	'95
Use permits for surface water only on an annual basis					v													
Until Lake Vegoritida reaches 515.5 masl, abstraction from Lake is prohibited	v																	
Until Lake Vegoritida reaches 515.5 masl, new wells are allowed only for irrigation of certain crops	80% multiannual (trees, grapevine) and up to 20% annual (except for maize, alfalfa, sugarbeets)				trees, asparagus, maize													
Until Lake Vegoritida reaches 515.5 masl, new wells are allowed only if drip irrigation is used	v	v																
New wells are allowed only:																		

389

---

[120] In 1996, the Vegoritida basin on the territory of PoK was divided into zones A and B. Zone A is dominated by tree crops while zone B by maize, sugarbeets and alfalfa. In 2000, zone A was restricted to a smaller area of the basin, while zone B was respectively extended.

Year of measures / Prefecture	2004/5 F	2004/5 P	2004/5 K	2003 F	2003 P	2003 K	2000/1 F	2000/1 P	2000/1 K	1998/9 F	1998/9 P	1998/9 K	1996/7 F	1996/7 P	1996/7 K	1994/5 F	1994/5 P	1994/5 K
- By local administration organisations, local land reclamation organisations, agricul-tural cooperatives								V	zA: V + farmer groups		V	zA: V + farmer groups		V	'97: zA: V + farmer groups		V	V + farmer groups
- If size of wells is: (L=large wells S=small wells)							L>4 inches	L>5 inches S<5 inches										
- If minimum distance of wells (m)[121] is:	350	350					300 L-S 350 L-L	100 S-S 300 L-L		300	500	zAz/B 350		500	zAz/B '97: 350 '96: zA - 350 zB - 300		300	350
- If minimum irrigated land property (ha) is:	10	10			2/S 15/L		15	2/S 15/L	zA: 15 zB: 6	10	15	zA: 15 zB: 6		15	zA: 15 zB: 6		15	15

F=Florina, P= Pella, K= Kozani

Source: Based on information collected from the archives of the Prefecture of Florina (Division of Florina (Division for Land Reclamation), Prefecture of Kozani (Division of Agriculture) and Prefecture of Pella (Division for the Management and Protection of Natural Resources).

[121] Mainly refers to wells for irrigation. In most cases, the distance of a new well from wells for public water supply was set at 500m (this restriction is not repeated in the table).

Annex VI: Prefectural water restrictive measures in the Mygdonian basin

Year of measures	2004	2000	1996	1995
Measures based on (among others):	2003 Water Law, CMD 2004 for Natural Wetland Park	1987 Water Law	1987 Water Law	1987 Water Law
EIA required for issuing permits	Yes	Yes		
Specific conditions for industries	From June to August, wells must reduce daily abstraction by 20%		Wells can be replaced only if industry operates legally and treats effluents	
Specific conditions for farmers	From June to August, not allowed to use any type of spray irrigation from 10:00-16:00. For all new irrigation works, drip irrigation or artificial rain multi-snouts used, wherever possible	For all new irrigation works, drip irrigation must be used wherever this is possible		
New wells allowed only:				
- If minimum distance of wells (m)[122] is more than:	300 m (in zone C of basin) between irrigation wells and other wells	200 m (in part of basin) between private irrigation and water supply wells and between private irrigation and industrial wells		
- If minimum irrigated land property (ha) is more than: (condition applicable in entire Prefecture)	5 ha for annual crops 2 ha for tree and wine crops	5 ha for annual crops 2 ha for tree and wine crops	5 ha	

---

[122] In most cases, the distance of a new well from wells for public water supply was set at 500m (this restriction is not repeated in the table).

Year of measures	2004	2000	1996	1995
New wells prohibited in:	In zones A and B of basin (by 2004 CMD for National Wetland Park). The 2004 CMD only allowed to maintain and modernize existing wells damaged mechanically but prohibited their replacement with new ones in zones A and B		In the entire basin of Lake Koronia until 30/6/97 (extended to 31/12/98), except for water supply use	Within a radius of 300 m from lake
Abstraction from Lake Koronia prohibited	Yes		Yes	Yes * Irrigation networks continued abstraction but had to reduce it by 20% annually

Source: Based on information collected from the archives of the Prefecture of Thessaloniki (Division for Water Resources and Land Reclamation) and of the Region Central Macedonia (Water Resources Management Directorate).

Annex VII: Effluent disposal permits in the Vegoritida basin

Main polluters in the basin	Status of disposal permits for effluents and wastewater	Remarks
Prefecture of Florina		
PPC Steam-electricity plant Amyndeo	Temporary, of 21-6-2005 (for 6 months)	
PPC Lignite-mine Ptolemaida-Amyndeo	Final permit, of 2-7-1991	There is WTTP for human wastewater
Public Slauther House Amyndeo	Final permit, of 5-7-2000	
EAS Amyndeo Wine production	Final permit, of 24-10-2003	
Vegoritids AE Wine Production	No permit*	* No information available in the Prefectural archives / assumption of lack of permit
Hatzis Wine Production	Final permit	
Fourkiotis Dairy production	Final permit	
Sideris AE (now FAGE AE) Dairy industry	Final permit, of 5-4-2004	
Public Corporation Xino Nero	Final permit, of 23-12-2004	
Municipality Amyndeo	Final permit	For wastewater from existing sewerage network
Prefecture of Kozani		
PPC Steam-electricity plant Agios Dimitrios	Temporary, 6 months	But additional works needed are nearing completion
PPC Steam-electricity plant Kardia	Temporary,	Urban wastewater and effluents of cooling towers treated, but not industrial effluents
PPC Steam-electricity plant Amyndeo Ptolemaida	Final permit	Both urban waste and industrial effluents treated (WTTP operates well, discharges into River Soulou)
PPC Lignite-mine South Field	No permit	Treatment plant in construction (request of MEPPW as a result of the EIA process)
PPC Lignite-mine North Field	Permit	Treatment/disposal on the land surface
PPC Lignite-mine Kardia	Permit	Treatment/disposal on the land surface
Public Slaugher House Ptolemaida	Final permit	
Tzobras Dairy Production	No permit	(seasonal operation December-early August) Sole registration of effluents with the sewerage network of local municipality Mourikiou, but municipality has not WWTP either, so effluents remain untreated

Main polluters in the basin	Status of disposal permits for effluents and wastewater	Remarks
PPC LIPTOL Lignite brick production and small steam-electricity plant	Final permit (for personnel building and WTTP)	But no permit for the medical treatment unit
Municipality Ptolemaida	Final permit (for UWWTP)	But UWWTP also receives wastewater from a local hospital, although the Health Directorate of the PoK opposes this because hospital wastewater should be pre-treated before disposal to the UWWTP. However, the elected Prefectural Council approved the continuation of this practice, against the objections of the Health Directorate

Status: November 2005.

Note: The Pella Prefecture is not included in this table, because there are no significant sources of industrial pollution on its territory of the Vegoritida basin. Water polluting municipalities of the Pella Prefecture are not served by significant sewerage networks and treatment plants yet.

Sources: Personal communication Kozani Prefecture, Division of Health, 1/12/2005; Personal communication Florina Prefecture, Division of Health, 30/11/2005; Phone communication PPC, Mining Directorate, 14/11/2005; Common Ministerial Decision 15782/1849 on a "Special pollution reduction programme of Lake Vegoritida and River Soulou from hazardous substances", Official Government Journal B 797, 25 June 2001.

Annex VIII: Effluent disposal permits in the Koronia subbasin (of the Mygdonian basin)

Main polluters in the basin (date of set up)	Status of disposal permits for effluents and wastewater		
Year	2005	1998	1996
Number of industries	16	22	20
Maxim textile production (1989)	Final	Temporary	Expired temporary
Astir fabric-dye (1990)	Final	Temporary	Temporary
Voulinos fabric-dye (1983)	Final	Temporary	Temporary
Protex fabric-dye (1976)	Final	Final	Final
Novaknit fabric-dye (1983)	Final	Final	Final
Kyknos fabric-dye (1988)	Closed	Final	Final
Eurovafi fabric-dye (1992)	Final	Final	Final
Ektor/Atlantis fabric-dye (ex- Finex) (1975)	Closed	Expired temporary	No permit
Rainbow fabric-dye (1983)	Closed	Final	Temporary
Apostolou fabric-dye (1992)	Final	Final	Temporary
Krinos fabric-dye (1993)	Closed	No permit	No permit
AGNO dairy products (1989)	Final	Expired temporary	Temporary
EGS food cans (1989)	Closed	Final	Expired temporary
ETENA vegetable processing (1988)	Final	Expired temporary	Expired temporary
FAGE dairy products	Final	Final	Final
Pyramis stainless utensils (1984)	Final	Expired temporary	Temporary
Nikolaidis galvaniser (1987)	Final	Final	Final
Vafiadis foundry (1995)	No permit	No permit	No permit
Triena fish processing (1991)	Closed	Final	Temporary
Slaughterhouse Tachmazidis Co. (1997)	Final	No permit	
Kapnidis de-dye (1996)	Final	No permit (underground disposal)	
Georgantas de-dye (1992)	Final	Final (underground disposal)	Final

Status: April 2005.

Sources: Arabatzi et al. (1996); Tsagarlis (1998); Knight Piesold et al. (1998); Data provided by the Prefectural Division for Environmental Protection (personal communication, April 2005).

Annex IX: Comparison of Greek and other basin regimes

Assignment of ordinal values for the comparison of different cases regarding change in their institutional water basin regimes:

In the comparative work of Bressers & Kuks (2004c), which compared 24 European cases with respect to water basin regime change, the 13 theoretical variables used were scored on a scale of ordinal values from (0) to (4). The 24 cases and the 13 variables per case were too many to be handled in a purely qualitative way. For this reason, the ordinal values were allocated to summarise information in a uniform format and to facilitate comparison. Apart from allocating the ordinal values to the theoretical variables, the researchers of the 24 cases were asked to provide few short statements per variable ('key facts'). In this comparative work, scores were not considered as facts but as judgements. The use of ordinal values, thus, allowed making explicit judgements on the rating of variables for the comparison.

In general, the ordinal value of (0) indicates that regime changes, sustainability changes and change agents are absent or even negative, and that conditions for regime change towards more coherence (and thus integration) are unfavourable. The ordinal value of (4) indicates the positive other extreme: much more integrated regimes, much more sustainable resource use, forceful change agents in the 'good' direction (of more integration) and very favourable conditions. The following two tables provide qualitative descriptions of the more precise meaning of the ordinal values (0) to (4) per variable[241]. The third and last table lists the ordinal values allocated to the theoretical variables in the two Greek cases examined in this dissertation and to the variables in the other 24 European cases already scored in the context of the comparison by Bressers & Kuks (2004c).

---

[241] Source of information is an internal explanatory project note of the EUWARENESS project, which was the research basis of the case comparison of Bressers & Kuks (2004b).

Description of ordinal values (0)-(4) for regime change variables, external change impetus and sustainability implications:

	Extent	Internal coherence of public governance	Internal coherence of property rights	External coherence between governance and property rights	Total regime change[124]	External impetus for change towards more integration	Implications of regime changes for sustainability
0	Not changed while incomplete or has even decreased	Not changed or has even decreased	Not changed or has even decreased	Not changed or has even decreased	Extent and coherence have not changed or have even decreased	*No new or changed* factors or even the contrary (evolving factors that further blocked such regime change)	Sustainability has not changed or has even decreased due to regime changes
1	Only increased on minor aspects	Only increased on minor aspects	Only increased on minor aspects	Only increased on minor aspects	Extent and coherence only increased on minor aspects, or an increased extent led to less coherence	*Only minor or rather weak* new or changed factors	Sustainability has only increased on minor aspects due to regime changes
2	Increased on only a few of the important aspects or only somewhat on more of the important aspects	Increased on only a few of the important aspects or only somewhat on more of the important aspects	Increased on only a few of the important aspects or only somewhat on more of the important aspects	Increased on only a few of the important aspects or only somewhat on more of the important aspects	Extent and coherence increased on only a few important aspects or only somewhat on more of the important aspects, or an increased extent partially led to less coherence	*Moderate* impetus of the combined new or changed factors	Sustainability has increased on only a few of the important aspects or only somewhat on more of the important aspects due to regime changes
3	Considerably increased on many important aspects	Considerably increased on many of the important aspects	Considerably increased on many of the important aspects	Considerably increased on many of the important aspects	Extent or coherence have considerably increased on many important aspects	*Strong* impetus of the combined new or changed factors	Sustainability has considerably increased on many important aspects due to regime change
4	Completed or was already complete	Completed or was already complete	Completed or was already complete	Completed or was already complete	Extent and coherence have been completed or was already complete	*Very strong* impetus of the combined new of changed factors	Sustainability has been completed or was already complete

---

[124] The value given to "total regime change" is not necessarily the average of the four previous values. It is quite possible that the three coherence indicators to a certain degree compensate each other, so that a decreased coherence in the property rights system is made unimportant, if the regime became more public governance-oriented or the other way around.

Description of ordinal values 0-4 for the contextual conditions influencing regime change:

	Set of conditions for regime change	Tradition of cooperation	Joint problem	Joint chances	Credible threat	Institutional interfaces
0	Very unfavourable for regime changes in the direction of more integration	There was and remained a tradition of thinking opposed to integration	There was and remained a fundamentally divergent (conflictual) understanding of the problem	There was and remained a dominant notion of possible losses from integration	There was and remained a credible threat of interventions by a dominant actor to discourage integration	There were and remained strong institutions that frustrated integration attempts
1	Rather unfavourable for regime changes in the direction of more integration, or somewhat more favourable ones only evolved at the latest phase of the process	There was no tradition of thinking in terms of integration, or only evolving at the latest phase of the process	There was a fragmented understanding of the problem, or a more common one evolved only at the latest phase of the process	There was no notion of possible joint gains from integration, or it only evolved at the latest phase of the process	No credible threat of interventions by a dominant actor to solve the disputes to his own benefit, or it only evolved at the latest phase of the process	There were no well functioning institutions that provide fertile ground for integration attempts, or they only evolved at the latest phase of the process
2	Somewhat favourable for regime changes in the direction of more integration, but only on a part of the relevant aspects	There was some tradition of thinking in terms of integration, but only on a part of the relevant aspects or only with a part of the relevant actors	There was some common understanding of the problem, but only on a part of the relevant aspects or only with a part of the relevant actors	There was some notion of possible joint gains from integration, but only on a part of the relevant aspects or only with a part of the relevant actors	Some credible threat of interventions by a dominant actor to solve the disputes to his own benefit, but only on a part of the relevant aspects	There were some institutions that provide fertile ground for integration attempts, but only on a part of the relevant aspects or not functioning very well
3	Rather favourable for regime changes in the direction of more integration, for most of the relevant aspects	There was some tradition of thinking in terms of integration, for most of the relevant aspects and most of the relevant actors	There was some common understanding of the problem, for most of the relevant aspects and with most of the relevant actors	There was some notion of possible joint gains from integration, for most of the relevant aspects and with most of the relevant actors	Some credible threat of interventions by a dominant actor to solve the disputes to his own benefit, for most of the relevant aspects	There were some well functioning institutions that provide fertile ground for integration attempts, for most of the relevant aspects
4	Very favourable for regime changes in the direction of more integration, for most of the relevant aspects and during most of the period of the process	There was a strong tradition of thinking in terms of integration, for most of the relevant aspects and most of the relevant actors	There was a strong common understanding of the problem, for most of the relevant aspects and with most of the relevant actors	There was a strong notion of possible joint gains from integration, for most of the relevant aspects and with most of the relevant actors	Strong credible threat of interventions by a dominant actor to solve the disputes to his own benefit for most of the relevant aspects	There were strong and well functioning institutions that provide fertile ground for integration attempts, for most of the relevant aspects and with most of the relevant actors

Allocation of ordinal values 0-4 to key variables in the two Greek cases and the 24-European case sample previously studied:

Variables	I1	I2	NL1	NL2a	NL2b	NL2c	NL2d	B1a	B1b	B1c	B1d	B2a	B2b	B2c	B2d	E1	E2	F1	F2a	F2b	CH1a	CH1b	CH1c	CH2	GR1	GR2
Regime change:																										
Extent	4	2	4	4	4	4	4	2	3	3	2	4	2	3	1	3	3	4	3	2	4	4	4	4	2,5	3
Internal coherence public governance	3	2	3	3	3	3	3	2	4	3	3	3	1	2	1	3	3	3	3	2	3	3	2	3	1	2
Internal coherence property rights	2	1	2	4	2	2	2	3	4	1	4	4	2	0	0	3	3	3	3	3	3	3	2	1,5	1	1
External coherence p.r & p.g.	3	1	2	4	3	3	3	2	2	2	3	3	1	2	1	3	3	2	3	3	3	3	3	4	1	1
Total assessment regime change	3	1,5	3	4	3	3	3	3	4	2	3	4	1	2	1	3	3	3	3	2	3	3	3	3,5	1	1,5
Implications for sustainability	3	1	3	2	3	3	2	3	3	3	3	3	1	2	1	3	3	3	3	1	3	3	2	2,5	1	1,0
Force of external change impetus	2	1	4	4	4	4	3	3	1	3	3	3	1	4	0	3	2	3	3	3	3	3	3	3,5	2	3
EU originated pressure	1	1	0	0	1	0	1	1	0	1	0	0	1	1	1	1	1	3	1	1	0	0	0	0	1	1
National regime other	1	1	1	1	1	0	1	0	1	1	1	1	0	1	0	1	0	1	1	1	1	1	1	1	0	0
Problem pressure	1	1	0	1	0	0	0	1	0	1	1	1	1	1	0	1	1	1	1	1	1	0	1	1	0	0
Other case circumstances	1	0	0	0	1	1	0	0	0	1	0	0	0	0	0	1	1	1	1	1	1	0	1	0	0	1
Conditions:																										
Tradition	1	1	3	4	2	3	3	0	2	2	2	3	2	2	1	1	1	2	3	2	1	1	1	2	0	0
Joint problem	2	1,5	2	3	0	3	3	1	3	3	3	4	2	4	1	3	2	1,5	2	2	2	3	2	1,5	2	2
Joint chance	2	0	2	4	4	2	3	2	3	3	1	2	1	1	1	1	3	3	2	1	1	1	2	3	2	3
Credible threat	4	1	2	4	4	4	4	2	2	1	0	1	3	3	1	3	1	3	3	2	2	1	3	2,5	1	1
Institutional interfaces	4	1	3,5	3	3	2	4	2	3	3	2	2	2	2	1	3	2	3	3	2	4	3	3	3	2	2
Total assessment conditions	3	1	3	4	3	3	3	2	3	2	2	3	1	2	2	3	2	3	3	2	2	3	2	2,5	1,5	1,5

Note: The ordinal values (0-4) were allocated for the two Greek cases, based on the assessment of the relevant variables in this dissertation. The ordinal values for the 24-European case sample previously studied were provided by the researchers of the respective cases for the earlier comparison of Bressers & Kuks (2004c) (refer to the relevant case study literature for more descriptive information in Bressers & Kuks (eds) (2004a))

Abbreviations of cases and subcases used in the figures of chapter 9

I1	Idro Lake and Chiese River (Italy)
I2	Marecchia-Conca basin
NL1	Regge river basin
NL2	IJsselmeer lake: a. Fisheries, b. Gas drilling, c. Land reclamation, d. Recreation and nature
B1	Vesdre river basin: a. Lead-poisoning Verviers, b. Chronic floods, c. River pollution, d. Industrial pollution Wayai
B2	Dender river basin: a. Agriculture vs. nature, b. Polders vs. water floods, c. Sewing and purification, d. Pollution vs. drinking water
E1	Mula river
E2	Matarraña river
F1	Audomarois basin
F2	Sèvre Nantaise basin: a. Agricultural hydraulic works, b. Pollution
CH1	Valle Maggia : a. Hydroelectric production, b. Gravel quarrying, c. Flood protection
CH2	Lake Baldegg and Lake Hallwil
GR1	Vegoritis basin
GR2	Mygdonian basin